IDEALIZATION VII:

STRUCTURALISM, IDEALIZATION AND APPROXIMATION

POZNAŃ STUDIES
IN THE PHILOSOPHY OF THE SCIENCES AND THE HUMANITIES

VOLUME 42

EDITORS

Tomasz Banaszak (assistant-editor)

Jerzy Brzeziński

Robert Egiert (assistant-editor)

Andrzej Klawiter

Krzysztof Łastowski

Leszek Nowak (editor-in-chief)

Katarzyna Paprzycka (Pittsburgh)

Marcin Paprzycki (Odessa, TX)

Piotr Przybysz (assistant-editor)

ADVISORY COMMITTEE

Joseph Agassi (Tel-Aviv)

Étienne Balibar (Paris)

Piotr Buczkowski (Poznań)

Mario Bunge (Montreal)

Nancy Cartwright (London)

Robert S. Cohen (Boston)

Francesco Coniglione (Catania)

Andrzej Falkiewicz (Wrocław)

Ernest Gellner (Cambridge)

Bert Hamminga (Tilburg)

Jaakko Hintikka (Boston)

Jerzy Kmita (Poznań)

Władysław Krajewski (Warszawa)

Theo A.F. Kuipers (Groningen)

Witold Marciszewski (Warszawa)

Ilkka Niiniluoto (Helsinki)

Günter Patzig (Göttingen)

Marian Przełęcki (Warszawa)

Jan Such (Poznań)

Jerzy Topolski (Poznań)

Ryszard Wójcicki (Łódź)

Georg H. von Wright (Helsinki)

Zygmunt Ziembiński (Poznań)

The address: prof. L. Nowak, Cybulskiego 13, 60-247 Poznań, Poland.

Fax: (061) 535-535

E-mail: epistemo at PLPUAM11.AMU.EDU.PL

IDEALIZATION VII:

STRUCTURALISM, IDEALIZATION AND APPROXIMATION

Edited by
Martti Kuokkanen

Rodopi

AMSTERDAM – ATLANTA, GA 1994

∞ The paper on which this book is printed meets the requirements of "ISO 9706:1994, Information and documentation – Paper for documents – Requirements for permanence".

ISBN: 90-5183-792-5 (Bound) (CIP)
©Editions Rodopi B.V., Amsterdam – Atlanta, GA 1994
Printed in The Netherlands

CONTENTS

IDEALIZATION, APPROXIMATION AND MEASUREMENT

IDEALIZATION, APPROXIMATION AND COUNTERFACTUALS IN THE STRUCTURALIST FRAMEWORK

*Poznań Studies in the Philosophy
of the Sciences and the Humanities
Vol. 42, pp. 3–24*

Theo A.F. Kuipers

THE REFINED STRUCTURE OF THEORIES*

Introduction

This contribution gives a systematic introduction to the structuralist reconstruction of empirical theories, which is presupposed in most of the other
contributions. It starts from a number of global ideas about the nature and
structure of empirical theories. To begin with, it is frequently useful to
distinguish a core of principles and a belt of auxiliary hypotheses.

A theory is said to be ontologically stratified when there are two or
more kinds of entities involved. In this case the principles usually concern
only one kind of entities and their properties, and may be called internal
principles, whereas the auxiliary hypotheses connect the different kinds of
entities and their properties, and may be called bridge principles. It may or
may not be plausible to speak of a lower, micro-level and a higher, macrolevel.

Besides ontological stratification there is epistemological stratification:
they frequently go together, but are essentially independent.

A proper theory T can be defined as an epistemologically stratified
theory in the sense that it contains terms, and hence statements, that are
laden with one or more of its principles, that are T-laden or T-theoretical,
for short. The other terms of T are called T-unladen or T-non-theoretical. In
contrast to proper theories, experimental hypotheses can be defined as
improper theories, containing no theoretical terms of their own. A set of

---

* This text is an adapted version of a chapter of an advanced textbook in philosophy
of science in preparation, provisionally entitled *Cognitive-heuristic patterns in science*.
Brief presentations of the examples of classical particle mechanics, of the periodic table
of chemical elements, and of the psycho-analytic theory have been deleted, as well as a
section on relations between theories and theory-nets. Instead of the latter, a section on
idealization and concretization has been inserted.

connected experimental hypotheses is called an experimental theory. It should be noted that being *T*-non-theoretical is a theory-relative qualification of a term or a statement: they may well be laden with underlying theories.

The main function of a proper theory *T* is the explanation and prediction of *T*-unladen, experimental laws, i.e., true general hypotheses containing no terms laden with themselves or *T*. For this function the distinction between experimental and proper theories is of course crucial.

A last general feature of theories to be mentioned beforehand is that the principles of a theory, whether ontologically and/or epistemologically stratified or not, can frequently be subdivided into main principles, claimed to be true for the whole domain concerned, and special principles, only claimed to be true for a certain subdomain.

Why the Structuralist Approach?

There are two main approaches to the fine-structure of empirical theories. The *statement*-approach conceives theories primarily as sets of statements. In the case of an axiomatized theory all these statements are logical consequences of a subset of the axioms. This approach has long been considered as the only and obvious approach, e.g. by Carnap and Popper.

The set-theoretic or *structuralist* approach was introduced by Suppes and refined by Balzer, Sneed, and Stegmüller. Its basic idea is that theories frequently specify classes of set-theoretic structures satisfying certain conditions. A set-theoretic structure is an ordered set of one or more domain- or base-sets and one or more properties, relations or functions defined on them, which satisfy certain conditions.

For example a biological family is a structure $<A,C>$ with A as the (base-)set of members of the family and C as a ternary relation on A, i.e., C is a subset of $A \times A \times A$, such that $C(x,y,z)$ indicates that z is a child of x and y. It is a proper two-generation biological family if there are precisely two members x and y in A such that for all other z in A it holds that $C(x,y,z)$.

According to the structuralist view an axiomatized theory defines such a class of structures and the conditions imposed on the components of the structures are the axioms of the theory. The link with reality is made by the claim, associated with the theory, that the set of set-theoretic representations of the so-called intended applications form a subset of the class of structures of the theory.

Unfortunately, there has been much disagreement about what the proper approach to theories is, whereas it is easy to see that the two approaches are not at all incompatible. At least for so-called first-order statement theories, i.e., theories formulated as a set of statements of a so-called first-order language, it is evident that the set of models of the theory, i.e., the structures for which the statements of the theory are true, is precisely a set of structures that might also have been introduced directly in the structuralist way. If we do not restrict ourselves to first-order languages both approaches are essentially intertranslatable. Hence, the choice is a pragmatic question.

The main advantage of the structuralist approach is that it is much more a bottom-up approach than the statement approach. It invites one as it were to represent and analyse theories and their relations as closely to the actual presentations in textbooks as formally possible. As in scientific practice all kinds of useful mathematics may be used for that purpose. It does not mean that structuralist reconstruction of theories is an easy task. However, the statement approach is certainly more difficult for specific reconstructions. It is primarily useful in talking about theories in general and in studying logical, in particular model-theoretic, questions about the relation between sentences and their models. These questions and their answers become very complicated as soon as substantial mathematics is involved, e.g. real numbers. However, as we will indicate, even theoretical questions concerning for instance idealization and concretization as a truth approximation strategy can be treated relatively easily in structuralist terms.

Given our intention to be as useful as possible for actual scientific research I will restrict the attention to the structuralist approach. The best textbooks presenting the structuralist view on theories are Balzer [1982], Diederich [1981], Stegmüller [1973/1986] and Balzer, Moulines, Sneed [1987]. I will present the main general aspects without going into technical details which are not of primary importance for actual practice. My main goal is to make clear what kind of entities one may be looking for in theory formation and how standard questions about these entities can be explicated. Moreover, in passing I will explicate some standard Popperian concepts in structuralist terms.

The Slide Balance

Consider the slide-balance of *Figure 1*. On either side there can be placed any finite number of objects of various weights at all possible distances from the fulcrum S. The balance is assumed to be completely symmetric, the equal arms are as long as necessary, and the objects are pointmasses, i.e., dimensionless particles. Our domain of interest is the equilibrium states, i.e., all possible distributions of objects leading to equilibrium.

S

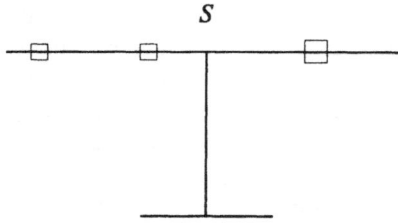

Figure 1

A plausible way to represent the equilibrium states, the intended applications, is as follows. We start by characterizing a possible or potential equilibrium state by a structure of the form $<P,Pl,d,w>$, where P is the finite set of particles involved and Pl the subset of P such that Pl and $P-Pl$ represent the particles to the left and to the right of S, respectively. For every particle p in P, $d(p)$ indicates the distance of p from S and $w(p)$ the weight of p. Technically speaking, d and w are positive real-valued functions on P. Let us call the set of structures $<P,Pl,d,w>$ satisfying all formal conditions the set of potential equilibrium models of our theory about the equilibrium states of the slide-balance, indicated by SBp.

Accordingly, our conceptual claim is that the equilibrium states can be represented as members of SBp, i.e., there is a subset E of SBp representing the real, that is, the empirically possible, equilibrium states: the SBp-set of intended applications.

The ultimate purpose of theory formation now is to try to characterize E explicitly by one or more additional conditions. As is well known, the adequate condition in the present case is specified by the so-called law of the balance: the sum of distance times weight of the objects on the left should be equal to the corresponding sum with respect to the objects on the right. Let us call the subset of SBp of members satisfying this condition the set of equilibrium models of our theory, indicated by SB. The proper empirical claim of the law of the balance can now be formulated as "$E=SB$", and this claim is true, *ceteris paribus*.

As we will see in other cases the relevant claim need not be as strong as in the present case. It might just have been the claim that E is a subset of SB. It will be helpful for later examples to add a more formal presentation of the naive theory of the slide-balance ($\mathbb{R}^+$ indicates the set of positive real numbers).

The naive theory of the slide-balance $<SBp,SB,D,E>$

$\longrightarrow$ contains $<P,Pl,d,w>$	iff
1) P is a finite set	(particles)
and Pl a subset	(the particles on the left of S)
2) d: $P \rightarrow \mathbb{R}^+$	($d(p)$: the distance of p from S)
SBp 3) w: $P \rightarrow \mathbb{R}^+$	($w(p)$: the weight of p)
4) *the law of the balance*	
SB $\Sigma p \in Pl\ d(p){\cdot}w(p) = \Sigma p \in P\text{-}Pl\ d(p){\cdot}w(p)$	

SBp-SB   *empirical content* (to be explained)

$E \subseteq SBp$ *conceptual claim*: the intended domain of applications can be represented as potential models, the intended applications

$E \subseteq SB$ *(naive weak) empirical claim*: the intended applications are equilibrium models (naive strong claim: $E=SB$)

Later we will see that the empirical claims are naive, because it appears to be impossible to test them in a non-circular way. But first I will present the general structuralistic set-up for unstratified theories.

Unstratified Theories

Let there be given a domain D of natural phenomena (states, situations, systems) to be investigated. D is supposed to be circumscribed by some informal, intensional description and may be called the *intended domain of applications*. Although D is a set, its elements are not yet mutually well distinguished. For this reason we do not yet speak of domain of intended applications.

In order to characterize the phenomena of D a set *Mp* of *conceptual possibilities or potential models* is construed. *Mp* is, technically speaking, a set of structures of a certain type, a so-called similarity type. In practice *Mp* will be the conceptual frame of a research program for D.

The confrontation of D with *Mp*, i.e., D seen through *Mp*, is assumed to generate a unique, time-independent subset $Mp(D)=I$ of all *Mp*-representations of the members of D, to be called the *Mp*-set of *intended applications*. Apart from time-independence, this is a conceptual claim. Of course, I will be a subset of the (*Mp*-)set of *empirical possibilities*, but it may be a proper subset, i.e., a more specific set of intended applications satisfying certain additional (more or less precise, but relatively observational) conditions.

Assuming that the set of empirical possibilities is a proper subset of Mp, i.e., not everything which is conceivable is empirically possible, I is also a proper subset of Mp. In certain cases I may be a one-element set, in particular when we want to describe the 'the actual world' in a certain context, that is, a realized empirical possibility, e.g. the description of conditions and results of a particular experiment.

A specific theory about D centers around an explicitly defined subset M of Mp, the *models* of the theory. More specifically, a specific unstratified theory is any combination of the form $UT = <Mp,M,D,I>$ with, beside the conceptual claims that M and I are both subsets of Mp, the (weak) empirical claim that I is a subset of M. Sometimes the strong empirical claim is made that I is equal to M, but we take the weak claim as standard. It is plausible to call UT true when its claim is true and false otherwise.

The general set-up of the structure of epistemologically unstratified theories will now be presented in a scheme. Such a theory is a meta-structure of the following form:

$<Mp,M,D,I>$ is an *epistemologically unstratified theory* iff

Mp:	potential models: a set of structures of a certain type
$M \subseteq Mp$:	models: the potential models that satisfy all axioms
$Mp-M$:	empirical content (to be explained)
D:	the intended domain of applications
$I \subseteq Mp$:	intended applications: conceptual claim: Mp-representation of D leads to the subset I of Mp
$I \subseteq M$:	empirical claim (strong claim: $I = M$)

$<Mp,M>$ is sometimes called the *theoretical core* of the theory, and $<D,I>$ may be called the *application target* of the theory.

The unstratified set-up of theories seems to be rather adequate for experimental theories, a combination of one or more experimental hypotheses, which contain by definition only terms that are understood independently of the theory concerned.

Basic Terminology

Before we go over to stratified theories, I we like to present some useful basic terminology, which can largely be seen as a structuralist explication of Popperian 'statement terminology' [Popper, 1959]. I will neglect all necessary provisos, in particular in regard to the complications arising from underlying theories. To use Lakatos's term [Lakatos, 1978], I explicate naive falsificationism, first unstratified, later stratified.

When the claim of theory $UT = <Mp,M,D,I>$ is false $I-M$ is by definition non-empty, in which case it is plausible to call its members instantial mistakes or (empirical) *counter-examples* of UT. Note that being a counter-example in this sense does not imply that it has been realized already and registered as such. The set of counter-examples $I-M$ is by definition a subset of $Mp-M$. Hence, $I-M$ can, whatever I is, only be non-empty when $Mp-M$ is non-empty. In other words, the members of $Mp-M$ may be called the *potential* counter-examples of the theory and, as was already announced a number of times, the set $Mp-M$ itself the *empirical content* of UT. From the present point of view Popper had this in mind with his notions 'potential falsifier' and 'empirical content', respectively.

Other plausible explications of Popperian terminology are easily found. UT is *falsifiable* (or empirical) if and only if $Mp-M$ is non-empty. UT^* is *better falsifiable* than UT when $Mp-M$ is a proper subset of $Mp-M^*$. The latter condition is equivalent to: M^* is a proper subset of M. In its turn, this is equivalent to stating that the claim of UT^* implies that of UT, and not conversely, that is, UT^* is stronger than UT.

The verification/falsification asymmetry also arises naturally in the present set-up. To *verify* theory UT it would be necessary to show that all members of I, that is all Mp-representations of D, belong to M. In interesting cases this will always be an infinite task, even in the case that I is finite, for the task is only finite when D is finite. To *falsify* UT it is 'only' necessary to show that there is *at least one* member of I not belonging to M. Hence, if a theory is true, verification will nevertheless not be obtainable if D is infinite. On the other hand, when a theory is false, falsification is attainable in principle, viz. by realizing one counter-example. Of course, if an attempt to falsify fails in such a way that the experiment provides an (*empirical*) *example* of UT, i.e., a member of M, this is called *confirmation* (or corroboration) of UT.

In the present set-up Popper's distinction between universal and existential statements is thus interpreted as the distinction between the general claim of the theory ($I \subseteq M$) that all intended applications are models of the theory, and the negation of this claim, the existential claim that at least one intended application is not a model ($I-M$ is non-empty).

A *basic statement* becomes a claim to the effect that a certain intended application x in I belongs to a certain subset F of Mp, defined by a certain condition on potential models, i.e., $x \in I \cap F$. An *accepted* basic statement presupposes of course that the relevant intended application has been realized.

The basic statement $x \in I \cap F$ is in conflict with theory UT if it can be demonstrated on conceptual grounds that $F \cap M$ is empty. Such basic statements may be seen as more direct explication of Popper's idea of 'potential falsifiers', compared to 'potential counter-examples'. However, it is easy to show that the suggested statement concept of potential falsifier has become essentially redundant. It is easy to check that a true potential falsifier of UT, i.e., "$x \in I \cap F$" is true, implies that I-M is non-empty and hence that x is a counter-example of UT. Conversely, the existence of counter-examples of UT is easily seen to imply that there must be true potential falsifiers. By consequence, demonstrating empirically the existence of a counter-example, i.e., realizing a potential counter-example, goes hand in hand with demonstrating that there is a true potential falsifier. Hence, the statement concept of potential falsifier is not needed in the face of the counter-example concept.

The Slide-Balance Reconsidered

The problem with the slide-balance is that it might be impossible to test the claim in a non-circular way without leading to infinite regress. For, to test the claim it is necessary to measure the distances and the weights. Whereas distance measuring does not require something like a slide-balance, weight measuring may not only actually be done by using a slide-balance, there might even be no other possibility. If the weight of a particle is measured by a slide-balance the law of the balance is obviously presupposed. Hence, assuming that the weight of a particle can only be measured by a slide-balance, the concept of weight is SB-theoretical and leads to the so-called *problem of theoretical terms*. A non-circular test of the claim of the theory would presuppose that the weights of the particles have been measured before with the same or another slide-balance. Hence, we get either an infinite regress or circular testing, if we stick to "$I \subseteq M$" as the empirical claim of the theory. There is, however, a way out of this dilemma, by restricting the empirical claim to SB-non-theoretical terms. Later we will see that the situation in the present example is for two reasons not as dramatic as suggested, but this did not exclude that the example could, by way of thought experiment, be transformed into an instructive example of genuine theoretical terms.

In order to formulate a new empirical claim we introduce the set of potential partial equilibrium models $SBpp$, being the structures of SBp

without the *SB*-theoretical weight component and the corresponding 'status-condition', viz. clause 3). By consequence, there is a restriction or *projection function* π from *SBp* onto *SBpp* projecting every potential model on the potential partial model arising from deleting *w* and clause 3). Hence, for $x = <P,Pl,d,w> \in SBp$ the projection of *x*, π(*x*), is equal to $<P,Pl,d> \in$ *SBpp*. For an arbitrary subset *X* of *SBp*, π*X*, the projection of *X*, is defined as the subset of *SBpp* containing precisely the projections of the members of *X*.

For stratified theories we assume that the set of intended applications *E* no longer represents the equilibrium states seen through *SBp*, but seen through *SBpp*. Hence, the corresponding conceptual claim that *E* is a subset of *SBpp* is not laden with our theory about *SB*.

Now it is easy to see that the claim that *E* is a subset of π*SB* is not laden with the weight term, in the sense that it does not presuppose that the weights of the particles have been empirically determined. Hence, this revised empirical claim can be tested in a non-circular way.

It is plausible to call the members of *E*-π*SB*, if any, counter-examples of the theory. It is clear that they have to come from *SBpp*-π*SB*. Hence, it is now plausible to call this set the empirical content and its members potential counter-examples. Note that the empirical content reduces to the empirical content of the unstratified theory (*SBp*-*SB*) when *SBpp* and *SBp* are identical and π is, by consequence, the identity-function.

Unfortunately, the new claim is not only non-circular, it is also vacuous, for the empirical content is empty. The claim says in fact that all intended applications can be extended to models of the theory. To be precise, the claim is that every $<P,Pl,d>$ in *E* can be supplied with a positive real-valued function *w* on *P* such that $<P,Pl,d,w>$ is in *SB*. But it is easy to check that this is possible for every member of *SBpp*. In other words the empirical content *SBpp*-π*SB* is empty.

However, the situation changes when we take so-called constraints into consideration: in the present case we have to require also that the weights assigned to the same particle, occurring in different applications, should be the same. In contrast to the distance of the objects from the fulcrum, our concept of weight is such that the weight of particles is constant in different applications. The formal treatment of constraints, however, will be postponed to a later section.

Before I return to the general exposition I will summarize the formal features of the theory of the slide-balance, leaving out the plausible specification of the projection function π:

The refined theory of the slide-balance $<SBp,SBpp,SB,\pi,D,E>$

$\longmapsto$	contains $<P,Pl,d,w>$	iff
$\|\mapsto$	contains $<P,Pl,d> = \pi <P,Pl,d,w>$ iff	
	1) P is a finite set	(particles)
	and Pl a subset	(the particles on the left of S)
$SBpp$	2) $d: P \rightarrow \mathbb{R}^+$	($d(p)$: the distance of p from S)
SBp	3) $w: P \rightarrow \mathbb{R}^+$	($w(p)$: the weight of p)
SB	4) *the law of the balance* $\Sigma p \in Pl\ d(p) \cdot w(p) = \Sigma p \in P\text{-}Pl\ d(p) \cdot w(p)$	
$SBpp$-πSB	*empirical content* (without w-constraint empty, with w-constraint non-empty)	
$E \subseteq SBpp$	*conceptual claim*: the domain of intended applications D can be represented as potential partial models	
$E \subseteq \pi SB$	*empirical claim*: the intended applications can be extended to models (strong claim: $E = \pi SB$)	

By way of digression, it is interesting to note that, assuming the weight-constraint, the SB-theory explains the following experimental, i.e., SB-unladen, *factor slide law*: if, starting from an equilibrium, the distances of all objects are multiplied by the same factor, there is again equilibrium. For it follows trivially from the law of the balance that it remains satisfied.

As a matter of fact, in the present case it is not difficult to formulate an experimental law such that the notion of weight can be explicitly defined, apart from a proportionality constant, on its basis. The law referred to states the following: given a unit object at a unit distance at one side of S, every other object p has a 'unique equilibrium distance' $d_u(p)$ at the other side. The weight $w(p)$ is then defined as $1/d_u(p)$, hence, such that in the relevant cases the law of the balance is satisfied by definition. Hence, for these cases the law cannot be tested in a non-circular way. But there is no regress, let alone infinite regress. For, given the definition, the rest of the law of the balance is a straightforward empirical claim which can be directly tested.

By consequence, on closer inspection the theory of the slide balance does not give rise to the problem of theoretical terms, when certain experimental laws are taken into consideration. Of course, this does not affect the instructiveness of the SB-theory as an almost proper theory. Moreover, it illustrates an interesting way in which a seemingly proper theory may be on closer inspection a sophisticatedly formulated experimental theory, in the present case: the conjunction of the 'unique equilibrium distance law', the weight-definition on its basis, and the law of the balance.

There is still one other reason why the problem of theoretical terms is not as dramatic in the case of the slide-balance: there are other ways of measuring the weight of objects than by using a slide-balance. But let us now turn to the general set-up of stratified theories, designed for proper theories.

Stratified Theories

The general set-up of the structure of epistemologically stratified theories can now directly be presented in a scheme. Such a theory is a meta-structure of the following form:

	$<Mp,Mpp,M,\pi,D,I>$ is an *epistemologically stratified theory* iff
Mp:	potential models: a set of structures of a certain type
Mpp:	potential partial models: the substructures of Mp restricted to non-theoretical components
$M \subseteq Mp$:	models: the potential models that satisfy all axioms
$\pi:Mp \rightarrow Mpp$:	the projection function (from Mp onto Mpp) $\pi X = \{\pi(x)/x \in X\}$, for $X \subseteq Mp$, implying $\pi X \subseteq Mpp$
πM:	projected models
$Mpp - \pi M$:	empirical content
D:	intended domain of applications
$I \subseteq Mpp$:	intended applications (non-theoretical): conceptual claim: non-theoretical representation of D leads to the subset I of Mpp
$I \subseteq \pi M$:	empirical claim (strong claim: $I = \pi M$)

Examples

In the literature there have been presented structuralist reconstructions of many well-known theories. Here is a short list with references:

Newton's classical (gravitational) particle mechanics: Sneed [1971], Zandvoort [1982] and Balzer, Moulines, Sneed [1987]. In this example the distinction between general and special principles of a theory is fundamental.

14

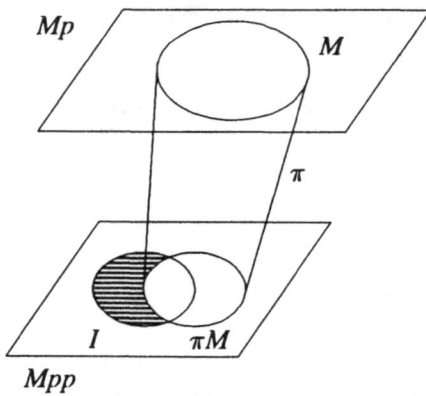

Now it is plausible to call $<Mp, Mpp, M, \pi>$ the *theoretical core* of the theory and $<D,I>$ remains the *application target*.

Figure 2 illustrates the revised empirical claim: the shaded area, representing $I\text{-}\pi M$, should be empty (on conceptual grounds!).

Figure 2

Simple equilibrium thermodynamics [Balzer et al., 1987]

Mendeleev's and the refined theory of the periodic table of chemical elements: Hettema and Kuipers [1988].

Freud's psycho-analytic theory: Balzer [1982] and Stegmüller [1986]. In this example the distinction between absolute and relative empirical content [see below] is very important.

Arrow-Debreu's neoclassical theory of individual and collective demand: Janssen & Kuipers [1989]. In this example there are other than epistemological reasons for the distinction between potential models and partial potential models.

Jeffrey's theory of decisions [Stegmüller, 1986]

Jakobson's theory of literature [Stegmüller, 1986]

In the Balzer [1982], Diederich [1981], Stegmüller [1973/1986] and Balzer, Moulines, Sneed [1987] one can find numerous other examples.

From the fact that the theories of Freud and Jakobson can be reconstructed in the structuralistic way it follows that this way of reconstruction is, like the statement approach, applicable to qualitative, non-mathematical theories. From the other examples it is evident that the present approach is also well suited for quantitative theories, a kind of theory for which the statement approach leads to all kinds of complications.

In a sense it is a trivial claim that every empirical theory can be reconstrued in structuralist fashion. Hence, there should be additional reasons to do so in particular cases. A general reason frequently is the desire to get a better insight in the theory; beside that, one may be interested in particular questions, such as whether the theory has empirical content, whether it is an experimental or a proper theory, what its precise relation is to another theory, etc. But the main function of getting acquainted with the structuralist approach in general, and by examples, is of course the heuristic role it may play in the construction of new theories.

Absolute and Relative Empirical Content

It is always possible to divide the axioms on the one hand into analytic (A) and substantial (S) axioms and on the other into non-theoretical (N) and theoretical (T) axioms, leading to four types of axioms: NA, TA, NS, TS. For both distinctions it holds that it is advisable in case of doubt to choose the cautious classifications, i.e., S and T, respectively.

The following survey will speak for itself.

Mpp	(the set of potential partial models):	NA
Mp	(the set of potential models):	$NA + TA$
Mpart	(the set of partial models):	$NA \qquad + NS$
M	(the set of models):	$NA + TA + NS + TS$

It is clear that $<Mpp,Mpart>$ is the (theoretical) core of a partial theory, i.e., an unstratified, hence experimental theory, constituting a substantial part of the full theory. The empirical content of the full theory was defined as $Mpp-\pi M$, let us call it more specifically the *absolute empirical content* (AEC). The empirical content of the partial theory, the *partial empirical content* (PEC), is of course $Mpp-Mpart$. Given the trivial fact that πM is a subset of *Mpart*, the partial empirical content is automatically a subset of the absolute empirical content. The interesting question is whether the full theory has something to add to the partial theory, i.e., whether the *relative empirical content* (REC), defined as $Mpart-\pi M$, is non-empty.

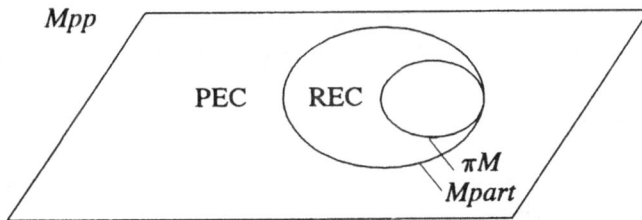

Figure 3

It is easy to check that the absolute empirical content $(Mpp\text{-}\pi M)$ is the union of the partial empirical content $(Mpp\text{-}Mpart)$ and the relative empirical content $(Mpart\text{-}\pi M)$. *Figure 3* depicts PEC and REC explicitly. AEC is the union of both.

As a consequence, if a theory has relative (and/or partial) empirical content it has absolute empirical content. Conversely, however, a theory may have absolute empirical content without having relative empirical content, in which case the absolute empirical content coincides with the partial empirical content.

Balzer [1982] claimed that the general psycho-analytic theory has partial, and hence absolute, but no relative empirical content. Stegmüller [1986], however, is able to prove that it also has relative empirical content. Stegmüller then continues with the interesting observations that for this proof it is not necessary to take constraints and/or special laws into consideration and that classical particle mechanics CPM has only (relative) empirical content when constraints and special laws are taken into considerations. Hence, according to the relative content criterion for empirical impact the theory of Freud is in a sense even superior to that of Newton.

But we like to add that the CPM example makes clear that a generic theory (including constraints) need not have relative empirical content in order to be useful. The important research question is whether a generic theory can be supplemented with special laws which have relative empirical content [Cf. Bunge, 1977].

Intended Applications Reconsidered

In this section I will start with some general remarks about the set of intended applications I, then I will formulate three different ways of determining I, and conclude with some elaboration of the problem of theoretical terms.

I was introduced as 'D seen through Mp' and later revised as 'D seen through Mpp', i.e., I represents the intended domain of applications with the conceptual means of Mpp. I will restrict my formulations to the revised situation, when not otherwise stated, for it includes the particular case that $Mp = Mpp$.

It is evident that I is Mpp-dependent, and Mpp is manmade. Hence we subscribe to a fundamental form of conceptual relativity. But this need not imply an extreme form of relativism: empirical claims are objectively true or false, for their truth or falsehood depends on nature.

In its turn, the objective character of empirical claims does not imply that D, Mpp (and hence I) and Mp are fixed beforehand, and that the task remains to formulate a subset M of Mp leading to a true empirical claim. As a matter of fact, in practice, the determination of D, Mpp, Mp and M is a complicated dialectical interaction process, guided by the desire to formulate informative and true empirical claims. Unfortunately, it seems difficult to discern general patterns in this interaction process, without making some important idealizations.

However, if we assume, by idealization, that Mpp is fixed, the determination of (D and hence) I can be governed by at least three different principles.

If we are interested in all relevant empirical possibilities, I coincides with the set of empirical possibilities at the Mpp-level. Let $EMPOSpart$ indicate this subset of Mpp. Although we may not have an explicit characterization of $EMPOSpart$, there is a clear empirical criterion for membership: x in Mpp belongs to $I=EMPOSpart$ iff x can be empirically realized. If we are interested in a well-defined subset of the set of $EMPOSpart$, i.e., empirical possibilities satisfying some explicit condition, membership determination is not fundamentally different. In both cases we will speak of *empirical determination* of I.

In this case, the obvious target of theory development is an explicit charaterization of I, i.e., a set of models M is sought for which the strong claim holds: $I=\pi M$, such that M may be called *the true (Mp-)theory about I*, or simply, *the (conceptually relative) truth*. In Kuipers [1992a/b] the formal structure of truth approximation by unstratified and stratified theories is studied (extensively/briefly).

The notion of an empirical possibility is here presented as an absolute qualification, but there may well be cases where it makes good sense to distinguish levels of empirical possibility, as e.g. suggested by the following sequence of 'lower' to 'higher' levels: the physical, chemical, biological, psychological, cultural-socio-economical level. Being empirically possible at a higher level then implies being empirically possible at a lower level, but not the converse. Another example concerns the idea of empirically possible states of an artifact, assuming that it remains intact, which means a severe restriction to its physically possible, including broken, states. Such refinements can easily be built into empirical determination of I, as long as the boundaries between the different levels of empirical possibility may be assumed to be sharp.

In many cases, however, the interest is directed to a proper subset of $EMPOSpart$, of which the membership is not sharply defined. One important way in which I can then have been circumscribed is by so-called *paradigmatic determination*.

> *Definition*: I is *paradigmatically determined* if there are *PAR* and *SIM* such that
> 1. I is a subset of *EMPOSpart*   the intended applications
> 2. *PAR* is a finite subset of I the paradigmatic examples
> 3. *SIM* is a binary relation on *Mpp*   a similarity relation
> 4. for all x in $I-PAR$ there is y in *PAR* such that $SIM(x,y)$

The elements of *PAR* may for instance be determined by the founding father of the theory and corresponds to one of the meanings Kuhn [1962] had in mind with the term 'paradigms' and which he later called 'exemplars'. Of course, the main source of vagueness is the notion of similarity, for it will as a rule not be possible to define this notion sharply, at least not at the beginning of the research process.

In both cases of determination, assuming that the theory has (at least absolute) empirical content, the empirical claim of a stratified theory will not be trivial. As is easy to verify, the empirical claim becomes trivial in the third way of determination of I: so-called *auto-determination*: for x in *EMPOSpart*, x belongs to I iff x belongs to πM, i.e., x is the projection of a model of a stratified theory. In the case of auto-determination the theory in question is typically not something to be tested, but it will have been designed for other purposes.

In case of an unstratified theory the set of intended applications is of course a subset of the set of empirical possibilities at the *Mp*-level, which is then on its turn a subset of *Mp*. For the further determination of I there are again the same three possibilities of empirical, paradigmatic and auto-determination. In the case that there are proper theoretical terms involved, all three forms of determination lead to problems.

Let us briefly restate and elaborate the background of (epistemological) stratification of a theory in T-theoretical and T-non-theoretical terms. Let there be an unstratified theory $UT = <Mp,M,D,I>$ and assume that UT has non-empty empirical content $Mp-M$. The term t occurring as component in Mp is said to be T-theoretical iff every known method of measuring of t in a specific intended application results in a model of UT. It is T-non-theoretical otherwise. Let UT contain at least one T-theoretical term and let us first assume that I is supposed to be empirically or paradigmatically determined, in which case the empirical claim "I is a subset of M" is non-trivial. However, testing this claim is impossible, for it leads either to circularity or to an infinite regress, as is not difficult to check. In the case of auto-determination the problem is that determination of the membership of M leads to circularity or infinite regress.

The remedy for these problems is the epistemological stratification of the theory in terms of a partial theory containing precisely all T-non-theoretical terms. Assuming that the stratified theory has non-empty (absolute) empirical content Mpp-πM, either the empirical claim or auto-determination is non-trivial, depending on whether I has or has not been fixed in advance, respectively.

The indicated definition of T-theoreticity is a pragmatic one, due to the "every known method of measuring"-clause and goes back to Sneed [1971]. Given the fact that the class of known methods can only increase, the definition is perfectly compatible with the advice to classify a term as T-theoretical in case of doubt. However, it is tempting to look for an intrinsic definition of theoreticity. Gähde [1983] has put forward an intrinsic definition. However, this proposal is not only highly technical and restricted to quantitative terms, it has also been criticised severely [Cf. Schurz, 1990].

Idealization and Concretization

Theories which are roughly about the same domain are frequently related. At each moment they may constitute a network of theories, i.e., a partially ordered set of theories that are directly or indirectly related. Such a network depicts the synchronic situation, the succession of networks indicates the diachronic development. The main relations studied in the structuralist approach are those of *specialization*, *theoretization* and *reduction*. For the purpose of this contribution, however, a fourth relation is more important: the relation of *concretization*, or its converse *idealization*.

In this section the relation of concretization will be defined and illustrated by the transition of the theory of ideal gases to that of Van der Waals. Some remarks will be made about concretization as a truth approximation strategy.

Concretization or factualization, as it has been presented by the Polish philosophers Wladislaw Krajewski [1977] and Leszek Nowak [1980], is basically a relation between real-valued functions. Hence, let us assume that the structures to be considered contain one or more base-sets and one or more real-valued functions defined on them, with or without one or more real constants. Structure y is called a *concretization* of x and x an *idealization* of y, indicated by $con(x,y)$, if y transforms, directly or by a limit procedure, into x when one or more constants or functions occurring in y uniformly assume the value 0. It is easy to see that it is a necessary condition for $con(x,y)$ that x and y have the same base-sets. Moreover it is easy to check that con is reflexive, antisymmetric and transitive.

The next task is to define the binary relation of concretization between theories. We will do this as weakly as possible: Y is a *concretization* of X and X an *idealization* of Y, indicated by $CON(X,Y)$, if and only if all members of X have a concretization in Y and all members of Y have an idealization in X. At first sight one might think that the second clause should be strengthened to: and all members of Y have a *unique* idealization in X. However, this would exclude e.g. 'inclusive' concretization triples $<X,Y,Z>$ with X as subset of Y and Y of Z and $CON(X,Y)$ and $CON(Y,Z)$.

It is trivial that CON is reflexive and transitive. However, it need not be antisymmetric, contrary to what one might expect. But sufficient for antisymmetry of $CON(X,Y)$ is that X and Y are convex. X is *convex* is defined as: for all x and z in X, if $con(x,y)$ and $con(y,z)$ then y is in X.

The transition from the theory of ideal gases to Van der Waals's theory of gases has frequently been presented as a paradigmatic case of concretization. So let us start by formulating the relevant models in naive structuralist terms, without the (theory-relative) distinction between theoretical and non-theoretical terms. $<S,n,P,V,T>$ belongs to the set of *potential gas models* (*GMp*) iff S represents a set of thermal states of n moles of a gas and P, V and T are real-valued functions defined on S representing pressure, volume and (empirical absolute) temperature, respectively.

Specific gas models are *GMp*'s satisfying an additional condition. *Ideal gas models* (IGM) satisfy in addition $P(s)V(s)=nRT(s)$ for all s in S, or simply $PV=nRT$, where R is the so-called ideal gas constant. For *Van der Waals gas models* (WGM) there are non-negative real constants a and b, within certain fixed intervals, such that $(P+(n^2a/V^2))\,(V-nb)=nRT$.

Note first that it is a necessary condition for $con(x,y)$ (x and y in *GMp*) that $S_x=S_y$. Note also that IGM and WGM are convex.

It is easy to check that WGM is a concretization of IGM (formally: $CON(\text{IGM},\text{WGM})$): each element of WGM transforms into an element of IGM by substituting the value 0 for a and b, and with each element of IGM there even can be associated several elements of WGM corresponding to arbitrary selections of values for the constants a and b within their respective intervals.

In Kuipers [1992a/1992b] I show among others that my refined definition of truthlikeness is such that "truth approximation by concretization" is perfectly possible. I illustrate this by indicating that WGM is closer to the truth than IGM, assuming the heuristic hypothesis that the true set of empirically possible gases is, in its turn, a concretization of WGM.

In these articles I also point out how concretization can also play a crucial role in not directly empirical scientific research directed at proving interesting theorems for certain sets of structures, as Hamminga [1983] showed for neo-classical economics. In another paper this type of concretization aiming at provable interesting truths is illustrated by the concretization, due to Kraus and Litzenberg, of the theory of Modigliani and Miller concerning the capital structure of firms [Cools, Hamminga and Kuipers, to appear].

Constraints

I have referred several times to so-called constraints. Whereas laws and axioms in the normal sense lay down restrictions to individual potential models, constraints impose restrictions to sets of potential models. A particular type of constraint is a so-called identity-constraint, guaranteeing that a function assigns in different potential models, with some common base-sets, the same value to the same individual. The weight-function in the case of the slide-balance as well as the mass-function in the case of classical particle mechanics are cases in point.

A constraint can be formally defined in a very general way.

> *Definition*: C is a *constraint* on the set S iff
> 1) C is a set of subsets of S
> 2) the union of the sets in C exhausts S ($\cup C = S$)
> C is a *transitive* constraint if it satisfies in addition:
> 3) if X is in C and Y is a subset of X then Y is in C
>   (subset-preservation, or transitivity)

Although there are interesting cases of non-transitive constraints the following remarks presuppose transitivity. To begin with, it is easy to prove that all singleton sets $\{x\}$, for x in S, belong to C, hence a (transitive) constraint does not exclude any individual potential model.

Let us now first concentrate on the typical role of a constraint C on Mp in a stratified theory $ST = \langle Mp, Mpp, M, C, \pi, D, I \rangle$. The standard empirical claim was "I is subset of πM", which could be paraphrased by saying that all members of I can be extended with theoretical components to genuine models, i.e., there is a subset X of M such that $\pi X = I$. Taking the constraint into consideration this claim is strengthened to: there is a subset X of M *belonging to* C such that $\pi X = I$. Hence, now both M and C restrict the degrees of freedom for the supplementation of theoretical components. In the corresponding versions of the strong claim the clause "X is a subset of M" is simply replaced by "X=M". It is clear that a stratified theory may even be a pure *constraint-theory*, in the sense that C is non-trivial and M is trivial, i.e., M=Mp.

The following reformulation of the standard claim with a constraint is instructive. Let A(ST), the *application space* of ST, be defined as the set of projections of all subsets of M satisfying C (formally: $A(ST) = \pi(P(M) \cap C)$). The standard claim comes now down to: I is in A(ST).

It is also plausible to define now the *(absolute) empirical content* AEC(ST) as P(Mpp)-A(ST), i.e., the subsets of Mpp which are excluded by M and C. Note that AEC(ST) reduces to $P(Mpp)-\pi(P(M))$ when C is trivial, i.e., when C=P(M), and to P(Mp)-P(M) when π is the identity function. It is easy to check that AEC(ST) is empty in these respective cases iff the originally defined empirical contents $Mpp-\pi M$, respectively, Mp-M are empty. Hence, the suggested new definitions of empirical content reproduce the original ones on the level of sets of sets of potential (partial) models.

Similar relations hold for the plausible definition of the *relative empirical content* REC(ST): P(Mpart)-A(ST). And again it follows almost trivially that non-empty REC(ST) implies non-empty AEC(ST), but not the converse.

Constraints make also sense in other cases, e.g. in partial theories and in unstratified theories. For the first case, let Cpart be a constraint on Mpp. Like Mpart, Cpart represents empirical restrictions. We may say that Mpart captures the standard empirical laws, whereas Cpart captures *constraint empirical laws*. Of course, the associated claim states that I is a subset of Mpart belonging to Cpart. It is important to note that many empirical laws are constraint laws, or mixtures of standard and constraint laws.

If the indicated partial theory with constraint is isolated from the full theory, it is clear that we have an unstratified theory with constraint. In general, an unstratified theory with constraint is of course of the form <Mp,M,C,D,I>, where C is a constraint on Mp.

Non-empirical Theories

Following Popper, non-empirical theories are by definition theories which are not (intended to be) falsifiable. One may distinguish at least four types of non-empirical theories:

metaphysical theories are supposed to make claims about reality without assuming any particular conceptualization or, equivalently, they make claims generalizing over conceivable conceptualizations of reality, *mathematical and logical theories* deal with defined abstract objects, i.e., mental constructs,

conceptual theories concern ways of looking (perspectives) to a certain domain,

normative theories deal with what is (supposed to be) ethically, juridically, esthetically (in)admissable.

It is evident that almost all technical ingredients presented for empirical theories are also useful for non-empirical theories. In fact, Suppes [1957] invented the structuralist representation of empirical theories by transferring, as far as possible, the standard way of presenting mathematical theories to empirical theories. The crucial difference is that non-empirical theories do not make general empirical claims. The claims which are associated or made by them typically are either conceptual (logical, mathematical etc.) or restricted to individual intended applications. A typical claim in a mathematical theory is a mathematical theorem to the effect that the models of the theory can be proven to have a certain explicitly defined property. A typical claim of a specific conceptual theory is that a certain intended application is (or is not) a model of that special theory. Of course, generic theories, i.e., theories with vacuous empirical claim, are conceptual theories.

The 'structuralist theory of the structures of empirical theories' is a perfect example of a theory which is primarily intended as a conceptual theory (although one may strengthen it to a genuine empirical theory). As a consequence, the foregoing exposition provides not only an elaborated example of a conceptual theory, it can also convince the reader of the usefulness of conceptual theories.

Department of Philosophy
University of Groningen
A-weg 30
9718 CW Groningen, Netherlands
T.A.F.Kuipers@Philos.rug.nl

REFERENCES

Balzer, W. [1982]. *Empirische Theorien: Modelle, Strukturen, Beispiele*. Braunschweig/ Wiesbaden: Vieweg.

Balzer, W., Moulines, C.U. & Sneed, J.D. [1987]. *An Architectonic for Science*. Dordrecht: Reidel.

Bunge, M. [1977]. The GST Challange to Classical Philosophies of Science. *Int.J. General Systems* **4**, 29−37.

Cools, K., Hamminga, B., Kuipers, T. [to appear]. Truth Approximation by Concretization in Capital Structure Theory. *Idealisation in Economics*. *Poznań Studies* (Ed. B. Hamminga).

Diederich, W. [1981]. *Strukturalistische Rekonstruktionen*. Braunschweig/Wiesbaden: Vieweg.

Gähde, U. [1983]. *T-Theoretizität und Holismus*. Frankfurt: Peter Lang.

Hamminga, B. [1983]. *Neoclassical Theory Structure and Theory Development*. Berlin: Springer.

Hettema, H. & Kuipers, T. [1988]. The Periodic Table: Its Formalization, Status, and Relation to Atomic Theory. *Erkenntnis* **28**, 387−408.

Janssen, M. & Kuipers, T. [1989]. Stratification of General Equilibrium Theory: A Synthesis of Reconstructions. *Erkenntnis* **30**, 183−205.

Krajewski, W. [1977]. *Correspondence Principle and Growth of Science*. Dordrecht: Reidel.

Kuhn, T. [1962]. *The Structure of Scientific Revolutions*. Princeton UP.

Kuipers, T. [1992a]. Naive and Refined Truth Approximation. *Synthese* **93.3**, 299−341.

Kuipers, T. [1992b]. Truth Approximation by Concretization. In: J. Brzeziński and L. Nowak (Eds.). *Idealization III: Approximation and Truth*. Poznan Studies. Vol. **25**, pp. 159−179.

Lakatos, I. [1978]. *The Methodology of Scientific Research Programs*. Cambridge UP.

Nowak, L. [1980]. *The Structure of Idealization*. Dordrecht: Reidel.

Popper, K. [1959]. *The Logic of Scientific Discovery*. London: Hutchinson.

Schurz, G. [1990]. Paradoxical Consequences of Balzer's and Gähde's Criteria of Theoriticity. *Erkenntnis* **32**, 161−214.

Sneed, J. [1971]. *The Logical Structure of Mathematical Physics*. Dordrecht: Reidel.

Stegmüller, W. [1973]. *Theorie und Erfahrung*. Band II, Teil D; [1986]. *Theorie und Erfahrung*. Band II, Teil H. Berlin: Springer.

Suppes, P. [1957]. *Introduction to Logic*. New York: Van Nostrand.

Zandvoort, H. [1982]. An Extension of Sneed's Reconstruction of Classical Particle Mechanics to Complex Applications, and an Alternative Approach to Special Force Laws. *Erkenntnis* **18**, 39−63.

*Poznań Studies in the Philosophy
of the Sciences and the Humanities
Vol. 42, pp. 25 – 48*

C. Ulises Moulines and Reinhold Straub

APPROXIMATION AND IDEALIZATION
FROM THE STRUCTURALIST POINT OF VIEW

Introduction

There are at least three things in philosophy of science we have learnt since
the heroic times of logical positivism, when scientific theories were considered simply as formal calculi related to brute empirical facts in a straightforward manner. We have learnt that this was an unbearably oversimplified
view of science because:

1) There are no such things as brute empirical facts a theory may be built
   upon or confronted with. We may still speak of the "empirical basis" of
   a given theory T but this expression has only a relative meaning. The so-
   called empirical basis of theory T is a collection of data already conceptualized from the point of view of an underlying theory T' or, more
   likely, from that of a loose assembly of theories T'_1, T'_2, etc. For example, the empirical facts for a dynamical theory are conceptual structures
   coming from kinematics, which in turn presuppose theories like physical
   geometry, chronometry, and mereology.
2) A scientific theory is a cultural product which, as any other cultural
   product (with the possible and doubtful exception of purely mathematical
   theories) essentially contains irreducibly pragmatic components, i.e.
   components that cannot be defined adequately in purely syntactic,
   semantic, or model-theoretic terms. Since there is presently no general
   formal theory of pragmatics applicable to all pragmatic notions, this
   implies that the philosophical analysis of science, both in general and in
   concrete cases, will have to make use of concepts that are not completely formalizable. The notion of explanation is a case in point and another
   example is the concept of application of a theory.

3) No scientific theory fits the "facts" it is supposed to systematize in a completely accurate way; almost no theory is related to the other theories it is supposed to be related to in an exact way; and, generally, no theory works unless a certain measure of "idealization" with respect to its "outer world" is admitted. These too are essential insights to take account of when trying to grasp the nature of scientific theories. That is, any empirical theory can be effectively applied or related to other theories only by allowing for some degree of idealization or approximation in the use of the theory and in the theoretical reconstruction of the data. The degree of inaccuracy with which a theory works may be expressed numerically or qualitatively, according to the type of theory and application. In so-called "mathematized" empirical theories, that is, in those theories that systematically use metrical concepts, there will be a tendency to express their degree of inaccuracy partially, though *not completely*, in numerical terms, whereas in so-called "qualitative" theories, that is, in those theories that only use classificatory or comparative concepts, the degree of inaccuracy assumed will be given, explicitly or more frequently implicitly, by topological, non-quantitative comparisons. Only in qualitative theories of a very simple kind it may be possible to avoid the (implicit or explicit) use of approximations, but only at the expense of producing quite trivial claims — claims that have almost no empirical content and lack technological applicability. In qualitative theories with a minimal amount of sophistication as well as in the more fully mathematized theories of the natural and social sciences, the use of idealizations and approximations will be absolutely essential.

Let us call these three results of the metascientific reflection of the last decades the principles of "theory-ladenness", of "praxis-ladenness", and of "approximation-ladenness" of science, respectively. The first two are, it seems, widely accepted and taken into account in present-day philosohy of science. The third one is not so popular, though it has increasingly become a matter of study from different perspectives in the last years. The approximation-ladenness of all of empirical science seems to us even more obvious than its theory- and praxis-ladenness and we suppose there is no need to make propaganda for it in the present text. Furthermore, the three principles mentioned are tied together in a deep, though not entirely obvious way.

In this paper we will concentrate on the approximation-ladenness of science. However, we hope that our analysis will also make clear several ways in which this aspect is intimately connected to theory- and praxis-ladenness. To make the point very briefly, our rough thesis is that, on the one hand, empirically meaningful approximation and idealization are only possible because of theory-ladenness and, on the other hand, the concepts of approximation and idealization we need for empirical theories are essentially constituted by some irreducibly pragmatic components in addition to the semantic ones. We hope these somewhat cryptic remarks will become clear in the course of the exposition.

I

Some Methodological Distinctions

The first distinction we would like to make is the well-known one between *idealization* and *approximation*. Different approaches in present-day philosophy of science have proposed different explications for these notions and for their mutual relationship. From the point of view of the structuralist metatheory we advocate, we would like to propose the following explication. Idealization is the more general, approximation the more particular notion. In a sense, every approximation is (or rather: rests on) a specific kind of idealization, though the converse is not true: Not every kind of idealization is (or rather: may be translated into) a form of approximation.

Idealization, under its most general interpretation, may just be characterized as any attempt to relate a given theory, as a conceptual unit, to something outside itself − be it the "empirical reality", or the "data", or perhaps also other theories as different conceptual units. To make such a relationship fruitful, one must "idealize" − that is, in general terms, one must bring the theory's *ideal conceptual structures* in touch with this "something" outside itself, which is always *alien* to the theory. This does not work adequately unless one is prepared to do some *force* on one of the two terms of the relationship (or on both at the same time). This force is "idealization".

These metaphors may be given a more controllable meaning by using technical notions of the structuralist approach. We remind that one of the essential components of a theory's identity according to structuralism is the class of its potential models, M_p. Roughly speaking, M_p represents a theory's *conceptual framework*. Now, idealization in general has to do with the use of the elements of M_p to say something about things outside the theory. If we idealize a physical system so as to be amenable to treatment by theory T, what we are doing is to conceptualize that system as a potential model of T. For example, when we say that, in order to apply Newton's gravitational theory to the planetary system, we have to idealize the latter as a set of particles moving on smooth paths, what we are doing is to "convert" or "reconstruct" that system into a potential model of Newton's theory. Whether it will *also* be an *actual* model, that is, a system satisfying Newton's law of gravitation, is a further question. To answer it, we generally need not only idealization in its general sense, but also its specific kind called "approximation". We may be successful in idealizing but fail in approximating.

The same goes for idealization viewed as a relationship between two different theories. When we say that we may get Kepler's theory out of Newtonian mechanics by a process of idealization, what we mean, in structuralistic terms, is that the potential models of Kepler's theory are being linked (in a certain formalizable way) to their counterparts − the potential models of Newtonian mechanics. Again, this by itself does still *not* guarantee that we may derive Kepler's theory from Newton's laws. In order to guarantee this we need something more specific − approximation.

These considerations about the distinction and relationship between idealization and approximation lead naturally to the second conceptual distinction which is relevant to our topic: a distinction between *intra-* and *intertheoretical* approximation.

The second step we have to make after we have "idealized" a physical system (or the "data") into a potential model of T is intratheoretical approximation − an internal business of T. The second step we have to make after we have "idealized" a theory T into another one T' is intertheoretical approximation − a relationship between two conceptually distinct theories. Intratheoretical approximation is intimately related to the notion of *application* of a theory and of its *empirical claim*. Intertheoretical approximation has more to do with the global structure of science and with such vexing issues as incommensurability and reduction of theories.

Under the general setting of idealization and approximation suggested here, many more interesting things could be said about idealization in general, especially about idealization which *does not* involve approximation, as well as about intertheoretical approximation. Lack of space prevent us from doing this. We will concentrate our investigation on intratheoretical approximation as a particular form of idealization. The results we present here are a continuation and revision of work already done on approximation within the structuralist program[1]. Some of the results, however, are com-

---

1 See, especially, Moulines [1976, 1980] and the most up-to-date systematic account of structuralism, *An Architectonic for Science*, by W. Balzer, C. U. Moulines, and J. D. Sneed; since we will frequently refer to this work, we will abbreviate it as *Architectonic*.

pletely new and imply a substantial shift of perspective, especially as far as the idea of the approximative empirical claim of a theory is concerned.

II

The Structuralist Apparatus for Approximation

In the following, we assume the reader is well-acquainted with the essentials of the structuralist apparatus to represent theories in general. We just remind that a theory, according to structuralism, is to be represented as an ordered net of structures called "theory-elements". Each theory-element consists of a core K and a domain of applications I, and the "(central) empirical claim" of each theory-element is just that K can be effectively applied to I; in set-theoretical terms, the standard symbolic notation for such a claim is "$I \in A(K)$", where A is a set-theoretical operator which can be defined precisely. Each core K is a complex entity constituted, in turn, by a number of different components: the class M_p of potential models, the class M_{pp} of partial potential models (the relative "non-theoretical" structures), the class M of actual models, the class C of "constraints" (inter-model connections) and the class L of "links" (inter-theory connections). M_p and M_{pp} are related by a many-one function r ("restriction"), which correlates potential models with their corresponding non-theoretical structures in M_{pp}.[2]

In order to simplify the exposition, we will forget about the components C and L. Nothing essential is lost by this simplification when dealing with approximation. Under this simplification, the empirical claim of a theory-element becomes more simple and it may be symbolized by the formula:

$I \in r(M)$.

Now, the metatheoretical problem we have to face is that, in any "really existing" empirical theory, the proposition expressed by this formula will normally be false (unless it is trivially true). The reason is, of course, that no interesting empirical theory applies exactly to the whole of its domain of intended applications. There will always be some "noise". But, though false, the empirical claim will be, in more or less "well-functioning" theories, "approximately true" — or, at least, this is what we would like

---

2 For more details on all this, see *Architectonic*, especially the first chapters.

intuitively to be able to say. That is, intuitively we may want to say that, although "$I \in r(M)$" is false, "$I \widetilde{\in} r(M)$" is true. Then, our metatheoretical problem is to look for a satisfactory explication of the dash " $\sim$ " in the latter formula.

In Chapter VII of *Architectonic* as well as in Moulines [1976, 1980] such an explication was offered. We briefly summarize the main ideas.

The most basic intuition we wish to explicate is essentially a metatheoretical claim of the sort: "potential (or partial) model x is an approximation of potential (or partial) model y in theory-element T". How can we render this idea precise?

Potential models of a theory-element T are all elements of the same class M_p of T. The particular form the elements of this class may take does not matter on the general level. They need not be structures of metrical concepts, nor do we need to suppose that the comparison rests on some special measurement methods. All we need is a well-defined class M_p (that is a class of structures determined by the same structural axioms) and an approximation relation between any two elements of that class. There is a well-known method for defining such a relation in topology, namely to introduce the concept of a *uniform structure*, or "*uniformity*" for short. Thus, let us explicate model-theoretic approximation by defining a uniformity on M_p.

For the formal definition of a uniformity, we need the following special notation: $Po(X)$ is the power-set of X; if M is a class, then $\Delta(M)$ is the "diagonal" of M, i.e., the class of all pairs of identical elements of M; if u is a set of ordered pairs then u^{-1} denotes the set of ordered pairs of u in which the order of the components is reversed, $u^{-1} = \{(y,x)/(x,y) \in u\}$; finally, if u_1 and u_2 are sets of ordered pairs then $u_1 \circ u_2$ denotes the usual product of relations:

$$u_1 \circ u_2 = \{(x,y) / \exists z: (x,z) \in u_1 \ \& \ (z,y) \in u_2\}.$$

The following axioms correspond essentially to those given by Bourbaki (1961), though applied to models as "points".

Definition 1: If M_p is a class of potential models, then U is a *uniformity on M_p* iff
  1) $\emptyset \neq U \subseteq Po(M_p \times M_p)$
  2) $\forall u_1, u_2: u_1 \in U \ \& \ u_1 \subseteq u_2 \rightarrow u_2 \in U$
  3) $\forall u_1, u_2: u_1 \in U \ \& \ u_2 \in U \rightarrow u_1 \cap u_2 \in U$
  4) $\forall u \in U: \Delta(M_p) \subseteq u$
  5) $\forall u \in U: u^{-1} \in U$

6) $\forall u_1 \in U \; \exists \; u_2 \in U \colon u_2 \circ u_2 \subseteq u_1$

The Problems with the Original Explication

The reader may find a detailed exposition of the rationale of the previous conditions as interpreted in the framework of empirical theories in *Architectonic*, Ch. VII. A model-theoretic uniformity U is the base to deal with approximation according to structuralism. However, this notion is not sufficient by itself to deal with the specific problems of *empirical* theories. Uniformities in this general sense have to be endowed with more restrictive conditions to deal especially with the following issues, which according to structuralism are quite basic when analyzing empirical theories and empirical approximation:

1. The relationship between T-theoretical and T-non-theoretical concepts: when considering approximation, we would like to have approximation at the T-theoretical level (M_p) closely related to approximation at the T-non-theoretical level (M_{pp}). In the structuralist literature, this problem has sometimes been called the "induction problem" (from T-theoretical to T-non-theoretical approximation) — though, of course, it has nothing to do with the classical "induction problem" of epistemology.
2. "Really admissible" approximation: not all "blurs" $u \in U$ will be acceptable to express genuine empirical approximation (otherwise we would trivialize the idea itself). Some further constrictions should be put on a subset A of admissible blurs in U.
3. The precise approximate version of the empirical claim: different alternatives to apply the dash " $\sim$ " in " $I \in r(M)$ " are conceivable, and we should choose the most plausible one.

In *Architectonic*, the first problem was solved by adding a further axiom to the above-stated axioms of a uniformity. This is the so-called condition of the "pseudo-diagonal". Define the pseudo-diagonal as $\psi(M_p) =: \{(x,x')/ x,x' \in M_p \& r(x) = r(x')\}$. Then the additional axiom says that: for all $u \in U$, $\psi(M) \subseteq u$. With this additional axiom, the general notion of a uniformity is restricted to a so-called "empirical uniformity". Under this condition, it may be shown that a uniformity on M_p induces a corresponding uniformity on M_{pp} (see *Architectonic*, pp. 339 and ff.).

The second issue was attacked by adding some necessary (though not sufficient) conditions of admissibility for blurs, in particular the existence of upper bounds (see *Architectonic*, VII.2.2).

Finally, the approximate empirical claim was explicated by defining "$\sim$" in terms of the existence of an admissible blur to which the two models belong and applying this dash to *sets* of models *twice* in the form of the exact empirical claim, $I \in r(M)$ (see *Architectonic*, VII.2.3).

Now, we maintain that the core ideas behind these answers to the three problems of approximation posed are, in principle, on the right track. However, their concrete form is still defective in certain important formal and epistemological respects, and should be revised. The main problems are essentially these: The pseudo-diagonal condition is just too "coarse"; the conditions of admissibility should be rendered more plausible from the empirical point of view by considering the effect of *local relativization* when settling the upper bounds; and the most plausible rendering of the approximate empirical claim is, both for technical and epistemological reasons, *not* the one proposed in *Architectonic*.

In the third part of this article we introduce the required modifications to solve the problems mentioned. The proposed revisions are not only interesting as an improvement of the original structuralist explication of approximation, but also provide, on their own, some deeper insights in the nature of approximation in general and are, we believe, of mathematical interest by themselves.

III

Blurs on Two Levels

If U is a uniform structure on M_p, $r(U)$ need not be a uniformity on M_{pp}. So the introduction of uniform structures as the general concept to describe approximate relations in theory-elements meets with a technical difficulty. The difficulty concerns the critical axiom for uniform structures that every blur contains the product of a blur with itself (*Def. 1−6*, above). The easiest way out of this difficulty is the simple requirement that it should not occur: that only such uniformities may be defined on M_p which induce a uniform structure on M_{pp}. Then the mathematical problem is left how to characterize such uniformities. This task will be partly fulfilled in the sequel by giving a fairly general sufficient condition. But additionally, we claim that this condition (if it is general enough to cover all of its intended applications) has an epistemological significance, too. We think that it is in both respects, the technical and the philosophical one, superior to the concept of an "empirical uniformity", as defined in *Architectonic*. But first we turn to the discussion of this older concept.

Pseudo-Diagonals

The requirement that every blur of the uniform structure on M_p must contain the pseudo-diagonal $\psi(M_p)$ was introduced for purely technical reasons in the first place. Some philosophical justification was tried, but the concept couldn't be applied in real case studies. One of several principal deficiencies of this concept is the trivialization of constraints. Take for instance the identity constraint for the mass function in classical particle mechanics *(CPM)*. Let X be an arbitrary subset of $M_p(CPM)$. If the blurs on M_p contain the pseudo-diagonal, then X satisfies the constraint almost exactly: for every $x \in X$ replace the mass function m_x by the constant function $m^*:P_x \to \mathbb{R}^+$ with $m^*(p)=1$ for every $p \in P_x$. Then a set X' of potential models is obtained which satisfies the constraint and for every blur $u, X \sim X'$ wrt u holds. (Remember $X \sim X': \Leftrightarrow \forall x \in X \exists x' \in X'(x \sim x') \& \forall x' \in X' \exists x \in X(x \sim x')$.)

Now we prove that such uniformities are uniquely determined by the uniformity they induce on M_{pp}. This means, that the approximation of the theoretical components is determined by the approximation of the nontheoretical components. To see this, consider the following construction: for a relation v in M_{pp} define $r^{-1}(v) := \{(x,y)/(r(x),r(y)) \in v\}$ and for a uniform structure V on M_{pp} define $r^{-1}(V) := \{u \subseteq M_p \times M_p / \exists v \in V: r^{-1}(v) \subseteq u\}$. This is a uniform structure on M_p; it can be shown that it is the coarsest uniform structure which induces V. It is easy to see that its blurs contain the pseudo-diagonal. It is called the "initial uniformity wrt r and V".

If a uniform structure U on M_p is "empirical" (in the sense of *Architectonic*, viz. every blur $u \in U$ contains the pseudo-diagonal), then $r(U)$ is a uniform structure on M_{pp}. The system $r^{-1}(r(U))$ is the initial uniformity wrt r and $r(U)$. The following proposition shows that $U=r^{-1}(r(U))$. The initial uniform structures on M_p wrt r and a uniform structure on M_{pp} are exactly the "empirical uniform structures" in the sense of *Architectonic*.

Proposition 1: The blurs of a uniform structure U on M_p contain $\psi(M_p)$ iff $r(U)$ is a uniform structure and U is the initial structure wrt r and $r(U)$.

Proof:
1. If $r(U)$ is a uniform structure and U is the initial structure wrt r and $r(U)$, then $U=r^{-1}(r(U))$. Since the blurs of the latter contain $\psi(M_p)$, U is an "empirical" uniformity.
2. If U is an "empirical" uniformity, then $r(U)$ is a uniform structure. We have to prove that $U=r^{-1}(r(U))$.

(a)　That U contains $r^{-1}(r(U))$ is clear from the fact that the initial structure is the coarsest uniformity which induces $r(U)$.

(b)　$r^{-1}(r(U))$ contains U: For $u \in U$ a $v \in U$ exists with $v^3 \subseteq u$. We claim that $r^{-1}(r(v)) \subseteq u$. (Proof: $(x,y) \in r^{-1}(r(v)) \Rightarrow (r(x),r(y)) \in r(v) \Rightarrow \exists x',y'(r(x') = r(x) \ \& \ r(y') = r(y) \ \& \ (x',y') \in v)$. Since v contains $\psi(M_p)$, $(x,x') \in v$ and $(y',y) \in v$. Hence $(x,y) \in v^3 \Rightarrow (x,y) \in u$.) From $r^{-1}(r(v)) \in r^{-1}(r(U))$ and $r^{-1}(r(v)) \subseteq u$ it follows $u \in r^{-1}(r(U))$.

So every blur of an "empirical" uniformity does not only contain the pseudo-diagonal $\psi(M_p) = r^{-1}(\Delta(M_{pp}))$, but the preimage $r^{-1}(v)$ of a blur from $r(U)$. This result rests essentially on the critical axiom.

The requirement that every blur should contain the pseudo-diagonal is by far too strong. It should be possible to select the coarsity of the approximation of theoretical components without changing the uniform structure for the non-theoretical components.

Homogeneous Relations and Bases

In order to find a more general condition for uniformities which implies induction from the theoretical to the non-theoretical level, consider a relation $v \subseteq M_p \times M_p$ and $(x,y) \in v$. Now change the theoretical components of x such that a potential model x' arises, viz. $x' \in M_p$ and $r(x') = r(x)$. In general, there is no potential model y' with $r(y') = r(y)$ and $(x',y') \in v$. The relation in M_p need not be homogeneous with respect to replacements of theoretical components. To settle this notion, we introduce the following definition.

Definition 2: A relation $v \subseteq M_p \times M_p$ is called *homogeneous*, iff for every $(x,y) \in v$ and $x' \in M_p$ with $r(x') = r(x)$ there exists $y' \in M_p$ with $r(y') = r(y)$ and $(x',y') \in v$.

Remark: A relation v in M_p is homogeneous iff one of the two following equivalent conditions is satisfied:

(a)　$\forall \ x,x' \in M_p(r(x') = r(x) \to r[v(x')] = r[v(x)])$

(b)　$\forall \ x \in M_p(r(v)(r(x)) = r[v(x)])$.

The following diagram depicts the difference between homogeneous and non-homogeneous relations.

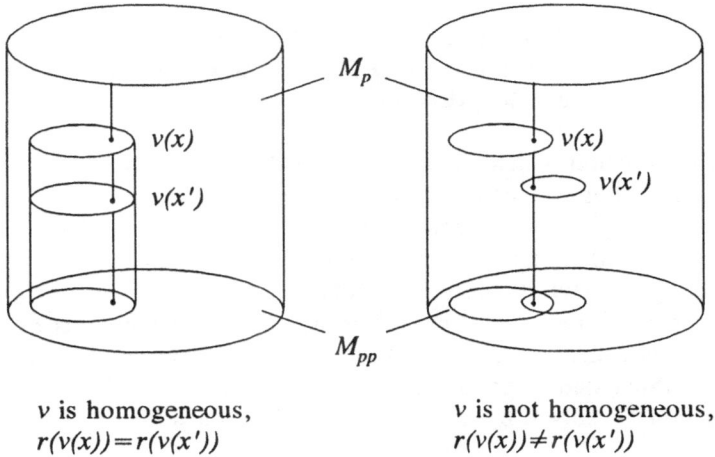

v is homogeneous, v is not homogeneous,
$r(v(x))=r(v(x'))$ $r(v(x))\neq r(v(x'))$

Fig. 1

Since the concept of a homogeneous relation is not defined only for symmetric relations, there is a distinct concept of "co-homogeneous" relations (for every $(x,y)\in v$ and $y'\in M_p$ with $r(y')=r(y)$ there exists $x'\in M_p$ with $r(x')=r(x)$ and $(x',y')\in v$). This concept is as suitable as the defined one. Obviously, v is co-homogeneous iff v^{-1} is homogeneous. But it does not hold that v is homogeneous iff v^{-1} is homogeneous.

The intersection of homogeneous relations need not be homogeneous, but the union of any set of homogeneous relations is homogeneous. Thus for a reflexive relation v there exists the homogeneous kernel, that is the union of all homogeneous relations contained in v. Finally, the product of two homogeneous relations is homogeneous. Thus far for the structural behaviour of homogeneous relations.

Definition 3: A uniform structure U is called *homogeneous*, iff it has a base of homogeneous blurs, viz. every blur in U contains a homogeneous blur.

This concept generalizes the concept of an empirical uniformity where every blur satisfies the pseudo-diagonal condition:

Proposition 2: If U is a uniform structure on M_{pp}, then the initial uniformity $r^{-1}(U)$ is homogeneous.

Proof:

Take $u \in r^{-1}(U)$. There is a $v \in U$ with $r^{-1}(v) \subseteq u$. Then $r^{-1}(v) \in r^{-1}(U)$ and $r^{-1}(v)$ is homogeneous: if $(x,y) \in r^{-1}(v)$ and $r(x')=r(x)$, then $(r(x),r(y)) \in v$ $\Rightarrow (r(x'),r(y)) \in v \Rightarrow (x',y) \in r^{-1}(v)$.

Remark: A relation which contains the pseudo-diagonal need not be homogeneous; a homogeneous relation need not contain the pseudo-diagonal. That an empirical uniformity is homogeneous follows from the first proposition which makes essential use of the critical axiom. Conversely, a homogeneous uniformity need not be an empirical one: for instance, the canonical uniformity on $\mathbb{R}^n$ is homogeneous with respect to any restriction function $r:\mathbb{R}^n \rightarrow \mathbb{R}^m$ $(m < n)$, but there are of course some blurs which don't contain the pseudo-diagonal. But the "problem of induction" is solved by homogeneous uniformities, too.

Proposition 3: If U is a homogeneous uniformity on M_p, then $r(U)$ is a uniform structure on M_{pp}.

Proof:

1. We start by proving a more general result: If $\forall u \in U \; \exists \; v \in U(v \subseteq u$ & $(r(v))^2 \subseteq r(v^2))$, then $r(U)$ is a uniform structure.
Proof:
$u' \in r(U) \Rightarrow \exists u \in U(r(u)=u') \Rightarrow \exists v \in U((r(v))^2 \subseteq r(v^2)$ & $v^2 \subseteq u) \Rightarrow r(v) \in$ $r(U)$ and $r(v^2) \subseteq r(u)$ and $(r(v))^2 \subseteq r(v^2) \subseteq r(u) = u'$.

2. If U is a homogeneous uniformity on M_p, then the general result of step 1 above is applicable: If v is homogeneous, then $(r(v))^2 \subseteq r(v^2)$.
Proof:
$(x,z) \in (r(v))^2 \Rightarrow \exists y((x,y) \in r(v)$ & $(y,z) \in r(v)) \Rightarrow \exists x', y', y'', z''((x',y') \in$ v & $(y'',z'') \in v$ & $r(x')=x$ & $r(y')=y=r(y'')$ & $r(z'')=z)$. Since v is homogeneous, there exists a z' with $r(z')=r(z'')$ and $(y',z') \in v \Rightarrow$ $(x',z') \in v^2 \Rightarrow (x,z)=(r(x'),r(z')) \in (r(v))^2$.

3. If U is homogeneous, then by step 2 above it has a base of the sort required by step 1, hence $r(U)$ is a uniform structure.

Remark: A uniform structure on M_p, which induces a uniform structure on M_{pp} need not be homogeneous, as can be shown by constructing a suitable counter-example. This fact could emerge as unfavourable, namely if the concept is too special to cover all of its intended applications. But otherwise its speciality would be advantageous. In the sequel, we give an interpretation of the property of being homogeneous and we show how homogeneous uniformities can be used as a criterion for induction in the case of non-homogeneous uniformities. In the proof we used a more general condition for induction: this condition is much more difficult to handle and it doesn't seem to have any epistemological interpretation.

That the canonical uniformity on $\mathbb{R}^n$ is homogeneous already indicates that this condition seems to be a natural one. The informal argument we raised against the pseudo-diagonal condition (that it trivializes the constraints) is blocked obviously in the case of homogeneous uniformities. We cannot obtain a similar subset of M_p simply by replacing the theoretical components in the potential models of the subset. By replacing theoretical components we do not in general obtain a similar model.

The Significance of Homogeneous Uniformities

A uniform structure on M_p which has a base of homogeneous blurs induces a uniform structure on M_{pp}, as we have seen previously. We ask now whether the existence of a base of homogeneous blurs has a philosophical significance or is merely of technical interest as a criterion for induction. Since the canonical uniformity on $\mathbb{R}^n$ has such a base, it does not seem to be an "exotic" condition at least. But what does it mean in the case of theory-elements?

Homogeneous uniformities are "homogeneous" in the following sense: consider two subsets X and X' of M_p with $r[X]=r[X']=M_{pp}$. These subsets represent two systems of theoretical extensions of M_{pp}. If we look at the uniformity on M_p in an entourage of X, we would expect it to be neither finer nor coarser than in any entourage of X'. To be precise: For $X \subseteq M_p$ and $w \in U$ we define the entourage $w[X]$ of X: $w[X] := \{y \in M_p \ / \ \exists \ x \in X((x,y) \in w)\}$; uniform spaces $(w[X],V)$ and $(w[X'],V')$ have the same coarsity wrt r if $r(V)=r(V')$.

Proposition 4: If U is a uniformity on M_p with a base of homogeneous blurs and (M_p',U') is a uniform subspace of (M_p,U), and if there is a $w \in U$ and $X \subseteq M_p$, such that $r[X]=M_{pp}$ and $w[X] \subseteq M_p'$, then $r(U')=r(U)$.

Proof:
1. $u \in r(U') \Rightarrow \exists \ \bar{u} \in U': r(\bar{u})=u \Rightarrow \exists \ v \in U: \bar{u}=v \cap (M_p' \times M_p') \Rightarrow v' :=$ $v \cap w \in U \Rightarrow \exists \ v^* \in U: v^* \subseteq v'$ and v^* is homogeneous. We prove that

$r(v^*) \subseteq r(\bar{u})$. (From this we infer $r(\bar{u}) = u \in r(U)$ because $r(v^*) \in r(U)$.) Let $(y,y') \in r(v^*)$. There is some $x \in X$ with $r(x) = y$. v^* is homogeneous, therefore there is some x' with $r(x') = y'$ and $(x,x') \in v^* \Rightarrow (x,x') \in v \cap w \Rightarrow x' \in w(x) \Rightarrow x' \in M'_p \Rightarrow (x,x') \in v \cap (M'_p \times M'_p) \Leftrightarrow (x,x') \in \bar{u} \Rightarrow (r(x), r(x')) = (y,y') \in r(\bar{u})$.

2. $u \in r(U) \Rightarrow \exists\, \bar{u} \in U: r(\bar{u}) = u \Rightarrow v := \bar{u} \cap (M'_p \times M'_p) \in U' \Rightarrow r(v) \in r(U')$ and $r(v) \subseteq r(\bar{u}) \Rightarrow r(\bar{u}) \in r(U')$.

Remark: The proposition can be slightly generalized: If (M'_p, U') is a uniform subspace of (M_p, U), and if there is a base B for U and $X \subseteq M_p$, such that $r[X] = M_{pp}$ and $\forall\, x \in X \; \exists\, w \in B(r[w(x) \cap M'_p] = r[w(x)])$, then $r(U') = r(U)$.

Any subspace M'_p of M_p which satisfies the condition mentioned in the proposition (or in the remark) induces the same uniformity. In this sense the uniform structure on M_p can be called homogeneous with respect to different theoretical extensions of M_{pp}.

The proposition may be of technical use, too. If we find a uniform space (M''_p, U'') with a base of homogeneous blurs, $M_p \subseteq M''_p$ and $r'': M''_p \to M_{pp}$ with $r''(x) = r(x)$ for all $x \in M_p$ such that (M_p, U) is a subspace which satisfies the condition mentioned in the proposition, then $r(U)$ is a uniform structure on M_{pp}. (Note that the concept of a homogeneous relation and the concept of a homogeneous uniformity can be defined for any uniform space (X, W) and surjective function $r: X \to Y$.) So we have found a criterion for induction. The uniform structure U need not have a base of homogeneous blurs: it only induces the same structure on M_{pp} than U'', viz. a uniform structure.

Several Restriction Functions

So far, we have only considered a single restriction function, the one which cuts off the theoretical components of the potential models. A uniform structure U may induce a uniform structure on the image of r, the canonical restriction function, but not on the image of some function r', which cuts off other components. Likewise U may have a base of homogeneous blurs with respect to r, but not with respect to r'.

In some sense we presuppose the distinction between theoretical and non-theoretical terms in the theory-element if we require induction only for the canonical restriction r. We cannot use the terms approximately until this distinction is established, since we do not know how a suitable uniformity looks like before we haven't established that the required induction property is actually fulfilled. Now, this seems to be a somewhat disadvantageous feature of our approximation approach within structuralism. Indeed, we would like to have an approximation concept which is as flexibel as possible − in particular, it should be more general than a quite particular way of "sorting out" the theoretical from the non-theoretical terms.

On the other hand, an approximation structure is independent of theoreticity if it induces a uniform structure on the image of any possible restriction function. Independence of theoreticity in this sense is a requirement which seems to be philosophically appealing though it is technically quite unpleasant. We mention three arguments in favour of this requirement.

First, the idealized and the approximate meanings of terms should not be separated. The extensions of empirical terms can only be determined approximately, and so the structure of their approximate application is clearly a part of their meaning. If a once T-theoretical term is becoming a T-non-theoretical one due to the emergence of a new theory-element (like mass in classical collision mechanics due to the emergence of CPM − cf. *Architectonic*, Ch.II.3.4), we would not expect a change in the uniform structure of $M_p(T)$, at most a change in the boundaries of admissible inaccuracy.

Second, if we didn't require independence, we would obtain a new necessary criterion for theoreticity: only such possible restriction functions which induce a uniformity are admissible. This reduces to triviality if the uniform structure has the induction property for every possible restriction function. This would avoid an unfounded strengthening of the criteria for theoreticity.

Third, the approximation structure has a bearing on the determination of the theoretical terms of T: a term is T-theoretical (according to the so-called "informal" structuralist criterion of theoreticity − cf. *Architectonic*, Ch.II), if any method of determination presupposes a model of T. If there are several methods of determination, some of them precise ones which presuppose a model of T, the rest inaccurate ones which are independent of any model of T, then we would nevertheless regard the term as T-theoretical.

At any rate, what we want to investigate here is the extent to which approximation may be compatible with different choices of the restriction function. But our results may be useful for the analysis of a single restriction function too.

Normally, there are too many possible restriction functions and we cannot examine each one of them. So we need some general results to simplify the task. For a given relation $v \subseteq M_p \times M_p$, we ask for which possible restriction functions v is homogeneous.

If it is not possible to replace the components to be cut off at all, then any $v \subseteq M_p \times M_p$ is homogeneous wrt every possible restriction function r. This is a trivial consequence of the definition of a homogeneous relation. For instance, if a potential model contains a base set B and a real-valued function on B, then it is impossible to replace B by a set B' distinct from B: this would violate the characterization of the function defined on B. The structure obtained by this replacement is no potential model. Hence, v is homogeneous wrt cutting off the base set.

In some cases where replacement of components is possible within M_p, the following proposition can be applied. We begin with some terminological clarifications. We assume that the potential models are tuples of length n (without auxiliary base sets). For $J \subseteq \{1,...,n\}$ let r_J be the restriction function which cuts off the members of $(x_1,...,x_n) \in M_p$ with indices in J. For r_J define its pseudo-diagonal $\Psi_J := \{(x,x')/r_J(x)=r_J(x')\}$.

It is easy to see that for all $J,J' \subseteq \{1,...,n\}: \Psi_J \circ \Psi_{J'} \subseteq \Psi_{J \cup J'}$. Unfortunately, the converse is not true: For instance, in the case described previously, where it is not possible to replace only the base set without violating the characterization of a function defined on the base set, it may be possible to replace the base set and the function simultaneously. Then the pseudo-diagonal which correponds to the cutting off operation of both components is irreducible to the pseudo-diagonals corresponding to the cutting off operations wrt to each component. But in the case where we can replace a set of components in two (or more) steps, each step leading to a potential model, we can prove that a relation is homogeneous step by step, too.

Proposition 5: Let $J,J' \subseteq \{1,...,n\}$ and $u \subseteq M_p \times M_p$. If $\Psi_J \circ \Psi_{J'} = \Psi_{J \cup J'}$ and u is homogeneous wrt r_J and wrt $r_{J'}$, then u is homogeneous wrt $r_{J \cup J'}$.

Proof:

Let $(x,y) \in u$ and $x'' \in M_p$ with $r_{J \cup J'}(x)=r_{J \cup J'}(x'')$. Hence $(x,x'') \in \Psi_{J \cup J'}$ $\Rightarrow \exists\, x' \in M_p: (x,x') \in \Psi_J$ & $(x',x'') \in \Psi_{J'}$, hence $r_J(x)=r_J(x')$ and $r_{J'}(x')=r_{J'}(x'')$. Since u is homogeneous wrt r_J there exists $y \in M_p$ with $r_J(y')=r_J(y)$ and $(x',y') \in u$. From $r_{J'}(x')=r_{J'}(x'')$ it follows that there is some $y'' \in M_p$ with $r_{J'}(y'')=r_{J'}(y')$ and $(x'',y'') \in u$, because u is homogeneous wrt $r_{J'}$. From $r_J(y'')=r_J(y')$ and $r_J(y')=r_J(y)$ it follows $r_{J \cup J'}(y)=r_{J \cup J'}(y'')$.

This result, together with the trivial remark formulated earlier, should considerably reduce the number of restriction functions to be examined in order to prove that a given relation is homogeneous wrt every possible restriction function. If there exists for a given uniformity a base of such relations, then every possible restriction function induces a uniformity on its image. The proposition may be also useful for the analysis of a single restriction function, namely if it cuts off more than one component and if it can be factorized by restriction functions which cut off less components. For instance, in *CPM* the replacement of both the mass function and the force function of a potential model may be done by replacing first the mass function and then the force function. So the proof that a given relation is homogeneous is simplified. Of course, all these results provide only sufficient, not necessary criteria.

IV

Locally Different Admissible Inaccuracies

The main problem in dealing with local differences of admissible inaccuracy is the connection of local features with global ones. In *Architectonic*, local differences are described by a set A of blurs, which has to be bounded from above in U; its members represent degrees of inaccuracy which are admissible somewhere in M_p. The globally admissible inaccuracy is defined by the reflexive and symmetric relation $x \sim y$ iff $\exists u \in A((x,y) \in u)$. So this relation is just the union of all admissible blurs.

But then, only a multitude of global relations is obtained, which does not amount to a definition of the locally admissible inaccuracy. For this purpose, the blurs would have to be fixed to their domains of applicability. It has to be said, which blur of A is admissible for a given $x \in M_p$. How can we fix a blur to a potential model?

The simplest idea is the introduction of a function $f: M_p \to U$ which is bounded by A. The value $f(x)$ is to be interpreted as the (maximal) blur which is admissible for $x \in M_p$. Given such a function $f: M_p \to U$ with values in A, define the boundary of admissible inaccuracy by $x \sim y$ iff $\exists z (x \in (f(z))(z)$ & $y \in (f(z))(z))$. (Note that $f(z) \subseteq M_p \times M_p$ and $(f(z))(z) = \{y/(z,y) \in f(z)\}$ by definition.) Instead of $f(x)$ we write u_x; let's call models x, y with $x \sim y$ "similar models". Obviously, this relation is reflexive and symmetric, and it is defined by the set of localized admissible blurs. If the function f is constant, and its value is a symmetric relation v, we obtain $x \sim y$ iff $(x,y) \in v^2$. This generalizes the situation in metric spaces, where balls with diameter δ are obtained by taking the radius $\delta/2$.

Yet such a function is not supposed to operate arbitrarily. Potential models which lie in close proximity should be endowed with similar blurs. Similar models should get similar blurs. These blurs should be locally similar, that is, $u_x(x)$ and $u_y(y)$ should be similar for similar x and y. But how can we define local similarity for blurs? We shall proceed by generalizing the situation in pseudometric spaces. It can be said that there the function f assigns similar blurs to similar models, if there is a non-expanding function $\varphi:M_p \to \mathbb{R}^+$, such that, given the "distance" d of the space, $f(z) = \{(x,y)/d(x,y) \le \varphi(z)\}$. ("Non-expanding" means: $|\varphi(x) - \varphi(y)| \le d(x,y)$). If x and y are similar, $\varphi(x)$ and $\varphi(y)$, and consequently $u_x(x)$ and $u_y(y)$ are even more so. The notion of a non-expanding function, however, cannot easily be generalized to the case of uniform structures. So we are looking for such a concept which is independent of the real numbers and nevertheless guarantees that there is a non-expanding function in the special case of pseudometric spaces.

Consider the following situation: it is clear by definition that $x \sim y$ for all $x,y \in u_z(z)$. Of course, there may be $t \in M_p$ with $t \sim z$ but $t \notin u_z(z)$. However, if $t \sim x$ for all $x \in u_z(z)$ but $t \notin u_z(z)$, we wouldn't feel that $u_z(z)$ really describes the admissible inaccuracy around z. So, intuitively, it seems reasonable to require that $u_z(z)$ should be a maximal set with pairwise similar members, which we call "similarity circles" (following Carnap's *Aufbau*). Only such similarity circles depict the local amount of admissible inaccuracy. Interestingly, this intuitive requirement solves the problem how to define an assignment of similar blurs to similar models, too: if the sets $u_z(z)$ are all similarity circles, and if the uniform structure is defined by a pseudometric, then there exists a non-expanding function φ which produces the circles $u_z(z)$. We explain this vague statement before proceeding to formal results.

Given a pseudometric space d on M_p and a function $\varphi:M_p \to \mathbb{R}_0^+$, define a uniform structure U by d and a function $f:M_p \to U$ by $f(z) := \{(x,y)/d(x,y) \le \varphi(z)\}$. The set $u(a,x) := \{y \in M_p/d(x,y) \le a\}$ is called entourage of $x \in M_p$ with radius $a \in \mathbb{R}_0^+$. Obviously, $(f(z))(z) = u(\varphi(z),z)$. Entourages can be the same for different radii. For instance, if there are no points in M_p which are farther from x then obviously $u(2,x) = u(3,x)$. For a given entourage $u(a,x)$ there is always a least radius b with $u(b,x) = u(a,x)$. Take $b := \inf\{a'/u(a',x) = u(a,x)\}$; then $u(b,x) = u(a,x)$ (we have defined the entourages using "$\le$" instead of "$<$" in order to get this property.) Intuitively, this least radius is the degree of admissible inaccuracy at $x \in M_p$. Now suppose that φ is the least function of all functions which determine the same entourages: for a given function $\Psi:M_p \to \mathbb{R}^+$ define $\varphi(x) := \inf\{a'/u(a',x) = u(\Psi(x),x)\}$. In the case of a pseudometric space we say that f assigns similar blurs to similar models if this least function $\varphi:M_p \to \mathbb{R}_0^+$ is non-expanding.

Proposition 6: If $u(\varphi(z),z)$ is a similarity circle for every $z \in M_p$, then $\varphi:M_p \to \mathbb{R}_0^+$ is a non-expanding function.

Corollary: φ is uniformly continuous.

Proof:

Suppose that there are x,y with $d(x,y) < |\varphi(x)-\varphi(y)|$. Hence $\varphi(x) \neq \varphi(y)$. Take $\varphi(x) < \varphi(y)$, so $d(x,y) < \varphi(y)-\varphi(x)$, hence:

1. $d(x,y) + \varphi(x) < \varphi(y)$.
2. $u(\varphi(x),x) \subseteq u(\varphi(y),y)$, because $z \in u(\varphi(x),x) \Rightarrow d(x,z) \leq \varphi(x)$, and from $d(z,y) \leq d(x,z)+d(x,y) \leq \varphi(x)+d(x,y) < \varphi(y)$ follows $z \in u(\varphi(y),y)$.
3. Since $u(\varphi(x),x)$ and $u(\varphi(y),y)$ are both similarity circles, they must be equal by 2. So there is no $z \in u(\varphi(y),y)$ which is not in $u(\varphi(x),x)$. But then the function $\varphi':M_p \to \mathbb{R}_0^+$ defined by $\varphi'(z)=\varphi(z)$ for $z \neq y$ and $\varphi'(y)=d(x,y)+\varphi(x)$ is smaller than φ (by step 1 above) but produces the same entourage of y as φ, contradicting the assumption that there is no smaller function than φ which produces the same entourages.

So we can find a non-expanding function which produces the entourages, that is, we can say that the entourages are changing smoothly, that is, similar models have similar entourages. For uniform spaces, we take the result of proposition 6 as a *definition* of a non-expanding assignment of locally admissible blurs to potential models. Now, one could ask whether there are enough assignments of blurs to potential models which satisfy the antecedent of proposition 6: for uniform spaces it can be shown that every blur contains a symmetric blur which can be built out of localized blurs such that similarity circles arise.

Proposition 7: For every $u \in U$ there is a symmetric blur $v^* \subseteq u$ and a function $\varphi:M_p \to U$, such that $(\varphi(z))(z)$ is a similarity circle (wrt $u \cap u^{-1}$) for all $z \in M_p$ and $v^*=\{(x,y)/\exists z(x \in (\varphi(z))(z) \ \& \ y \in (\varphi(z))(z))\}$.

Proof:

There is a $v \in U$ with $v^2 \subseteq u \cap u^{-1}$ and $v=v^{-1} \Rightarrow \forall x \in M_p(v(x) \times v(x) \subseteq u \cap u^{-1})$, so $v(x)$ is a set of pairwise similar members (wrt $u \cap u^{-1}$, which is a reflexive and symmetric relation) for every $x \in M_p$. By the lemma of Zorn we can prove that $v(x)$ is contained in a maximal set of pairwise similar members. Select for every $x \in M_p$ a maximal set C_x of pairwise similar (wrt $u \cap u^{-1}$) members with $v(x) \subseteq C_x$. Then $v \cup (\{x\} \times C_x) \in U$. Define $\varphi:M_p \to U$ by $\varphi(x):= v \cup (\{x\} \times C_x)$.

1. $(\varphi(x))(x)=(v \cup (\{x\} \times C_x))(x)=v(x) \cup C_x=C_x$, that is, a similarity circle.

2. $v^* := \{(x,y)/\exists z(x \in (\varphi(z))(z) \ \& \ y \in (\varphi(z))(z))\}$ is symmetric and is contained in $u \cap u^{-1}$, since $(\varphi(z))(z) \times (\varphi(z))(z) = C_x \times C_x \subseteq u \cap u^{-1}$, because C_x is a similarity circle wrt $u \cap u^{-1}$.

3. $v \subseteq v^*$: $(x,y) \in v \Leftrightarrow y \in v(x) \Rightarrow y \in C_x \Rightarrow (x,y) \in C_x \times C_x \subseteq v^*$. It follows that $v^* \in U$, since $v \in U$.

(Note that the proof uses the critical axiom for uniform structures.)

We conclude, that a symmetric relation v^* according to the last proposition is a suitable boundary of admissible inaccuracy. A function which builds up v^* by assigning blurs to potential models is an admissible localization of the admissible inaccuracy.

V

The Approximate Empirical Claim

Given a boundary of admissible inaccuracy, there are a lot of possible formulations of the approximate empirical claim. The starting point is the idealized empirical claim $I \subseteq r(M)$. In principle, we can blur the set of intended applications I, the class of actual models M, its restriction $r(M)$ and finally the restriction function r (or, equivalently, both I and $r(M)$ at the same time). We may render these four possibilities formally as follows:

(EC1) $\widetilde{I} \subseteq r(M)$, i.e., $\exists X(X \sim I \ \& \ X \subseteq r(M))$,

(EC2) $I \subseteq \widetilde{(r(M))}$, i.e., $\exists X(X \sim r(M) \ \& \ I \subseteq X)$,

(EC3) $I \subseteq r(\widetilde{M})$, i.e., $\exists X(X \sim M \ \& \ I \subseteq r(X))$,

(EC4) $\widetilde{I} \subseteq \widetilde{(r(M))}$, i.e., $\exists X,Y(X \sim I \ \& \ Y \sim r(M) \ \& \ X \subseteq Y)$.

In *Architectonic*, (Ch. VII.2.3), it was argued that (EC4) should be preferred for methodological reasons. However, formal considerations below will show that this is *not* the most plausible choice to express the approximate empirical claim of a theory.

To reduce the number of alternatives, we follow two principles: a technical one, which says that we should not distinguish between set-theoretically equivalent formulations, and an epistemological one, which says that we should not blur the same components of the claim twice. The first step towards reducing the number of alternative formulations is to show that (EC1) and (EC2) are equivalent. This is settled in the following proposition.

Proposition 8: The following formulations of the approximate empirical claim are equivalent:

(EC1) $\widetilde{I} \subseteq r(M)$,

(EC*) $\forall i \in I \exists m \in r(M): i \sim m$,

(EC2) $I \subseteq (\widetilde{r(M)})$.

Proof:

1. (EC1) $\Rightarrow$ (EC2) is well-known.

2. (EC2) $\Rightarrow$ (EC*): $\widetilde{I} \subseteq r(M)$: $\leftrightarrow \exists X(X \sim I \ \& \ X \subseteq r(M))$. For $i \in I$ there exists $x \in X$ with $i \sim x$ and $x \in r(M)$.

3. (EC*) $\Rightarrow$ (EC1): From (EC*) we infer $(r(M) \cup I) \sim r(M)$. Define $Y: = r(M) \cup I$. Hence $Y \sim r(M)$ and $I \subseteq Y$, viz. (EC1).

So (EC1) and (EC2) are just two possibilities to express (EC*) in a more compact way. The third formulation, $I \subseteq r(\widetilde{M})$, is stronger than these two, and equivalent to them if the relation $\sim$ is homogeneous. Define $u(x): = \{y/ y \in M_p \& y \sim x\}$ and $r(u)(x): = \{y/y \in M_{pp} \& y \sim x\}$. Note that $\forall x \in M_p (r[u(x)] \subseteq r(u)(r(x)))$.

Proposition 9: (EC3) $I \subseteq r(\widetilde{M})$ implies (EC*) $\forall i \in I \exists m \in r(M): i \sim m$. If $\forall x \in M_p(r[u(x)] = r(u)(r(x)))$, then (EC3) $\leftrightarrow$ (EC*).

Proof:

A. $I \subseteq r(\widetilde{M}) \leftrightarrow \forall i \in I \exists x \in M_p \exists m \in M(i = r(x) \& x \sim m)$. Proof:

1. $I \subseteq r(\widetilde{M}) \leftrightarrow \exists X(X \sim M \& I \subseteq r(X))$ by definition. If $i \in I$, then $i \in r(X) \Rightarrow \exists x \in X(i = r(x))$. Because of $X \sim M$ there is an $m \in M$ with $x \sim m$.

2. $\forall i \in I \exists x \in M_p \exists m \in M(i = r(x) \& x \sim m)$: hence there is a function $f: I \to M_p$, such that for all $i \in I$: $r(f(i)) = i$ & $\exists m \in M$ with $f(i) \sim m$. Then $(f[I] \cup M) \sim M$. Define $X: = f[I] \cup M$. Then $X \sim M$ and $I \subseteq r(X)$, because of $I \subseteq r(f[I])$ $(i \in I \Rightarrow f(i) \in f[I] \Rightarrow i = r(f(i)) \in r(f[I]))$ and $r(f[I]) \subseteq r(X)$.

B. $I \subseteq r(\widetilde{M})$, $i \in I \Rightarrow \exists \ x \in M_p \exists m \in M(i = r(x) \ \& \ x \sim m) \Rightarrow r(m) \in r(u)(r(x)) = u(i)$, hence $i \sim r(m)$.

C. $\forall i \in I \exists m' \in r(M): i \sim m' \ \& \ \forall x \in M_p(r[u(x)] = r(u)(r(x)))$. If $i \in I$ then there is an $m \in M$ with $i \in r(u)(r(m)) \Rightarrow i \in r(u(m)) \Rightarrow \exists x \in M_p(r(x) = i \& x \in u(m))$.

We have required that the uniform structure U must have a base of homogeneous blurs, but this does not imply that every blur is homogeneous. Normally, the admissible inaccuracy for the non-theoretical components will depend on the theoretical components, so the boundary of admissible inaccuracy will normally not be a homogeneous relation. Therefore (EC3) will normally be stronger than (EC*).

But we may conclude that $I \subseteq r(\widetilde{M})$ (formulation (EC3)) is preferable to $I \subseteq \widetilde{(r(M))}$ (formulation (EC2)) independently of this conjecture: for if the relation $\sim$ is homogeneous, then (EC3) is equivalent to (EC*) and if it is not homogeneous, then (EC3) seems to be more adequate than (EC*). If on the theoretical level the admissible inaccuracy of the non-theoretical components depends on the theoretical components, it is quite implausible that we should use the restriction of $\sim$ for the formulation of the approximate claim. Furthermore, the class M_p is the appropriate space for the formulation of approximate claims when the links and the constraints are to be included.

We didn't give any reasons so far for our dismissal of all formulations of the approximate empirical claim which blur the same components twice. If x and y are potential models and $\sim$ is a reflexive and symmetric relation describing the boundary of admissible inaccuracy and if we find that $x \not\sim y$, we would plausibly conclude that x and y are too far apart, that no one does approximate the other. Even if there is a third model z with $x \sim z$ and $z \sim y$, we wouldn't change our judgment, since $\sim$ is the boundary and not the product of $\sim$ with itself. So blurring models twice runs against the concept of a boundary of the globally admissible inaccuracy.

Prima facie, this doesn't seem to be the case in (EC4), i.e., $\widetilde{I} \subseteq \widetilde{(r(M))}$. It seems to express that it is allowed to blur the intended application and (the restriction of) an actual model. But as the following proposition shows, (EC4) just says that there is a "middle" model between i and $r(m)$, viz. the intended applications are blurred twice in order to attain (a restriction of) an actual model.

Proposition 10: Define for M_{pp}: $x \approx y$ iff $\exists z(x \sim z \& z \sim y)$. Then $\widetilde{I} \subseteq \widetilde{(r(M))} \Leftrightarrow I \subseteq \widetilde{(r(M))}$.

Proof:
1. $\widetilde{I} \subseteq \widetilde{(r(M))} \Leftrightarrow \exists X, Y(X \sim I \& Y \sim r(M) \& X \subseteq Y)$ by definition. This implies: $\forall i \in I \exists z \in X(i \sim z)$ and $X \subseteq Y$ yields $\forall z \in X \exists m \in r(M)(z \sim m)$. Therefore $\forall i \in I \exists m \in r(M)(i \approx m)$. From proposition 8 it follows that this is equivalent to $I \subseteq \widetilde{(r(M))}$.
2. $I \subseteq \widetilde{(r(M))} \Leftrightarrow \forall i \in I \exists m \in r(M)(i \approx m)$ (proposition 8). So $\forall i \in I \exists m \in r(M) \exists z \in M_{pp}(i \sim z \& z \sim m)$. Hence there is a function $f: I \to M_{pp}$ with $f(i) \sim i$ for all $i \in I$ such that there is an $m \in r(M)$ with $f(i) \sim m$. Then $I \sim f[I]$ and $r(M) \sim f[I] \cup r(M)$. Define $X := f[I]$ and $Y := f[I] \cup r(M)$. Then $X \sim I \& Y \sim r(M) \& X \subseteq Y$, that is $\widetilde{I} \subseteq \widetilde{(r(M))}$.

In sum, (EC1), (EC2) and (EC3) claim that for every $i \in I$ there is an $m \in r(M)$ in close proximity, where the closeness of the proximity is judged by a reflexive and symmetric relation. Blurring twice seems to be superfluous. All other possible formulations are even worse in this respect, because some components are blurred twice or threefold whereas others are only blurred once or twice.

At this point the objection might be raised, that (EC4) is misunderstood by the interpretation we gave to it. If we take $\sim$ as the admissible inaccuracy, then the intended applications are indeed blurred twice. But if $\approx$ denotes the admissible inaccuracy then there need not exist a blur $\sim$ such that $x \approx y$ iff $\exists z(x \sim z \& z \sim y)$. In this perspective, (EC4) appears as a strengthening of $I \subseteq \widetilde{(r(M))}$! The admissible inaccuracy is to be evenly distributed on the intended applications and the actual models. That is, every intended application can be described by a partial model close to it which can be extended approximately to an actual model. This partial model describing the intended application can be regarded as an approximate determination of the non-theoretical components of the intended application.

So with regard to the admissible inaccuracy in the sense of closeness of intended applications and actual models, (EC3) seems to be the best explication of the empirical claim. But there may well be a substructure in the claim due to the approximate determination of models, which can be divided into the approximate determination of the T-non-theoretical components (which does not require the laws of T) and the determination of the T-theoretical components (requiring the application of the laws). In some sense the empirical claim is strengthened, but it seems to be more adequate to impose an additional condition on the admissible inaccuracy: it has to be a relation $\approx$ which can be factorized by $x \approx y$ iff $\exists z(x \sim z \& z \sim y)$, where $\sim$ is a blur (or a relation defined by localized blurs).

The approximate determination of models has not been examined systematically so far. It is not clear whether (EC4) or even a further possibility (EC5) with $\widetilde{I} \subseteq (r(\widetilde{M}))$ will emerge as the adequate explication of it. In any case, (EC3) seems to be a promising starting point. Determinability of models is an additional claim strengthening the claim of approximate extendability of intended applications to actual models. It should be reconstructed by adding structures to the boundary of admissible inaccuracy.

Institut für Philosophie, Logik und
  Wissenschaftstheorie
Universität München
Ludwigstrasse 31
80539 München, Germany

Institut für Philosophie

Freie Universität Berlin
Habelschwerdter Allee 30
D-1000 Berlin 33, Germany

REFERENCES

Balzer, W., Moulines, C.U., and Sneed, J.D. [1987]. *An Architectonic for Science. The Structuralist Program.* Dordrecht: Reidel.

Bourbaki, N. [1961]. *Topologie générale.* Paris: Hermann.

Moulines, C.U. [1976]. Approximate Application of Empirical Theories: A General Explication. *Erkenntnis*, **10/II**, 201−227.

Moulines, C.U. [1980]. Intertheoretic Approximation: The Kepler−Newton Case. *Synthese*, **45**, 387−412.

*Poznań Studies in the Philosophy
of the Sciences and the Humanities
Vol. 42, pp. 49−52*

Ilkka A. Kieseppä

A NOTE ON THE STRUCTURALIST ACCOUNT ON APPROXIMATION

Recently, C. U. Moulines and Reinhold Straub have criticized the structuralist account of approximation presented in Balzer, Moulines and Sneed [1987] (referred to as BMS in this note) and proposed several modifications to it. However, in what follows we shall see that the account of approximation presented in BMS is flawed in a much more dramatic way than the various criticisms of Moulines and Straub indicate. The theorem below shows that there are no interesting examples of *classes of admissible blurs*, on which the account of approximation given in BMS is based.

For the sake of completeness, a few definitions will be recapitulated.[1]

Definition 1. A *uniformity* U on a set X is a family of sets of satisfying the
   following axioms 1.1. − 1.6.:
   1.1 $\emptyset \neq U \subseteq \mathrm{Po}(X \times X)$
   1.2 If $u \in U$, $u \subseteq v$ and $v \subseteq X \times X$, then $v \in U$.
   1.3 If $u_1, u_2 \in U$, then $u_1 \cap u_2 \in U$.
   1.4 If $u \in U$, then $\{(x,x) \mid x \in X\} \subseteq u$.
   1.5 If $u \in U$, then $u^{-1} = \{(y,x) \mid (x,y) \in u\} \in U$.
   1.6 If $u \in U$, there is a $v \in U$ such that
        $v \circ v = \{(x,z) \mid (\exists y)((x,y) \in v \wedge (y,z) \in v)\} \subseteq u$.

Of course, the uniformities discussed below will be defined on the set of potential models M_p of some theory-element

$$T = <M_p, M, M_{pp}, GC, GL, A, I> .$$

---

1 Definitions 1, 2 and 3 below are essentially the same as DVII-1, DVII-5 and DVII-6, respectively, in Balzer − Moulines − Sneed (1987).

Next, I shall recapitulate the definition of the restriction relation in the form relevant for the purposes of this note.[2] First, one defines

$$r : M_p \to M_{pp}$$

as a function which "cuts away" the theoretical components of each potential model, turning it into a partial potential model. Then one further defines $\bar{r}(u)$ as

$$\bar{r}(u) = \{(y,y') \mid \text{there are } x,x' \text{ such that}$$
$$y=r(x),\ y'=r(x') \text{ and } (x,x') \in u\}$$

for each $u \in Po(M_p \times M_p)$. Obviously, for any $u,v \in Po(M_p \times M_p)$,[3]

(1) $\bar{r}(u \cup v) = \bar{r}(u) \cup \bar{r}(v)$.

The definitions of the set of admissible blurs and of the *bound* of a set must still be recapitulated.

Definition 2. Suppose that $U \subseteq Po(M_p \times M_p)$ is a uniformity and that A is a subset of U. Define *Bound(A)* as
$Bound(A) = \{\ u \in U \mid (\forall u' \in U)((u \subseteq u' \to u' \notin A) \wedge$
$\qquad\qquad (u' \subset u \to u' \in A))\}$.

Here, of course, the symbol $\subset$ denotes the proper subset relation (*i.e.*, $u' \subset u$ iff $u' \subseteq u$ and $u \neq u'$).

Definition 3. If U is a uniformity on M_p, A is a class of admissible blurs in U if and only if it satisfies the following conditions 3.1 – 3.4.
3.1. $A \subseteq U$ and $A \neq \varnothing$
3.2.   For all $u \in U$: if $u \in A$, then $u^{-1} = \{(y,x) \mid (x,y) \in u\} \in A$.
3.3.   For all $u,u' \in U$: if $u \in A$ and $\bar{r}(u)=\bar{r}(u')$, then also $u' \in A$.
3.4.   For all $u \in A$ there is a $u_s \in Bound(A)$ such that $u \subseteq u_s$.

The following additional definition is also needed:[4]

---

2 *Ibid.*, pp. 338–9.
3 This is theorem TVII-4 (1) in *ibid.*
4 I have changed this definition to the form proposed by C. U. Moulines after he made an objection to my original formulation.

Definition 4. The restriction relation
$$r : M_p \to M_{pp}$$
is called *non-trivial* if for every $x \in M_p$ there is a $x' \in M_p$, $x' \neq x$, such that $r(x) = r(x')$.

Obviously, all the relations r corresponding to the examples of theory-elements given in chapter 3 of BMS are non-trivial in this sense. However, interesting sets of admissible blurs cannot exist for non-trivial relations r, as the following theorem shows, since there cannot be two admissible blurs such that one of them is a subset of the other.

Theorem. Suppose that
$$T = < M_p, M, M_{pp}, GC, GL, A, I >$$
is a theory-element in the sense of BMS and that the restriction relation r corresponding to it is non-trivial. Then there are no two admissible blurs $u_1, u_2 \in A$ such that $u_1 \subset u_2$.

Proof. The proof proceeds by contradiction. Suppose that U is the uniformity corresponding to A, and that $u_1, u_2 \in A$ are both different and such that $u_1 \subset u_2$. Then there is a $(x,y) \in u_2 - u_1$. By the non-triviality of r, there is $x' \in M_p$, $x' \neq x$, such that $r(x) = r(x')$ and

(2) $\quad \bar{r}(\{(x,y)\}) = \bar{r}(\{(x',y)\})$.

Thus, if u_3 is defined as

(3) $\quad u_3 = u_2 \cup \{(x',y)\})$

then $u_3 \in U$ by 3.1. and 1.2., and it follows from (1) and (2) that

$$\bar{r}(u_3) = \bar{r}(u_2 \cup \{(x',y)\}) = \bar{r}(u_2) \cup \bar{r}(\{(x',y)\}) =$$
$$\bar{r}(u_2) \cup \bar{r}(\{(x,y)\}) = \bar{r}(u_2 \cup \{(x,y)\}) = \bar{r}(u_2),$$

because $(x,y) \in u_2$. Because $u_2 \in A$, by 3.3. also $u_3 \in A$. Thus, by 3.4., a $u_s \in Bound(A)$ can be so chosen that $u_3 \subseteq u_s$. Let us put

(4) $\quad u = u_s - \{(x,y)\})$.

Because $u_1 \subseteq u_2 \subseteq u_3 \subseteq u_s$ and $(x,y) \notin u_1$, $u_1 \subseteq u_s - \{(x,y)\} = u$. Because $u_1 \in U$, it follows from 1.2. that $u \in U$. Because $(x,y) \in u_2 \subseteq u_3 \subseteq u_s$, it follows that $u \subset u_s$. Applying definition 2, it follows that $u \in A$. But by (3), $(x',y) \in u_3 \subseteq u_s$, so that because of (4) and the fact that $(x,y) \neq (x',y)$,

$(x',y) \in u,$

and it is possible to compute that

$$\bar{F}(u_s) = \bar{F}((u_s - \{(x,y)\}) \cup \{(x,y)\}) =$$
$$\bar{F}(u_s - \{(x,y)\}) \cup \bar{F}(\{(x,y)\}) = \bar{F}(u) \cup \bar{F}(\{(x',y)\}) =$$
$$\bar{F}(u \cup \{(x',y)\}) = \bar{F}(u).$$

It follows from 3.3. that $u_s \in A$. But this contradicts the assumption according to which $u_s \in Bound(A)$. $\square$

Department of Philosophy
University of Helsinki
P.O. Box 24
00014 University of Helsinki, Finland

REFERENCES

Balzer, W., Moulines, C.U. and Sneed, J.D. [1987]: *An Architectonic for Science. The Structuralist Program*. Dordrecht: Reidel.

Moulines, C.U. [1976]: Approximate Application of Empirical Theories: A General Explication. *Erkenntnis* **10**, 201−227.

Moulines, C.U. and Straub, R. [1994]: Approximation and Idealization from the Structuralist Point of View, in this volume.

Poznań Studies in the Philosophy
of the Sciences and the Humanities
Vol. 42, pp. 53–55

C. Ulises Moulines and Reinhold Straub

A REPLY TO KIESEPPÄ

Kieseppä points out some real difficulties in the original treatment of admissible approximation. Proposition 4 below implies that there is even a further difficulty, which is akin to his line of argument. The following remarks show how all of them can be superseded.

1. Definition 4 must be slightly amended:
The restriction function $r: M_p \to M_{pp}$ is *non-trivial*, if for every $x \in M_p$, there is a $x' \in M_p$ such that $x \neq x'$ and $r(x) = r(x')$.

2. For $v \subseteq M_{pp} \times M_{pp}$, the relation $r^{-1}(v) \subseteq M_p \times M_p$ is defined as
$r^{-1}(v) = \{(x,x') \mid (r(x), r(x')) \in v\}$.

3. *Proposition*: If $u \in Bound(A)$, then $u = r^{-1}(r(u))$.
Proof: Let $u \in Bound(A)$. Since r is onto, $r(u) = r(r^{-1}(r(u)))$. From TVII-8 (*Architectonic*, p. 349) it follows that $u \subseteq r^{-1}(r(u)) \in Bound(A)$. So u cannot be a proper subset of $r^{-1}(r(u))$, since this would imply $u \in A$, in contradiction to $u \in Bound(A)$. $\square$

4. *Proposition*: If the restriction function r is non-trivial, then there is no system of admissible blurs on M_p.
Proof: Suppose that A is a system of admissible blurs. $A \neq \emptyset$, so let $u \in A$. From $r(r^{-1}(r(u))) = r(u)$ it follows that $v := r^{-1}(r(u)) \in A$. Hence there is a $v_s \in Bound(A)$, $v \subset v_s$. By Proposition 3 above, $v_s = r^{-1}(r(v_s))$.

There is a pair $(x,y) \in v_s - v$. Then $v_s - \{(x,y)\} \in U$ because $v \subseteq v_s - \{(x,y)\}$ and $v \in U$. Further, $v_s - \{(x,y)\}$ is a proper subset of v_s. Hence $v_s - \{(x,y)\} \in A$.

Since r is non-trivial, there is some $x' \in M_p$ such that $x \neq x'$ and $r(x) = r(x')$. From $v_s = r^{-1}(r(v_s))$ and $(x,y) \in v_s$ it follows that $(x',y) \in v_s$. Therefore $r(v_s-\{(x,y)\})=r(v_s)$, hence $v_s \in A$, in contradiction to $v_s \in$ $Bound(A).\square$

5. We correct the definition of $Bound(A)$ in order to avoid this unpleasant result. This might be done in either of two ways:

a) $Bound(A) = \{u \in U \mid \forall u' \in U: (u \subset u' \rightarrow u' \notin A) \& (u' \subseteq u \rightarrow u' \in A)\}$.
b) $Bound(A) = \{u \in U \mid \forall u' \in U: (r(u)) \subseteq r(u') \rightarrow u' \notin A) \& (r(u') \subset r(u) \rightarrow u' \in A)\}$.

Intuitively, $Bound(A)$ in the sense of 5.a) represents quite the same idea as $Bound(A)$ in the sense of *Architectonic*. But now, neither Kieseppä's Theorem nor Proposition 4 can be obtained, since every member of $Bound(A)$ is an admissible blur by definition. Perhaps, version 5.b) is even more conservative than 5.a). By DVII-6(3) (*Architectonic*, p. 348), theoretical changes do not affect the admissibility of a blur. So one might propose that blurs should only be bounded with respect to the non-theoretical components of the potential models. Kieseppä's conclusion is blocked by definition 5.b), too.

6. Theorem TVII-7 (*Architectonic*, p. 349) can still be proved, if definition 5.a) (or 5.b)) is used for $Bound(A)$.

7. Theorem TVII-8 is no longer valid, if we change the definition of $Bound(A)$. Instead, it follows that $r^{-1}(r(u')) \in Bound(A)$ for every $u,u' \in U$ with $u \in Bound(A)$ and $r(u)=r(u')$. Proposition 3 above remains valid.

8. Theorem TVII-9 remains valid, if $Bound(A)$ is defined as in 5.a) or in 5.b). The proof refers to version 5.a).
Proof: It suffices to show that, for all $v \in B(A)$, there is a $v_s \in Bound(B(A))$ so that: $v \subseteq v_s$. Let $v \in B(A)$, $u \in A$ with $r(u)=v$ and $u_s \in Bound(A)$ such that $u \subseteq u_s$. Define $v_s := r(u_s)$. Clearly, $v \subseteq v_s$. It remains to show that $v_s \in Bound(B(A))$.

(1) $v_s=r(u_s)$, hence $r^{-1}(v_s)=r^{-1}(r(u_s))=u_s$. If $v' \in r(U)$ and $v_s \subset v'$, then $u_s=r^{-1}(v_s) \subset r^{-1}(v')$. Suppose $v' \in r(A)$. Then there is some $v \in A$: $r(v)=v'$ and $u_s' \in Bound(A)$ with $v \subseteq u_s'$. It follows that $u_s=r^{-1}(v_s) \subset r^{-1}(v')=r^{-1}(r(v)) \subseteq r^{-1}(r(u_s'))=u_s'$. Therefore $u_s \subset u_s'$, in contradiction to $u_s,u_s' \in Bound(A)$. So $v' \notin r(A)$.

(2) If $v' \in r(U)$ and $v' \subseteq v_s$, then there is $u' \in U$ with $r(u')=v'$. From $r(u') \subseteq v_s$ it follows that $r^{-1}(r(u')) \subseteq r^{-1}(v_s)=u_s$, therefore $r^{-1}(r(u')) \in A$. Now, $v'=r(u')=r(r^{-1}(r(u')))$ is a member of $r(A)$. $\square$

9. In sum, most of the concepts and results in *Architectonic* can be rescued from Kieseppä's attack. However, this account has been revised recently on the ground of informal considerations.

(1) Arbitrary changes of theoretical components should not be viewed as admissible. Usually, constraints connect potential models with regard to their theoretical components, too. Inaccuracy in this respect must be bounded in order not to trivialize the constraints.

(2) "With regard to different applications of the same theory-element usually different degrees of inaccuracy will be accepted." (*Architectonic*, p. 348)

This idea can be formalized in a straightforward manner: Instead of using a bounded set of admissible blurs, we introduce an approximation relation in M_p (which should be reflexive and symmetric) and a function $f : M_p \rightarrow U$, which attributes to every potential model "its" boundary of admissible inaccuracy. The approximation relation $\sim$ is defined as $x \sim y$ iff there is a z such that $x \in (f(z))(z)$ and $y \in (f(z))(z)$. This relation defines the admissible inaccuracy, whereas the blurs $f(x)$ $(x \in M_p)$ define the locally applicable degrees of admissible inaccuracy. A blur $u \in U$ is admissible for $z \in M_p$ if $u(z) \subseteq (f(z))(z)$, thus $f(z)$ is admissible for z, too.

10. Both amendments, 9.(1) and 9.(2), ensure that no argument, analogous to Kieseppä's theorem, can be raised against the new account of admissible inaccuracies.

Institut für Philosophie, Logik und
   Wissenschaftstheorie
Universität München
Ludwigstrasse 31
80539 München, Germany

Institut für Philosophie

Freie Universität Berlin
Habelschwerdter Allee 30
D-1000 Berlin 33, Germany

Poznań Studies in the Philosophy
of the Sciences and the Humanities
Vol. 42, pp. 57—79

Wolfgang Balzer and Gerhard Zoubek

STRUCTURALIST ASPECTS OF IDEALIZATION[1]

Introduction

The basic picture of concretization and idealization seems to be that there is 'the' real world which is described by human theories to more or less satisfactory degree. As the real world is very complex the theories describing it in the beginning provide very rough pictures, many real features are left out of account. Theories therefore are idealized. In the historical process more and more of the features originally ignored are incorporated into the theories which in this way become more accurate pictures of the world. This is the process of concretization.

Beginning with Marx, the scheme of concretization has been taken into account by social scientists. Idealized theoretical models are introduced in a first step, together with the claim that succeeding concretizations will lead to interesting or true insights into the phenomena. In the natural sciences the way of introducing idealized models is much older; it originated together with the experimental method in Galilei's times. In contrast to the situation in the social sciences subsequent concretization here is not crucial because the idealized models can be made to fit to real systems which are artifically constructed.[2] Recently, the scheme of idealization and concretization has become the subject of properly methodological investigation, mainly through the work of Nowak and Krajewski.[3] Attempts are made to

---

1 We are indebted to M. Kuokkanen for penetrating comments on an earlier version which led to substantial revision.

2 We do not intend to discuss the origin and the historical development of these ideas.

3 See Nowak [1980], Krajewksi [1977], the references in these works, and the volumes of the Poznań Studies over the last years.

explicate the concepts, to apply them to cases from the history of science, and to evaluate them in a context of other methodological issues.

It is tempting to associate the issue with materialism and metaphysical realism because 'the' real world seems to be necessary as a guide for concretization. Without the real world, concretization has no 'direction' and becomes arbitrary, or so it seems. However, not much can be said in favour of the existence of a fully determined real world independent of human description.[4] Therefore, idealization should be studied without the presupposition of an independent, fully fledged real world. Indeed, idealization and concretization should be regarded as ordinary intertheoretical relations between theories.

Starting from neutral ground, in the present paper we want to investigate the notion of idealization, and that of concretization which is just the converse of the former, as intertheoretical relations, and in structuralist[5] terminology. Krajewski's and Nowak's explications use a rather narrow syntactic format in which a theory's law is represented by a numerical equation and concretization by the introduction of a real-valued parameter. We generalize their account by using the structuralist format while sticking to their mode of modelling concretization by means of a real-valued parameter. Our generalization makes the notions applicable to various reconstructed examples as they can be found in the literature,[6] in contrast to the original syntactic accounts which would involve further, substantial effort in subsuming existing theories under the narrow syntactic format.[7] On the other hand, we do *not* generalize the treatment of concretization in terms of real-valued parameters for this would lead us to a general notion which has been studied in detail under the label of approximative reduction.[8] Therefore, sticking to real-valued parameters seems to be necessary in order to keep concretization distinct from approximative reduction. On our account, idealization and concretization are seen as a special case of approximative reduction. They can be clearly distinguished from reduction, and may be studied on their own.

In addition to this general point, we address three more special ones. First, we elaborate on a feature neglected by Nowak and Krajewski, namely

---

4 Typical for the issue is the development of H. Putnam who tried to start as a hard realist but ended in Putnam [1981].

5 See Balzer, Moulines and Sneed [1987], but also, for an easier approach to theoreticity, Balzer [1985].

6 Compare Balzer, Moulines and Sneed [1987] and the references in that work.

7 It is not even clear whether such subsumption is feasible in all cases.

8 See Part III below.

that the two theories' laws (the idealized and the concretized one) have the same form.[9] Second, we stress that in the examples from the natural sciences an additional requirement of continuity is satisfied.[10] Third, we suggest *quasi-metrical* spaces to be used for the treatment of 'real-life' examples of idealization and concretization.

I

Conceptual Preliminaries

Let Θ be some many-sorted, higher order similarity type for structures of the form

$$\langle D_1,\ldots,D_k,A_1,\ldots,A_m,R_1,\ldots,R_n\rangle$$

where $D_1,\ldots,D_k$ are sets, $A_1,\ldots,A_m$ are sets of mathematical objects,[11] and $R_1,\ldots,R_n$ are functions, relations or constants over $D_1,\ldots,D_k,A_1,\ldots,A_m$. We assume that one of the *auxiliary* base sets A_i is the set IR of real numbers, that all the relations $R_1,\ldots,R_n$ are proper relations[12] and we write $STR(\Theta)$ to denote the class of all structures of type Θ. Let Θ^e be an extension of Θ, obtained by simply adding a constant for a real number, so that structures of type Θ^e have the form

$$\langle D_1,\ldots,D_k,A_1,\ldots,A_m,R_1,\ldots,R_n,\alpha\rangle$$

---

9 An alternative attempt at explicating this feature was made in Kuokkanen and Tuomivaara [1992]. These authors' analysis, however, rests on the *assumption* that the functions expressing the law like connection (*f* on p. 81 and p. 89 in Kuokkanen and Tuomivaara [1992]) are the same in both the idealized and concretized theory (for instance, in (26.1) and (26.2) on p. 81). In our view, this assumption is not satisfied in all examples; it is too strong, and so the problem of explicating 'sameness of form' still is open.

10 In fact, if this requirement is added in general, idealization becomes a special case of approximative reduction.

11 The distinction between mathematical and "'non-mathematical'" objects is taken over from Bourbaki [1968], p. 259. For reasons of simplicity we do not impose the requirement of invariance under canonical transformtions used there to obtain a clean separation, however. A detailed account in model-theoretic notation is found in Balzer [1985a].

12 As contrasted to proper individuals. Thus each R_i is a set (or some more complex construct) of objects occurring in $D_1,\ldots,D_k,A_1,\ldots,A_m$, and not simply an element of $D_1,\ldots,D_k,A_1,\ldots,A_m$.

where α is an element of one of the sets $A_1,...,A_m$.

If Θ is a many-sorted similarity type of finite higher order then by a *theory T for* Θ we mean a pair[13] $\langle M,I \rangle$ such that 1) $M \subset STR(\Theta)$, 2) $I \subset STR(\Theta)$, 3) M is characterized by at least one cluster law. In this definition by a cluster law we mean a formula **B** containing at least two (proper) relation symbols **R,R'** such that, in the language of **B**, **B** cannot be written equivalently as a conjunction of $\mathbf{B}_1$ and $\mathbf{B}_2$ where $\mathbf{B}_1$ contains only **R** and $\mathbf{B}_2$ contains only **R'**. M is called the class of *models* and I the set of *intended applications* of T.

By explicitly stating the mathematical sets $A_1,...,A_m$ we may concentrate on the 'empirical' part of a theory and avoid inclusion of axioms for the mathematical entities. They may be presupposed as having been character-ized elsewhere. This way of defining models was proposed by Bourbaki [1968], and has proved its value in the formulation of mathematical theories — as shown by the volumes of Bourbaki — as well as in the reconstruction of numerous empirical theories.[14]

If $x = \langle D_1,...,D_k,A_1,...,A_m,R_1,...,R_n,\alpha \rangle \in STR(\Theta^e)$ then by *red(x)* we denote the *reduct* $\langle D_1,...,R_n \rangle \in STR(\Theta)$ of x. If x and y are two structures of type Θ we say that x is a *substructure* of y *in the narrow sense* if each component of x is a (possibly empty) subset of the corresponding com-ponent of y. For two structures x' and $y' \in STR(\Theta^e)$, x' is a *substructure* of y' if $red(x')$ is a substructure of $red(y')$ in the narrow sense, the last com-ponent (α) is the same in x' and in y', and it is an element of the base sets $D_1^{x'},...,A_m^{x'}$ of x'. We write $x' \sqsubseteq y'$ to express that x' is a substructure of y'.[15] Obviously, each substructure again is a structure of the respective type.[16]

---

13 For further details we refer to Balzer, Moulines and Sneed [1987]. The distin-guished role which potential models play in that book is suppressed here for reasons of simplicity. Working with structures instead of potential models the importance of theoretical terms which has been made a crucial point of the structuralist approach by some proponents and opponents is played down to what we think it's proper role.

14 See the examples in Balzer, Moulines and Sneed [1987], and the references there, as well as in Stegmüller [1986].

15 Note that this notion is slightly weaker than the usual model theoretic notion of a restriction of a structure.

16 Therefore it is not necessary to introduce the classes of substructures officially, as would be necessary in case we started with a class of potential models instead of the class of all structures, compare Balzer [1985] or Balzer, Lauth and Zoubek [1993].

II

Idealization is an Intertheoretic Relation

Intuitively, the idea of concretization is that of getting a more concrete picture of 'the' real system(s). Every theory being abstract to some degree, it can be made more concrete by taking into account more and more features or dimensions of the real system(s). The inclusion of one such new feature roughly is modelled by the introduction of a new parameter plus corresponding adjustment of the theory's (cluster) law. Conversely, each theory contains some element of idealization with respect to the real system(s) it describes. The history of physics is rich of examples of concretizations and idealizations, and both notions recently played some role in the discussion of Marxian methodology.[17]

Clearly, the notion of concretization involves two theories — and therefore it is a binary relation among theories. Idealization on the other hand need not necessarily be construed as a binary relation. Informally, we may say that a theory is just ideal, or contains an element of idealization. Informal language does not require that a theory be an idealization of *another theory*, the obvious reason being that in general such a binary expression would refer to some hypothetical theory which is not yet known. If T at some time is the best theory known to apply to certain phenomena and if we say that T is an idealization of theory T^* then T^* is unknown. For if T^* were known then T^* being more concrete — and in this sense better — than T would have replaced T, and T would not be the best theory available.

Informally, we may say that there are two notions of idealization: an unary and a binary one. The former is used to express that theory T 'is' an idealization (contains some idealization), the latter to state that T is an idealization of some theory T^*.

A moment's reflection shows that the unary notion is of little systematic use. The real system presupposed in it's use exists only in the heads of naive metaphysical realists; we need not repeat here the arguments against that kind of philosophy. To say that a theory is an idealization *simpliciter* is true but, trivially so. Ancient philosophers found out that knowledge is knowledge by ideas, i.e., ideal. This finding has developed to some extent; today we say that theories are abstract or approximately true. To say that a theory is an idealization does not convey more information than that.

---

17 See Krajewski [1977], Nowak [1980], and, for example, Hamminga [1989].

For this reason we will consider only the binary notion of idealization in the following. Still, there is a basic ambiguity even in the meaning of the binary notion. This ambiguity is rooted in the problem of identifying a scientific *theory*. Even in the precise, structuralist frame there are − at least − two different notions of a theory. One of these, described in the previous section, centers on single 'laws' or law-like, systematic statements. Each cluster law on this account gives rise to a 'theory' or *theory-element*. A second notion takes theories to consist of bundles or families of laws which are interrelated in some definite way to form a *theory-net*.[18] It is clear that in this terminology a relation between two theory-elements which both occur in one theory-net at the same time is an *inter*theoretical relation between theory-elements *and* a kind of *intra*theoretical relation between components of a 'theory' (a theory-net). With respect to this ambiguity, the *inter*theoretical aspect is primary because the *intra*theoretical relation can be defined in terms of the intertheortical relation. The intratheoretical relation between two 'nodes' (theory-elements) of a theory-net is just an intertheoretical relation between these nodes.

A still more ambiguous meaning of 'theory' obtains when different 'variants' or versions of 'the same' law are said to belong to the same theory. Sometimes, a binary notion of concretization or factualization is used in order to describe a feature of a theory's being applied to a real system. In this process of application, further assumptions motivated by the particular real system under investigation are added to the theory's law in order to obtain or 'derive' statements that may be fitted to the data, and the resulting statement again is called a 'law'. In this way two 'laws' belonging to the same theory may be related by the *intra*theoretical relation of concretization. We do not think that it is good metatheoretic strategy to apply the notion of a law to such statements for they may contain ad-hoc features which are not law-like at all. We think the notion of a law should be used to refer only to statements which can clearly be identified as such by having some standing in their respective scientific community, and we prefer to call the other statements 'application specializations'.

The binary, *inter*theoretical relation of concretization between theories (or theory-elements) has clear cut real-life examples, like the relations between van der Waals-ideal gas law, free fall-Newtonian theory of gravitation, or the ballistic equation without and with air resistance. If we extend its meaning to cover relations between laws and their application specializa-

---

18 In Balzer, Moulines and Sneed [1987] these are called 'theories' while the smaller units are called 'theory-elements'.

tions as well, confusion is hardly avoidable. Given the priority of the first, *inter*theoretical meaning in terms of the examples we prefer to reserve the term 'concretization' to cases of that kind.

To put the point differently, there are two dimensions or ways of looking at concretization. First, we can see concretization as a feature of the process of applying a given theory to a real system (one of its 'intended applications'). We start with a theory (model, law) which is 'concretized' in order to bring it 'nearer to the data' as obtained from a particular, real system. Second, we can see concretization as a relation between two theories which both are at hand: one theory is a concretization of the other. We have argued for the second view. Concretization as a relation between theories can be stated in precise terms because we have a precise notion of a theory, and this kind of concretization is clearly illustrated by examples. In order to take 'concretization' as a feature of application we have to clarify the nature of the entity which is created by concretization (what we called application specialization before). If this entity is regarded as a 'law' then − formally − there is no difference between concretization in application and concretization between theories, and it is sufficient to study the latter notion. If, on the other hand, application specializations are not treated as laws (or as giving rise to theories) the term concretization should be avoided in the context of application of a theory. In any case we may concentrate on the study of idealization and concretization as an intertheoretical relation.

In this paper we use the term 'theory' in the sense of 'theory-element' and consequently we are dealing with *inter*theoretical relations of idealization and concretization. Although this notion sometimes historically may involve a theory not yet known, there is quite some evidence now that such a theory will emerge at some time in the future. Besides, there are numerous cases from the history of science in which both theories: the idealization and it's (future) concretization, *are* known.

By an intertheoretical relation we mean a relation involving all the components of a theory. Besides the models, which clearly are involved in view of the previous discussion, idealization also has to refer to the intended applications. This is necessary to state that both theories describe 'the same' real systems.[19] Without such a condition, the formal relation

---

19 We admit that the use to be made in this respect of the intended applications in the following definition formally does not give much content to the notion of two set theoretic structures capturing the same real system. However, at least *some* formal representation of this relation is needed. If it is left entirely implicit misunderstandings are likely to occur, witness the recent discussion about incommensurability and Kuhn's

might obtain between two classes of models of rather unrelated theories like thermodynamics and equilibrium economics.

According to Nowak [1980], Chap. 2.3, and Krajewski [1977], Chap. 2.5, initial idealizing assumptions are removed by means of concretization, and the concretization of the initial, idealized theory is obtained by the introduction of some non-zero real number α and some modification of the law of the original theory referring to this number. This number is a value of some quantity or 'magnitude' or 'feature' of the system under investigation which may or may not have been taken into account beforehand. Nowak treats all these 'magnitudes' as being present in the beginning, those which are neglected in idealization having zero values. It must be emphasized, however, that usually not all the relevant magnitudes are known in the beginning.[20] Only during the development of a theory does it become clear which factors are relevant and have to be taken into account. The converse of concretization is idealization. If T is a theory and T^* is a concretization of T then T is an idealization of T^*.

More precisely, Nowak describes the transition from an 'idealized' theory given by equation

(1a) $F(x) = f_k(H(x))$

to a 'more concrete' one given by equation

(1b) $F(x) = f_{k-1}(H(x), p_k(x))$

where x denotes a system or an object, $F(x)$ a quantity or property of x, $H(x)$ some undisputed 'parameters' on which $F(x)$ depends, p_k the 'new' parameter taken into account by the concretization, and f_k, f_{k-1} denote the respective law-like connections existing between property F and the relevant parameters. These connections may change, and will change, during the transition, while the value of property F remains *identical*.[21] Equations (1a) and (1b) are assumed to hold under conditions expressed by '$G(x)$ and $p_k(x) = 0$' and '$G(x)$ and $p_k(x) \neq 0$', respectively, where $G(x)$ represents some neccessary condition for system x to belong to the theory's universe, and

---

move to stress identities. See, for example, Kuhn [1983].

20 This is obvious in the social sciences where the 'essential' features or factors relevant to describe a real system simply are unknown, some well established theories, like general equilibrium theory in economics, not withstanding.

21 This identity, compare Nowak [1980], p. 29, is somewhat puzzling, and seems to make sense only for a metaphysical realist.

$p_k(x)=0$ is called the *idealizing assumption*. Obviously, this scheme does not cover all forms of scientific laws; for instance it does not cover purely qualitative laws. Nowak's account therefore is somewhat narrow.

Krajewski does not commit himself to an identity of the values of F.[22] He models concretization as the transition from a 'law' of the form $C_F(x) \wedge p(x)=0 \rightarrow F(x)=0$ to a law of the form $C_F(x) \wedge p(x)>0 \rightarrow F'(x)=0$, where x denotes a system or object, $p(x)$ the value of one of its 'parameters', C_F a set of factual assumptions, and $F(x)=0$, $F'(x)=0$ denote the two 'laws' holding about x before and after the transition.[23] In this scheme, the quantities F and F' may be completely unrelated with each other as long as both have the same value in system x. In order to reduce this arbitrariness he uses his 'correspondence principle' CN which states that, 'asymptotically', equation $F'(x)=0$ passes into equation $F(x)=0$ when $p(x)$ goes to zero. This principle, however, is not further analyzed formally. Krajewski's approach also lacks generality, for a theory's laws will not always take the form of an equation of the above kind.[24]

Taking F and F' to be identical (as Nowak) or as unrelated (as Krajewski, without CP) corresponds to the two horns of a dilemma. On the one side, if F and F' are identical it is difficult to see what is meant by F. If the quantity does not change in the course of concretization it must be some kind of 'true' feature of the system, which is a rather metaphysical kind of entity. If, on the other hand, F and F' are completely unrelated, why should we say that the law expressed by equation $F'(x)=0$ has anything to do with the other law expressed by $F(x)=0$? The two laws just may be very different laws, having nothing at all to do with each other. The requirement that the system x satisfies a fixed set of conditions C_F *formally* does not change the situation.

It is rather obvious how to generalize these accounts in order to make them applicable to all kinds of laws occurring in empirical theories: just replace Nowak's equations $F(x)=f_k(H(x))$, $F(x)=f_{k-1}(H(x),p_k(x))$, by classes of models M and M^* the models x of which are made up from entities of the kinds denoted by F, f_k, f_{k-1}, H, p_k, and similarly for Krajewski's equations $F(x)=0$ and $F'(x)=0$. The essential point in the transition then is that the 'new' models $x^* \in M^*$ contain the 'new', non-zero number α while in the 'old' models this number is replaced by zero. In this terminology the

---

22 Krajewski [1977], Chap. 2.5.

23 Actually, Krajewski uses the term 'law' to denote the whole implications of the above form, and labels $F(x)=0$ as an 'expression' of a law.

24 Of course, both Nowak and Krajewski are aware of the somewhat restricted form of their metatheoretic formalism.

transition from an idealized theory with model class M to a more concrete one consists in replacing M by a new class M^* such that, if models in M are of type Θ then those in M^* are of the extended type Θ^e introduced in Part I above.

In the notation of Part I we obtain the following definition. If $T=\langle M,I \rangle$ and $T^*=\langle M^*,I^* \rangle$ are theories we say that T is an *inhomogenous idealization* of T^* iff there exist similarity types Θ and Θ^e such that the following five conditions are satisfied.[25]

1) T is of type Θ, T^* is of type Θ^e,
2) the structures of type Θ^e have the form $\langle D_1,...,D_k,A_1,...,A_m,R_1,...,R_n,\alpha \rangle$ where $\langle D_1,...,A_m,R_1,...,R_n \rangle$ is a structure of type Θ,
3) the last component of these structures, α, is a positive, real number, i.e. if $x=\langle D_1,...,D_k,A_1,...,A_m,R_1,...,R_n,\alpha \rangle$ is a structure of type Θ^e then there is some $j \leq m$ such that $0 < \alpha \in A_j = \mathbb{R}$,
4) $I \subset red(I^*)$,
5) $red(M^*) \cap M=\emptyset$.

By condition 4) the concretized theory has more intended applications, and the intended applications of the idealized theory are 'identical' to those of the concretized theory in the sense that their descriptions differ only in the constant α. T 'abstracts' from the complication arising by taking $\alpha > 0$ into account, and in this sense is an idealization of T^*. Condition 5) expresses that M and M^* are dramatically different, and that this difference arises from α. No 'old' model in M can be extended to a 'new' model of M^* simply by adding a positive α.

There are cases in which concretization involves more than one parameter, like in the transition from the ideal gas law to van der Waals' law, or that from Galileo's law of free fall to Newton's. In some such cases — like that of the gas law — the transition may be 'broken up' into one or more steps of the above form, each step involving just one new parameter. In other cases the parameters are related to each other and must not be varied independently. In the example of free fall, two 'Newtonian parameters' have to go to infinity (and their inverses to zero): the radius R of the earth, and the mass m of the earth, but in such a way that their proportion m/R^2 remains constant. The previous definition may easily be generalized to deal with several parameters at once. We just have to replace the number α by

---

25 We do not try to formulate the definitions fully formally but try to keep them legible at the cost of some formal sloppyness.

an n-tuple of such numbers, and adjust the 'rest' of the definition. By adding a formal constraint − as a relation among the real parameters − we could obtain a fully general version of the previous definition.[26]

It has been noted by many authors that in a relation of this kind the hypothesis of the idealized theory contains some counterfactual element. There, the parameter α is suppressed or set equal to zero while 'really' − i.e. in the better theory T^* − it is different from zero.[27]

On Nowak's account the two functions f_k and f_{k-1} in (1a) and (1b) above must be different in order to accomodate his two equations with the *same* $F(x)$. Krajewski on the formal level admits different forms as long as they are bound together by the correspondence principle. Now if we would admit completely different laws in the theories to be compared the notions of idealization and concretization would become inadequately weak for the definitions would apply to cases of completely different theories − maybe even from different disciplines − which intuitively have nothing to do with each other. It is obvious from the examples that some additional condition has to be satisfied: the two theories' laws have to be of the same, or similar, form.

The form of a law depends on the contingent syntactical formulation of the theory, and changes when the theory is formulated differently. This difficulty cannot be completely overcome simply by avoiding syntax, for our model classes of course contain some syntactic element in the form of their similarity types. Admitting the fundamental philosophical difficulty, we address the question of similarity of the form of laws by considering two 'suitable variants' of the two theories under comparison. As a necessary condition for idealization both theories have to have variants which satisfy a condition of similarity of their laws. Informally, we explicate this condition by requiring that the idealized theory's model-class M is obtained from that of the concretized theory, M^*, by setting the constant α to zero in models of the latter class. The problem arising here is of course that, in M^*, α is required to be non-zero, so 'setting α to zero' makes no sense inside M^*. In order to circumvent this problem we cannot avoid extending M^* to a superset M_0^* in which α may be zero. Thus M_0^* is a class of structures of the form $\langle D_1,\dots,R_n,\alpha\rangle$, where $\alpha \geq 0$ and $M^* \subseteq M_0^*$. If M^* is extended to such a superset M_0^*, and M is equal to the reduct of that part of M_0^* in which α is zero then the laws characterizing M indeed are similar to those characterizing M^*.

---

26 We note that the present definition could also be generalized if we replace α by a real-valued, non-negative function with at least one positive function value.

27 Compare Rott [1991] for a comprehensive discussion.

Adding this assumption to the previous definition we obtain a more homogenous notion of idealization.

If T and T^* are theories then T is a *homogenous idealization* of T^* wrt. M_0^* iff

1) T is an inhomogenous idealization of T^*
2) M_0^* is a class of models of type Θ^e such that
    2.1) in all $x = \langle D_1, \ldots, R_n, \alpha \rangle \in M_0^*$, $\alpha \geq 0$
    2.2) $M^* \subseteq M_0^*$
3) $M = red(M_0^* - M^*)$.

Condition 3) may be rewritten as follows: $\forall x (x \in M \leftrightarrow \exists x'(x' = \langle D_1, \ldots, R_n, \alpha \rangle \in M_0^* \wedge \alpha = 0 \wedge red(x') = x)$.

This definition also may be generalized to cover more than one parameter, and such that the parameters are interrelated.

The *empirical claim* of theory T states that every intended application can be extended to a model of M, i.e. $\forall x \in I \, \exists y (y \in M \wedge x \sqsubseteq y))$. If T is a homogenous idealization of T^* then the two theories' empirical claims are strongly related to each other as stated by the following theorem. We write I^0 to denote the set $\{\langle x, 0 \rangle \, / \, x \in I\}$.

Theorem 1
    a) If $I^0 \sqsubseteq I^*$ then $I^0 \sqsubseteq M_0^*$.
    b) If $I^* \sqsubseteq M_0^*$ and if $\exists Y (Y \subseteq I^* \wedge Y \sqsubseteq M_0^* - M^* \wedge I \subseteq red(Y))$
       then $I \sqsubseteq M$.

Proof: See appendix.

By part a), the empirical claim of the concretized theory 'restricted' to the intended applications of the idealized theory is implied by the latter's empirical claim. b) states that the idealized empirical claim follows from that of the concretized theory, provided an additional requirement is satisfied. The point of this requirement is that the concretized intended applications can be extended to, and thus explained by, a set of models Y in which the new parameter α is zero. This condition usually will not be satisfied, it points to the counterfactual nature of the idealized theory, or to the 'Kuhn loss'.

III

Idealization and Approximation

The previous definition of idealization involves a jump of the parameter from α to zero. In most examples − in the natural sciences in all examples − the transition is more continuous. Krajewski in Sec. 4.1 of his book requires that the expression of the law of the idealized theory should be deducible from that of the concretized theory in the limit when parameter α is zero. This may be made precise as follows.

Consider a sequence (α_i) of real numbers such that $\alpha_i \to 0$. For each α_i let x_i be some model in M_0^*, and let (x_i) be the sequence of these models. The condition of convergence now may be stated by requiring that this sequence of models (more precisely: the corresponding sequence of reducts) converges to some model in M: $red(x_i) \to x \in M$. Note that this kind of convergence cannot be achieved by successive substitution of α_i by α_{i+1} in one initial model x^*, and thus cannot be written in the form $x^*(\alpha_i) \to x^*(0)$.

All we need in order to complete this requirement is a suitable topology on the class M_0^* of extended models. Sometimes it will be convenient to use the broader class of mere structures $STR(\Theta^e)$ instead. In concrete cases such topologies can be defined rather easily. We will state some examples in Part IV below.

There remains the question of how a notion of convergence for sequences of models or structures should go into the final formulation of the condition of approximation between T^* and T. Basically there are two variants of such a formulation. In the first variant, we consider arbitrary sequences (x_i) of models containing real numbers (α_i) such that $\alpha_i \to 0$, and we require that every such sequence should converge to some model in M. This version however does not guarantee that each model of M will be obtained as a limit of some sequence of models with non-zero parameters. Moreover, there is no natural condition which would guarantee this in general. Such conditions may be formally investigated only when the topology is given.

In a second version we may require that for each model x in M there are sequences (α_i) in $\mathbb{R}^*$ and (x_i) such that $\alpha_i \to 0$, $x_i \in M_0^*$, and $red(x_i)$ converges to x. This second version has the advantage of guaranteeing that 'all of' T is approximated by T^* and thus that T^* really is an improvement of T. In the absence of further formal investigations of how these two conditions compare with each other, we prefer the second one, for the reason just stated.

Even at the general level where the topology is left undefined we have to say a bit more about it. Just requiring that the condition of approximation holds for some topology is too weak. Such a condition might be satisfied by choosing the topology very coarse such that every (or every interesting) sequence converges. The easiest way out of this problem at the general level is to relativize the general definition of idealization to a given topology.

We thus arrive at the following definition of continuous idealization. If $T=\langle M,I \rangle$ and $T^*=\langle M^*,I^* \rangle$ are theories, M_0^* is an extension of M^* as above and U is a topology on $STR(\Theta)$ then T is a *continuous idealization* of T^* wrt. M_0^* and U iff

1) T is an idealization of T^* wrt. M_0^*
2) for all $x \in STR(\Theta)$: $x \in M$ iff there exists sequences (x_i), (α_i) such that

    2.1) each x_i has the form $\langle D_1,\dots,R_n,\alpha_i \rangle \in M_0^*$
    2.2) $\alpha_i \to 0$ wrt. the natural topology in $\mathbb{R}$
    2.3) $red(x_i) \to x$ wrt. U.

Finally, T is a continuous idealization of T^* wrt. U iff there exists an extension M_0^* of M^* such that T is a continuous idealization of T^* wrt. M_0^* and U.

If T in addition is a homogenous idealization of T^* then for each model x of T there exists a sequence x_i in $M_0^*-M^*$ satisfying conditions 2.1) to 2.3) above. The reverse direction in this case does not hold in general. It's being satisfied will depend on the particular theory T^* and the topology.

We mention that by using the formal apparatus to be developed in Part IV we can prove that in case of continuous (homogenous) idealization the model class M is approximated by some subset X^* of M^* for any threshold $\alpha > 0$ in such a way that $d^*(M,red(X^*)) < \alpha$.[28] Then two cases are possible. First, X^* is part of an extension of I^*. In this case there exists a set $Y \subseteq red(I^*)$ of real systems for T such that $Y \sqsubseteq red(X^*)$. This may be called *factual* approximation. In the second case X^* is not part of an extension of I^*. In this case approximation cannot be achieved by means of real systems but only by means of 'ideal', theoretical systems. We might speak of 'counterfactual' or 'theoretical' approximation here.

As already noted in connection with inhomogenous idealization it is not difficult to extend this kind of definition to cover cases where several

---

28 d^* here denotes a quasi-metric d on structures lifted to the power set, compare Part IV.

parameters are set to zero simultaneously, and such that some given constraint is satisfied. We also note that it is of course possible to imbed the previous definitions into a dynamical perspective in which the idealized theory occurs at two succeeding points of time, t_1, t_2, $t_1 < t_2$, and the concretized theory at the later instant t_2. We then might state that the intended applications of T at t_1, I_1, satisfy the empirical claim[29] of T at t_1 only with an $\in$ larger than that needed for I^* and T^*, that I^* is much larger than I is at t_1 (compare requirement 4 of our first definition), or that T^* 'explains' or describes the more narrow range of intended applications, I_2, for T at t_2. While the second statement has been incorporated into our definitions the first one is in need of confirming historical data, and the third one is difficult to state in a precise way.

The notion of continuous idealization may be compared with that of approximative reduction. As there are several different versions of the latter notion in the literature[30] we have to make some choice. Let us look at Mayr's notion.[31] According to Mayr, in the above terminology a theory $T = \langle M, I \rangle$ approximately reduces to a theory $T^0 = \langle M^0, I^0 \rangle$ iff, with respect to some suitable uniformity on M^0 (or on the corresponding class of structures), 1) M is contained in the uniform completion of M^0, and 2) I and I^0 are appropriately related.[32] Condition 1) may be rephrased as saying that for each model x of M there is a Cauchy-sequence (x_n) of models in M^0 which converges (in the completion) to x. If we call the elements of M and M^0 the 'old' and 'new' models, respectively, the condition says that each model of the old theory can be approximated by, or is the limit of, a sequence of new models. In this sense the 'old' theory can be obtained in the limit from the 'new' one.

Obviously, Mayr's limit condition is aslo satisfied by our notion of continuous idealization: 'Each model of the old theory can be approximated by a sequence of models of the new theory'. However, in idealization, the models of both theories have the same form in the sense that they differ only in the value of α (and, in the homogenous case, are related in the form $M = red(M_0^* - M^*)$). In idealization, therefore, taking the limit is a rather local affair affecting the other 'parts' of the models in minor ways. In approximative reduction, on the other hand, the two theories' models may be rather different in their overall structure. Also, there is no reference to a real parameter (though in the examples such a parameter usually is present).

---

29 Compare Balzer, Moulines and Sneed [1987], Chap. 7.

30 Compare Rott [1991] for a comprehensive discussion.

31 Mayr [1981], echoed in Stegmüller [1986].

32 The exact nature of this relation was controversial but is not important here.

In fact, with some additional notation we could show that continuous idealization is a special case of approximative reduction. By adding further requirements to the latter notion's definition we obtain the notion of continuous idealization. Uniform completion is not central in this context for the models of T^* are likely to be complete already.

IV

Topological Notions

In order to complete the picture of continuous idealization the topology presupposed in the precious section has to be specified. We noted that the concrete definition of the topology will depend on the particular form of the theories involved, and will vary from case to case. We want to finish by looking more closely at this issue and see how far we can get in general as well as in terms of concrete examples.

In structuralist writings approximation usually is covered by means of a uniform space.[33] This notion has the advantage of yielding 'uniform' neighbourhoods, i.e. neighbourhoods independent of the point of which they are neighbourhoods. It's disadvantage is abstractness. In applications usually numerical definitions are employed.[34] Another disadvantage − which uniform spaces share with topological spaces − is that there is no natural way of lifting them to the level of power sets. This operation has turned out quite important, at least in the structuralist approach. For these reasons we prefer a slightly stronger notion, namely that of a quasi-metrical space.[35]

By $\mathbb{R}^*$ we denote the set of real numbers extended by an 'infinite' element ∞ plus the corresponding stipulations: $a < \infty$, $a + \infty = \infty$, $\infty + \infty = \infty$, for all $a \in \mathbb{R}$. $\langle X, d \rangle$ is a *quasi-metrical space* iff X is a non-empty set and $d: X \times X \to \mathbb{R}^*$ is such that for all $x, y, z \in X$:

1) $d(x,y) \geq 0$

---

33 The first reference in which this notion is systematically used in a meta-theoretic context seems to be Ludwig [1978]. For a precise definition, see Balzer, Moulines and Sneed [1987], Chap. 7, for instance.

34 Actually, we do not know any concrete example in which the uniformity of an empirical theory is not defined in a numerical way.

35 Niiniluoto uses distances, for instance in Niiniluoto [1984], Chap. 7, but is not explicit whether these should satisfy all the axioms for a metric which cause some problems in connection with 'lifting' to power sets.

2) $d(x,x) = 0$
3) $d(x,y) = d(y,x)$
4) $d(x,y) \leq d(x,z) + d(z,y)$.

If d, in addition to 1)$-$4), also satisfies 5) $d(x,y) \in \mathbb{R}$, and 6) if $d(x,y)=0$ then $x=y$, then d is called a *metric*.

Theorem 2 Every quasi-metrical space $\langle X,d \rangle$ induces a topology and a uniformity on X in a natural way.

Proof: Standard theorem in topology.[36] The sets $u(\delta)=\{\langle x,y \rangle \in X^2 \ / \ d(x,y) < \delta\}$ and $u(z,\delta)=\{x \in X \ / \ \langle z,x \rangle \in u(\delta)\}$ with $\delta \in \mathbb{R}^+$ and $z \in X$ are bases for a uniformity, and a topology, respectively. $\square$

Theorem 3 Let $\langle X,d \rangle$ be a quasi-metrical space, let $Po(X)$ be the power set of X, and let d^*: $Po(X) \times Po(X) \rightarrow \mathbb{R}^*$ be defined by

$$d^*(A,B) = max(\sup_{x \in A} \inf_{y \in B} d(x,y), \sup_{y \in B} \inf_{x \in A} d(x,y))$$

then $\langle Po(X),d^* \rangle$ is a quasi-metrical space.
The proofs of theorems 3, 5, 6 and 9 are given in the appendix.

Theorem 4 Let $\langle X_i,d_i \rangle$, $i=1,2,3...$ be a countable family of quasi-metrical spaces, $\pi(X_i)$ the product of all X_i (i.e. $\pi(X_i)=\{\Phi/\Phi: \mathbb{N} \rightarrow \cup\{X_i \ / \ i \in \mathbb{N}\}$ *and for all i:* $\Phi(i) \in X_i\}$), and let d_π: $\pi(X_i) \times \pi(X_i) \rightarrow \mathbb{R}^*$ be defined by $d_\pi(\Phi,\Phi')=\Sigma_i d_i(\Phi(i),\Phi'(i))$, then $\langle \pi(X_i),d_\pi \rangle$ is a quasi-metrical space.

Proof: The triangle inequality is preserved in the transition to the sum. Infinite values are treated by cases. $\square$

As a special case of Theorem 4 we obtain the 'product' of two spaces $\langle X,d \rangle$, $\langle Y,d' \rangle$ as the space $\langle X \times Y,d_\pi \rangle$ with $d_\pi(\langle x,y \rangle, \langle x',y' \rangle)=d(x,x')+d(y,y')$.

Theorems 3 and 4 allow for a fine grained construction of quasi-metrical spaces on the class of potential models (or proper models) in a structuralist setting. Let M_p be a class of structures of the form $\langle D_1,...,D_k,A_1,...,A_m, R_1,...,R_n \rangle$ as described above. Then each R_i is an element of a corresponding echelon set[37] E_i defined over $D_1,...,D_k,A_1,...,A_m$ by iteration of taking Cartesian products and power sets, starting with suitable elements of

---

36 See, for instance, Schubert [1964], p. 115.
37 See Bourbaki [1968], p. 259 and Balzer [1985], p. 8, for precise definitions.

$\{D_1,...,D_k,A_1,...,A_m\}$. Therefore, if on each 'base set' $D_1,...,D_k,A_1,...,A_m$ a quasi-metric d_j^0 is given $(1 \leq j \leq k+m)$, we can construct by means of Theorems 2 and 3 corresponding quasi-metrics d_i on each echelon set E_i.

Theorem 5 If M_p is a class of potential models in the sense of Balzer, Moulines, Sneed [1987] and if on each base-set occurring in a structure of M_p a quasi-metric is given then a quasi-metric can be constructed on M_p in a natural way.

Let us now consider several concrete examples of quasi-metrics which are of relevance in meta-scientific application.

First, let X be a set and $Po(X)$ the power set of X. Define d_1: $Po(X) \times Po(X) \to \mathbb{R}^*$ by

$$d_1(A,B) = \|A \Delta B\|$$

where $\|Y\|$ denotes the cardinality of set Y, and $A \Delta B$ the symmetric difference of sets A,B: $A \Delta B = (A \cup B) - (A \cap B)$.

Theorem 6 $\langle Po(X),d_1 \rangle$ is a quasi-metrical space.

A second family of examples covers Euclidean distances on real spaces. Define d_2: $\mathbb{R} \times \mathbb{R} \to \mathbb{R}$ by $d_2(\alpha,\beta) = |\alpha-\beta|$ where $|v|$ denotes the usual absolute value of a number v. For any natural number $n > 1$ define d_3: $\mathbb{R}^n \times \mathbb{R}^n \to \mathbb{R}$ by[38] $d_3(\langle \alpha_1,...,\alpha_n \rangle,\langle \beta_1,...,\beta_n \rangle) = (\Sigma_{i<n} |\alpha_i-\beta_i|^2)^{\frac{1}{2}}$, and d_4: $\mathbb{R}^n \times \mathbb{R}^n \to \mathbb{R}$ by $d_4(\langle \alpha_1,...,\alpha_n \rangle,\langle \beta_1,...,\beta_n \rangle) = max\{|\alpha_i-\beta_i| \ / \ i \leq n\}$.

Theorem 7 $\langle \mathbb{R},d_2 \rangle,\langle \mathbb{R}^n,d_3 \rangle$ and $\langle \mathbb{R}^n,d_4 \rangle$ are quasi-metrical spaces.
Proof: Standard analysis. $\square$

Let D be a set, $D' \subseteq D$, $D' \neq \emptyset$, and G be a family of functions $f:D \to \mathbb{R}$. On $G \times G$, define $d_5[D']$ by $d_5[D'](f,g) = \sup_{a \in D'} |f(a)-g(a)|$.

Theorem 8 $\langle G,d_5[D'] \rangle$ is a quasi-metrical space.
Proof: Standard analysis. $\square$

---

38 In general, '2' may be replaced by 'k' here.

An important extension of d_5 covers sets of functions whose domain is not necessarily the same. Let F be a set of real-valued functions whose domains may be arbitrary. For $f \in F$ let $dom(f)$ denote f's domain. On $F \times F$ we define

$$d_6(f,f') = \begin{cases} \infty & if\ dom(f) \neq dom(f') \\ d_5(f,f') & if\ dom(f) = dom(f'). \end{cases}$$

Theorem 9 $\langle F,d_6 \rangle$ is a quasi-metrical space.

We note that all the metrics occurring in mathematical literature are of course quasi-metrics, like for instance the different metrics on Hilbert space.

Finally, let us look at some examples from empirical theories. In classical mechanics one has to compare position functions ('paths'), mass functions and force functions of types s: $P \times T \to \mathbb{R}^3$, m: $P \to \mathbb{R}$, f: $P \times T \to \mathbb{R}^3$, respectively.[39] For instance, if s: $P \times T \to \mathbb{R}^3$ and s': $P' \times T' \to \mathbb{R}^3$ we set

$$d(s,s') = \begin{cases} \infty & if\ P \times T \neq P' \times T' \\ sup\{d_3(s(p,t),s'(p,t))\ /\ \langle p,t \rangle \in P \times T\} & if\ P \times T = P' \times T'. \end{cases}$$

m and f are treated in the same way. The same scheme applies to similar cases in other theories. In exchange economics, for instance, there are price functions, commodity bundles and utility functions of types p: $G \to \mathbb{R}$, q: $J \times G \to \mathbb{R}$, U: $J \times \mathbb{R}^n \to \mathbb{R}$ the quasi distance of which can be defined as that for s and s' above.

Often, the types of two functions to be compared are not strictly identical. For instance in the comparison of classical and relativistic mechanics we have to compare classical mass functions of the type m_c: $P \to \mathbb{R}$ with relativistic ones which are of the type m_r: $P \times T \to \mathbb{R}$. In this case we replace m_c by m_c^* where m_c^*: $P \times T \to \mathbb{R}$ is defined by $m_c^*(p,t) = m_c(p)$, and then take the quasi distance $d(m_r, m_c^*)$.

---

39 Compare Balzer, Moulines and Sneed [1987] for details.

V

Appendix

Proof of Theorem 1: a) Let $\langle x,0 \rangle \in I^0$, so $x \in I$. By assumption, $I \sqsubseteq M$, and by 3) of the definition of homogenous idealization, $M = red(M_0^*-M^*)$, so there is $y \in red(M_0^*-M^*)$ such that $y \sqsubseteq y$. By the definition of *red*, there is $z \in M_0^*$ such that $y = red(z)$. By the definition of M_0^*, z has the form $z = \langle u,0 \rangle$, so $y = red(z) = u$. Altogether, we have $\langle x,0 \rangle \sqsubseteq \langle y,0 \rangle = z \in M_0^*$. So $I^0 \sqsubseteq M_0^*$.

    b) Let $x \in I$, so $x \in red(Y)$, i.e. there is $y \in Y$ with $x = red(y)$. By assumption, $Y \sqsubseteq M_0^*-M^*$, so there is $z \in M_0^*$ with $y \sqsubseteq z$. By 3) of the definition of homogenous idealization, $M = red(M_0^*-M^*$, so $red(z) \in M$. But $y \sqsubseteq z$ implies $red(y) \sqsubseteq red(z)$, so $x = red(y) \sqsubseteq red(z) \in M.$ $\square$

Proof of Theorem 3: Define $d': X \times \text{Po}(X) \to \mathbb{R}^*$ by

$$d'(x,A) = inf\{d(x,y) \mid y \in A\}, x \in X, A \subseteq X.$$

It is easy to verify that d' satisfies the following conditions:

(1) $d'(x,\emptyset) = \infty$
(2) $d'(x,A) \le d(x,y)$ for all $y \in A$
(3) $d'(x,\{y\}) = d(x,y)$ for all $y \in X$
(4) $d'(x,A) = 0$ if $x \in A$
(5)   If $0 \le c \le \infty$ and $c \le d(x,y)$ for all $y \in A$ then $c \le d'(x,A)$
(6)   If $\delta \in \mathbb{R}^+$ and $\emptyset \ne A \subseteq X$ then there is some $y \in A$ such that $d'(x,A) \le d(x,y) \le d'(x,A) + \delta$.

Now let $d^*(A,B) = max\{\sup_{x \in B} d'(x,A),\ \sup_{y \in A} d'(y,B)\}$, then d^* is the function defined above. From the definitions of d', d^* and infimum we obtain for $x \in X$ and $A, B \subseteq X$, $A \ne \emptyset$:

(7)   If $x \in B$ then $d'(x,A) \le d^*(B,A)$
(8) $d^*(A,B) = 0$ if $A = B = \emptyset$
(9) $d^*(A,B) = \infty$ if $A \ne \emptyset$ and $B = \emptyset$
(10) $d^*(A,B) = d^*(B,A)$
(11) $d^*(\{x\},\{y\}) = d(x,y)$
(12)   If $A, B \subseteq X$ and $A = B$ then $d^*(A,B) = 0.$

Condition (8), for instance, is proved as follows. If $A = B = \emptyset$ then $\sup_{x \in B} d'(x,A) = min\{r \in [0,\infty] \mid \forall x \in B: d'(x,A) \le r\} = min[0,\infty] = 0$. Analogously,

$\sup_{y \in A} d'(y,B) = 0$, and these two equations yield $d^*(A,B) = 0$.

The triangle inequality for d^*:

(13) $\quad d^*(A,B) \leq d^*(A,C) + d^*(C,B)$, for all $A,B,C \subseteq X$

is proved by cases, using the well known techniques for handling infima and suprema. We demonstrate two cases in detail.

(i) $\quad A = B = \emptyset$. (8) yields $d^*(A,B) = 0 \leq d^*(A,C) + d^*(C,B)$.

(ii) $\quad A \neq \emptyset$, $B \neq \emptyset$, $C \neq \emptyset$.

For $x \in A$, $y \in B$, $z \in C$ we have $d(x,y) \leq d(x,z) + d(z,y)$ by assumption. By means of (2) this yields $d'(x,B) \leq d(x,y) \leq d(x,z) + d(z,y)$, so, by (5) and (7) for all $y \in B$: $d'(x,B) - d(x,z) \leq d(z,y)$ and $d'(x,B) - d(x,z) \leq d'(z,B) \leq d^*(C,B)$. This implies, by (5) and (7): $d'(x,B) - d^*(C,B) \leq d(x,z)$ for all $z \in C$ and $d'(x,B) - d^*(C,B) \leq d'(x,C) \leq d^*(A,C)$. So for all $x \in A$: $d'(x,B) \leq d^*(A,C) + d^*(C,B)$, which yields $\sup_{y \in B} d'(x,B) \leq d^*(A,C) + d^*(C,B)$. Analogously we obtain with (10) $\sup_{y \in B} d'(y,A) \leq d^*(B,C) + d^*(C,A) = d^*(A,C) + d^*(C,B)$. The last two inequalities and the definition of d^* imply (13). This, together with (10) and (12) are the defining conditions for a quasi metrical space. $\square$

Proof of Theorem 5: Proof: Let M_p contain structures of the type[40] $\langle k,m, \tau_1,\ldots,\tau_n \rangle$ where each τ_i is of the form $po(\tau_i')$. For reasons of simplicity we assume that there is some $l \leq n$ so that each $x \in M_p$ has the form $x = \langle D_1,\ldots, D_k, A_1,\ldots,A_m, R_1,\ldots,R_l, F_{l+1},\ldots,F_n \rangle$ and $R_1,\ldots,R_l$ are proper relations while $F_{l+1},\ldots,F_n$ are functions. That is, for each i with $l+1 \leq i \leq n$ there exist $k+m$-types τ_{i1}, τ_{i2} such that $\tau_i = po(\tau_i') = po(\tau_{i1} \times \tau_{i2})$, and F_i is a function with domain $dom(F_i) \subseteq \tau_{i1}(D_1,\ldots,A_m)$, and range $ran(F_i) \subseteq \tau_{i2}(D_1,\ldots,A_m)$. So $F_i: dom(F_i) \to ran(F_i)$, $F_i \subseteq \tau_{i1}(D_1,\ldots,A_m) \times \tau_{i2}(D_1,\ldots,A_m)$, $F_i \subseteq \tau_i'(D_1,\ldots,A_m)$, $F_i \in \tau_i(D_1,\ldots,A_m)$.

By assumption, for each j with $1 \leq j \leq k+m$ there is a d_j^0 such that $\langle D_j, d_j^0 \rangle$ (for $j \leq k$) and $\langle A_i, d_{k+i}^0 \rangle$ (for $j = k+i$, $1 \leq i \leq m$) are quasi-metrical spaces.

(1) Let $X = \tau(D_1,\ldots,A_m)$, $Y = \tau'(D_1,\ldots,A_m)$ be echelon sets over $D_1,\ldots,A_m$ of types τ,τ', respectively, and $d: X \times X \to \mathbb{R}^*$, $d': Y \times Y \to \mathbb{R}^*$ be quasi metrics on X and Y. By means of Theorem 2 we can construct quasi metrical spaces

---

40 Compare Balzer, Moulines and Sneed [1987], pp. 8, 9, for details.

$\langle po(\tau)(D_1,\ldots,A_m),d^*_X\rangle$ and $\langle po(\tau')(D_1,\ldots,A_m),d^*_Y\rangle$. By means of Theorem 3 we can construct the product space $\langle \tau \times \tau'(D_1,\ldots,A_m),d_\pi\rangle$.

Let $\Phi = \Phi[\tau,\tau',D_1,\ldots,A_m]$ be the class of functions $\{F/F\colon A \to B \wedge A \subseteq \tau(D_1,\ldots,A_m) \wedge B \subseteq \tau'(D_1,\ldots,A_m)\}$. On Φ we define a quasi-metric d'' as follows.

$$d''(F_1,F_2) = \begin{cases} \infty, & \text{if } dom(F_1) \neq dom(F_2) \\ \sup_{a \in dom(F_1)} d'(F_1(a),F_2(a)), & \text{if } dom(F_1) = dom(F_2). \end{cases}$$

(2) By means of (1) we may recursively define quasi-metrics $d_1,\ldots,d_l$ and $d'_{l+1},\ldots,d'_n$ on the basis of $D_1,\ldots,A_m$ and $d^0_1,\ldots,d^0_{k+m}$. Thus, for $1 \leq i \leq l$: $\langle \tau_i(D_1,\ldots,A_m),d_i\rangle$, and for $l+1 \leq l+i \leq n$ $\langle \tau'_{l+i}(D_1,\ldots,A_m),d'_{l+i}\rangle$ are quasi-metrical spaces.

(3) By (2), for each i, $1 \leq i \leq l$, a quasi distance $d_i(R,R')$ is defined for relations $R,R' \in \tau_i(D_1,\ldots,A_m)$. And for each j with $l+1 \leq l+j \leq n$ a quasi distance $d_{l+j}(F,F')$ is defined for any two functions $F,F' \in \Phi[\tau'_{l+j,1},\tau'_{l+j,2}, D_1,\ldots,A_m]$. A quasi-metric on M_p now is defined by

$$d(x,x') = \begin{cases} \infty, & \text{if } \langle D_1,\ldots,A_m\rangle \neq \langle D'_1,\ldots,A'_m\rangle \\ \Sigma_{i \leq l} d_i(R_i,R'_i) + \Sigma_{l+1 \leq l+j \leq n} d_{l+j}(F_{l+j},F'_{l+j}). \end{cases} \square$$

Proof of Theorem 6: Proof: Conditions 1)−3) of the definition are trivial. In order to prove condition 4) we show that (1) $A \Delta B \subseteq (A \Delta C) \cup (C \Delta B)$ for any three sets A, B, C. From this and $\|U \cup V\| \leq \|U\| + \|V\|$ condition 4) follows immediately. (1) is proved by cases. For instance, let $x \in A \cup B$ be such that $x \in A$ but not $x \in C$. Then $x \in A \Delta C$. The other cases are treated similarly. $\square$

Proof of Theorem 9: Let $f,g,h \in F$. We show that $d_6(f,g) \leq d_6(f,h) + d_6(h,g)$. If $dom(f) \neq dom(g)$ then $dom(h)$ must be different either from $dom(f)$ or from $dom(g)$. So the inequality holds with ∞ on both sides. If $dom(f) = dom(g)$ and $dom(h) \neq dom(f)$ the inequality holds with ∞ on the right hand side. Finally, if $dom(h) = dom(f) = dom(g)$ the proposition follwos from Theorem 8. $\square$

Institut für Philosophie, Logik und
    Wissenschaftstheorie,
University of Munich
Ludwigstrasse 31
80539 München, Germany

REFERENCES

Balzer, W. [1985]. *Theorie und Messung*. Berlin etc.: Springer.

Balzer, W. [1985a]. On a New Definition of Theoreticity. *Dialectica* **39**, 127−45.

Balzer, W., Moulines, C.-U. and Sneed, J.D. [1987]. *An Architectonic for Science*. Dordrecht: Reidel.

Balzer, W., Lauth, B. and Zoubek, G. [1993]. A Model for Science Kinematics. *Studia Logica*, **52**, 519−48.

Bourbaki, N. [1968]. *Theory of Sets*. Paris: Hermann.

Hamminga, B. [1989]. Sneed versus Nowak: An Illustration in Economics. *Erkenntnis* **30**, 247−65.

Krajewski, W. [1977]. *Correspondence Principle and the Growth of Science*. Dordrecht: Reidel.

Kuhn, T. [1983]. Commensurability, Comparability, Communicability. In: P. Asquith et al. (Eds.). *PSA 1982*. East-Lansing: APS.

Kuokkanen, M. and Tuomivaara, T. [1992]. On the Structure of Idealizations. *Poznań Studies in the Philosophy of the Sciences and the Humanities* **25**, 67−102.

Ludwig, G. [1978]. *Die Grundstrukturen einer physikalischen Theorie*. Berlin etc.: Springer, 2nd edition 1990.

Mayr, D. [1981]. Investigations of the Concept of Reduction II. *Erkenntnis* **16**, 109−129.

Niiniluoto, I. [1984]. *Is Science Progressive?*. Dordrecht: Reidel.

Nowak, L. [1980]. *The Structure of Idealization*. Dordrecht: Reidel.

Putnam, H. [1981]. *Reason, Truth and History*. Cambridge: UP.

Rott, H. [1991]. *Reduktion und Revision, Aspekte des nichtmonotonen Theorienwandels*. Frankfurt/Main etc.: Peter Lang.

Schubert, H. [1964]. *Topologie*. Stuttgart: Teubner.

Stegmüller, W. [1986]. *Theorie und Erfahrung*. Dritter Teilband. Berlin etc.: Springer.

Poznań Studies in the Philosophy
of the Sciences and the Humanities
Vol. 42, pp. 81–94

Andoni Ibarra and Thomas Mormann

COUNTERFACTUAL DEFORMATION AND IDEALIZATION IN A STRUCTURALIST FRAMEWORK

Introduction

Idealization is a central aspect of scientific knowledge formation, its explication is an important task of philosophy of science. Even if recently a lot
of work has been done in this field there is no unanimously accepted
account of idealization [cf. Niiniluoto 1986: 257]. However, as a rather
uncontroversial common point of departure one might take the assertion that
the idealizing character of science has something to do with its applicability.
Typically, scientific theories cannot be applied directly, rather their relation
to the actual world is mediated and indirect. For instance, the laws of
nature can hardly be interpreted as universal statements that involve a
material implication — rather they have to be conceived of as counterfactual
conditionals that apply to certain unrealized or even unrealizable situations.

One way to explain the idealized character and the indirect and mediated
applicability of scientific theories is provided by possible worlds semantics
[cf. Lewis 1986]: to explain in this framework that empirical theories such
as mechanics or thermodynamics are idealizing theories is to say that they
apply to some appropriate possible worlds where such entities as frictionless
planes, point masses, ideal gases, etc. exist. Thereby, the framework of
possible world semantics enables us to understand the counterfactual
character of scientific laws. However, it remains incomplete as long as the
relation between the actual and the "ideal" worlds is not elucidated.

Another approach that has contributed to a deeper understanding of the
problem of applicability of empirical theories is structuralism [cf. Balzer,
Moulines and Sneed 1987]. This has been done by developing a highly
sophisticated description of the structure of empirical theories. However,
until today, structuralism has hardly taken any notice of possible world
semantics: except for a side remark of Stegmüller [1979: 12] on the modal

character of constraints and a paper of Miroiu that intended to relate the structuralist concept of theoretical function to the modal concept of Kripke frames [cf. Miroiu 1984] we know of no attempt to bring possible world semantics and structuralism in closer contact to each other.

Finally, as the approach of philosophy of science that has most explicitly dealt with the problem of idealization, the "Poznań School" [cf. Nowak 1980, 1989, 1991 and Krajewski 1977] has to be mentioned. Although Nowak repeatedly emphasizes the counterfactual character of economic laws [cf. Nowak 1980, 1991], he never refers to any kind of possible world semantics or to any other account of modal logic. Moreover, his approach is rather syntactically minded and he makes no mention of the structuralist approach.

Only very recently, did structuralism and the idealization approach of the Poznań School come into closer relation. As far as we know, Kuokkanen [1988] and Hamminga [1989] are the first who have explicitly attempt to discuss the relationship between the structuralist approach and the Poznań school. Without doubt the work of these authors provides important steps for establishing a mutually fruitful relation between both strands of thought. We think, however, that more can be done.

The outline of this paper is as follows: In part I we introduce counterfactual (or idealizing) deformation procedures following some recent ideas presented in Nowak [1989, 1991]. As a concrete example we consider the various idealizing deformation procedures performed in the elementary theory of the simple pendulum as they have been studied by Laymon in a series of articles [cf. Laymon 1982, 1985, 1987]. In part II we study counterfactual deformation operators in the structuralist approach. Part III deals with the complementary concepts of idealization and concretization of theories. In part IV we apply the framework of structuralism *cum* idealization structure to the elucidation of the counterfactual character of empirical laws.

I

Counterfactual Deformations

Nowak [1989, 1991] discusses the topic of idealization in the framework of *counterfactual deformations*. The fundamentals of his approach can be succintly described as follows. We start with a set O of possible objects and a set U called the universe of properties. The state of affairs that an object o has a property $u \in U$ is denoted by $<o, u, U>$.

(1.1) Definition. Let U, U' be universes of properties and o a possible object. A *potentialization* or *counterfactual deformation* of $<o,u,U>$ is a triple $<o,u',U'>$. It is called a *soft* counterfactual deformation iff $U=U'$, and a *hard* counterfactual deformation iff $U \neq U'$.

Counterfactual deformations are not arbitrary, a few make sense, and the huge majority does not. An all-important task of a theory of counterfactual deformations that is worth its salt is to distinguish between "good" and "bad" ones. For this purpose we have to take into account the structure of the universes of properties that are involved. A universe of properties U usually is not simply a set, i.e., a heap of unrelated properties, but a set endowed with further structure. Moreover, in the case of hard counterfactual deformations the universes U and U' typically are structurally related. We mention some important cases:

U has the structure of a Cartesian product $U=U_1 \times U_2$: In this case we write $<o,u_1,u_2,U_1,U_2>$ instead of $<o,<u_1,u_2>,U_1 \times U_2>$. The product structure of the universe of properties enables us to define two important cases of hard counterfactual deformation: *reduction* and *transcendentalization*. *Reduction* consists in that a given object is counterfactually postulated not to have some properties it actually has. *Transcendentalization*, as the counterpart of reduction, is the counterfactual deformation that an object has properties that it actually does not possess at all. The precise definition of these counterfactual deformation procedures is the following:

(1.2) Definition. Let $U=U_1 \times U_2$ be a universe of properties with projection maps p_i: $U_1 \times U_2 \to U_i$ defined by $p_i (u_1,u_2)=u_i$, $i=1,2$. A *reduction* of $<o,u_1,u_2,U_1,U_2>$ defined by p_i is the counterfactual deformation defined by:

$$<o,u_1,u_2,U_1,U_2> \Rightarrow <o,u_i,U_i>, \ i=1,2.$$

A *transcendentalization* defined by p_i is an "inverse" deformation defined by $<o,u_i,U_i> \Rightarrow <o,u_1,u_2,U_1,U_2>$, $i=1,2$.

Another important example of a structured universe of properties U is the case if there is a distinguished *extremal* element $u_0 \in U$. For instance, if U has the structure of a vectorspace, the base point 0 is often chosen as an extremal element:

(1.3) Definition. Let U be a universe of properties with distinguished extremal element $u_0 \in U$. An *ideation* of $<o,u,U>$ is a counterfactual deformation $<o,u,U> \Rightarrow <o,u_0,U>$.

The various kinds of counterfactual deformations described so far may be combined with each other in various ways yielding a bunch of counterfactual deformations. In the following we shall concentrate on one special case of counterfactual deformation, to wit, *idealization* defined by:

(1.4) Definition. The counterfactual deformation of *idealization* is defined as the combination of *reduction* and *ideation*.

To give a concrete example of idealization let us consider the idealizations involved in an elementary treatment of the simple pendulum [cf. Laymon 1987: 204]. We can treat realistically the bob as being physically extended, or we can perform a counterfactual idealization treating it as a point mass. We can treat the sine of the angle w of displacement as exactly $sin(w)$, or we can perform a soft counterfactual deformation treating it as w. Moreover, we can realistically treat the medium as having hydrostatic effects on the pendulum, or we can perform the idealizing deformation as if it were vacuum. We may conceive these counterfactual deformations of the physical system "symple pendulum" P as caused by counterfactual deformation *operators* b,w, and m applied to P in the following way:

b: P (extended bob) $\Rightarrow P$ (point mass bob)	(ideation)
w: P $(sin(w)) \Rightarrow P(w)$	(soft deformation)
m: P (hydrostatic) $\Rightarrow P$ (vacuum)	(reduction)

The operators b,w, and m may be combined with each other. For instance, $bw(P)$ is to be read as: first the system P is subjected to the counterfactual deformation that the sine of the angle w of displacement is taken to be w, and then the system $w(P)$ is subjected to the counterfactual deformation of treating the pendulum's bob as a point mass.

We observe that the combinations of the operators satisfy the following properties:

(1)	$bb = b$, $ww = w$, $mm = m$	(idempotence)
(2)	$bw = wb$, $bm = mb$, $mw = wm$	(commutativity)
(3)	$(bw)m = b(wm)$	(associativity)

Thus, if we add the trivial deformation operator *id* that does not change anything at all, the set *(b,w,m,id)* can be endowed with the structure of a complete *(semi)lattice* [cf. Laymon 1987: 204] that can be displayed in the following diagram:

(1.5) Lattice of deformation operators of the pendulum.

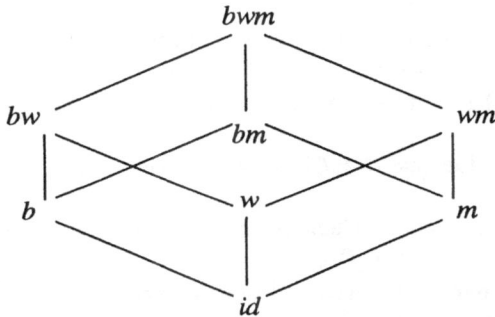

In the following we want to argue that this lattice structure of the set of counterfactual deformation operators is typical for the role idealization plays in the application of empirical theories to phenomena of the real world.

II

Counterfactual Deformations in the Structuralist Approach

We now want to reformulate the sketch of counterfactual deformations given above in the framework of the structuralist philosophy of science. This leads to a better understanding of how idealization works, or so we want to argue. In reformulating the theory of counterfactual deformation in the structuralist framework, we take the above operator approach seriously. This means, we conceive counterfactual deformations as operators defined on the class M_p of possible models of a structuralist theory element.

Recall that the general format of a structuralist (partial, potential) model is the following [cf. Balzer, Moulines and Sneed 1987]:

(2.1) $x = \langle A_1,...,A_n,f_1,...,f_p \rangle$

where the A_j's are the base sets, i.e., sets of empirical entities like particles, molecules, fields, genes, commodities, persons, the theory is about, and possibly auxiliary mathematical entities e.g. the real numbers R etc. In the following we'll be somewhat more explicit about the relations f_i, mentioning explicitly their carriers U_i, i.e., instead of (2.1) we write

$$(2.2) \quad x = <A_1,...,A_n,f_1,...f_p,U_1,...,U_p>$$

where each f_i is a subset of U_i. When appropriate we abbreviate (2.2) by $x=<A,f,U>$. Then we may generalize Nowak's counterfactual deformations for (partial) (potential) models of structuralist models as follows:

(2.3) Definition. Let $x=<A,f,U>$ be a structuralist (partial) (potential) model.

(a) x' is a soft counterfactual deformation of x if and only if x' is of the form $x'=<A,f',U>$.

(b) x' is a hard counterfactual deformation of x if and only if x' is of the form $x'=<A,f',U'>$ and $U\neq U'$.

According to structuralism the application of an empirical theory is not a monolithic all-or-nothing affair, a theory has not one single universal application to the world as a whole, rather it possesses a variagated and extended family I of intended applications. However, different applications usually are not independent of each other but interrelated. This interrelation can be explicated by the concept of structuralist constraints [cf. Balzer, Moulines and Sneed 1987]. With respect to the problem of idealization this means that we should define counterfactual deformations not for a single isolated model $x\in M_p$ but for all elements of M_p simultaneously. This is made precise in the following definition:

(2.4) Definition. Let $T=<K,I>$ be a structuralist theory element, $K= <M,M_p,M_{pp},r,C>$. A *counterfactual deformation operator d* of T is a map $d: M_p \rightarrow M_p$ of the following form:

(1) $d(<A,f,U>) = <A,f',U'>$

(2) $dd = d$

(3) $d(M) \subseteq M$.

In other words, d *is a projection of M_p onto itself that preserves the subset M of M_p*. The first condition doesn't need any further comment, it is just Nowak's definition applied to structuralist models. The second condition may be explained by the mass point idealization for the pendulum already mentioned above. If we counterfactually assume that the bob of a pendulum is a mass point, a second application of the same counterfactual deformation procedure to that system does not yield anything new, i.e., we have $d(d(x)) = d(x)$. The third condition describes, so to speak, the *direction* of the counterfactual deformation operators: Their *raison d'être* is to eliminate certain factualities that hinder potential models from being actual ones. The purpose of counterfactual deformation is to transform a "good" potential model into an actual one. Of course, this cannot be done for any potential model but we should require that the procedure of counterfactual idealizing does not lead us astray, i.e., if x is already a model of T any idealizing deformation $d(x)$ of x should also be a model of T. As a reasonable generalization of the lattice of counterfactual deformation operators of the pendulum theory we now propose the following assumption concerning the structure of the deformation operators of a structuralist theory element:

(2.5) Assumption. Let $T = <K,I>$ be a structuralist theory element, and D its set of counterfactual deformation operators. Then the set D is a (complete) *semilattice*, i.e., the concatenation symbolized by ⊛ of counterfactual deformations enjoys the following properties:

(1) $d \circledast d = d$ (idempotence)
(2) $d \circledast d' = d' \circledast d$ (commutativity)
(3) $(d \circledast d') \circledast d'' = d \circledast (d' \circledast d'')$ (associativity).

These conditions are read off directly from Laymon's example (1.5). One could ask more specific questions about the specific structure of D, e.g. is D a special kind of semilattice, e.g. a distributive or even Boolean lattice? However, there is not the space here to pursue these kinds of questions any further. We are content with the general observation that the semilattice of counterfactual deformation adds a further element of concretization to the structuralist notion of a theory element:

(2.6) Definition. A *theory element T with an idealization structure D* is a structuralist theory element $<K,I>$ endowed with a semilattice D of counterfactual deformation operators defined on M_p. It is denoted by $<K,I,D>$.

The completion of a theory element by a semilattice of counterfactual deformation operators leads to a refining of the empirical claim of a theory:

(2.7) Definition. Let $T = <K,I,D>$ be a theory element with idealization structure. Denote the fibre of potential models over x by $r^{-1}(x)$, i.e.,

$r^{-1}(x) := \{y \ / \ y \in M_p$ and $r(y)=x\}$. Then *the empirical claim of T (without constraints)* is the assertion that for all $x \in I$ there is an appropriate $d_x \in D$ such that the following holds:

$$d_x(r^{-1}(x)) \ \cap \ M \neq \emptyset.$$

Stated informally, this definition requires that all intended applications of T can be extended to potential models of the theory in such a way that appropriate counterfactual deformations of these potential models yield actual models of T. Obviously, if D is trivial, i.e., $D=\{id\}$, (2.7) boils down to the familiar definition of the empirical claim of a theory element. It might be useful to explain the steps of this counterfactual path from data to theory [cf. Laymon 1982] in some detail. For a single intended application x this runs as follows:

First Step: Embedding of the data into the theoretical framework.
  In the structuralist framework this means to expand a partial model x to the set $r^{-1}(x) := \{y/r(y)=x\}$ of potential models that are projected by r onto x. Usually, this set has more than one element, i.e. theoretical expansion in the framework of T is not unique.
Second Step: Applying appropriate counterfactual deformation operators to the set $r^{-1}(x)$: $r^{-1}(x) \Rightarrow d_x(r^{-1}(x))$.
  The application of d might result in a reduction of the number of elements, i.e., there might be y_1, y_2 with $d_x(y_1)=d_x(y_2)$, but usually the set $d_x(r^{-1}(x))$ is not a singleton, i.e., theoretical expansion *plus* counterfactual deformation of x does not yield a uniquely determined result. The most important point, however, is whether the intersection $d_x(r^{-1}(x)) \cap M$ is non-empty. If this set is non-empty the empirical claim of T is true for x.

An analogous 2-step-procedure has to be carried out if we take structuralist constraints into account thereby treating the whole set I of intended applications instead of a single element x of I. Further modifications of (2.7) are necessary when considerations of approximations come into play.

III

Idealization and Concretization

Counterfactual idealizing procedures might be useful to create exact laws and theories that deal with idealized objects and relations like mass points and economic men, however, these laws and theories do not describe the actual world but tell how physical systems would behave under some counterfactual conditions [cf. Niiniluoto 1986: 255]. Thus, idealization is not an end in itself. Rather, with some qualification, it might even be characterized as a necessary evil or a makeshift solution for the problem of applying theories to the actual world which is definitively not solved by idealization alone. Thus, since we hardly can get rid of idealization *ceteris paribus* we may conclude: the less idealization the better. In order that such a maxim makes sense we have to presuppose that idealization is not a yes-or-no affair but comes in degrees. Hence, in this section we engage in the task to define degrees of idealization.

Having described the set idealizing deformation operators as a semilattice there is a canonical partial order on D defined as follows:

(3.1) Definition. Let D be a semilattice. A *partial order between the elements of D* is defined as follows:

$$d \leq d' \text{ iff there is a } d'' \text{ such that } d'' \circledast d = d'.$$

The partial order $(D, \leq)$ defines in a natural way a partial order on M_p in the canonical way:

(3.2) Definition. Let $T = <K,I,D>$ be a theory element with idealization structure. Then the order relation $\leq$ defined on D by (3.1) induces an *order relation on M_p* in the following way:

$$x \leq y \text{ iff there is a } d \in D \text{ with } d(x) = y.$$

Hence, if $d \leq d'$, we get $d(x) \leq d'(x)$ because there is a d'' such that $d''(d(x)) = d'(x)$. This corresponds to the intuitive idea that $d'(x)$ is a stronger counterfactual deformation of x than $d(x)$ since it amounts to the deformation d *plus* another deformation d''.

Having at our disposal the notion of degrees of idealization we now introduce the concept of *concretization* as the inverse of idealization, i.e., concretization = de-idealization, we can describe the order relation defined in (3.2) informally in the following two complementary ways:

$$x \leq y \text{ iff} \begin{cases} x \text{ is a concretization of } y. \\ y \text{ is an idealization of } x. \end{cases}$$

According to the general "holistic philosophy" of the structuralist approach we should not be content with the definition of idealization and of concretization for single potential models, rather we should strive to apply these concepts to theory elements as wholes. This leads to genuinely structuralist versions of idealization and concretization:

(3.3) Definition. Let $T= <K,I,D>$ and $T'= <K',I',D'>$ be structuralist theory elements with idealization structures. T is called a *concretization* of T' iff the following conditions hold:

(1) $K= <M,M'_p,M'_{pp},r'>$, $K'= <M',M'_p,M'_{pp},r'>$

(2) $D \subseteq D'$

(3) for all potential models x of M'_p for which there is a counterfactual deformation operator $d' \in D'$ such that $d'(x) \in M'$ there is a counterfactual deformation operator $d \in D$ with $d \le d'$ such that $d(x) \in M$

(4) $I=I'$.

If T is a *concretization* of T' then T' is called an *idealization* of T. This is denoted by $T' \le T$.

Conditions $(1)-(4)$ are natural requirements for concretizations. They ensure that the counterfactual deformations of a concretization T are not stronger than the counterfactual deformations of the more idealized theory T'. More vividly, this can be expressed by the statement that for a concretization T of a theory T' the counterfactual paths from data (i.e. theoretically expanded intended applications) to theory (i.e. actual models of the theory) [cf. Laymon 1982] are shorter than for the idealization T' of T, or, *vice versa*, that for an idealization T' of T those paths are longer. This is illustrated in the following diagram:

Obviously, the relation $\le$ between theory elements with idealization structure is reflexive and transitive but not symmetric in general. Hence, the "logic of idealization" is a *S4*-logic.

(3.4) Definition. Let $N_L(T_0)=\{T_l \ / \ T_l= <K_l,I,D_l>;\ l\in L\}$ be a set of theory elements with idealization structures. $N_L(T_0)$ is an *idealization net with base* T_0 if and only if the following conditions hold:

(1) there is a $l_0 \in L$ such that for all $l \in L$: $T_l \le T_{l_0}$

(2) If $T_l \le T_{l'}$ and $T_{l'} \le T_l$ then $T_l = T_{l'}$

In the next section we'll show that the idealization procedures that take place in empirical science can be described with the help of idealization nets. The example we'll discuss is the paradigmatic example of the Poznań school, to wit, the Marxian law of value.

IV

The Counterfactual Character of Empirical Laws

As has often been observed, most laws of most scientific theories are counterfactual laws, i.e. they do not directly apply to any actual objects in any actual situation, rather they tell us how physical or social systems would behave under idealized counterfactual conditions [cf. Niiniluoto 1986: 255]. A famous case in question is Marx's law of value that asserts that under certain counterfactual conditions C_i the price ratio of any two commodities equals the ratio of their values. These conditions have been carefully explicated in Nowak [1980: 3−22]. Following Hamminga [1989] we can state a highly idealized version of the law of value as follows:

(4.1) $\quad C_1 \& C_2 \& \ldots \& C_8 \Rightarrow (p(x) = w(x))$,

where $\Rightarrow$ is to be interpreted as a counterfactual conditional. To recast (4.1) in the framework of possible world semantics let us first recall a piece of jargon of possible world semantics. If x is a possible world where a proposition p holds this x is called a p-world. Let us assume that there is a certain "neighborhood" system [cf. Lewis 1986] for the actual world a_0 such that it makes sense to speak about nearness. Then (4.1) is rendered as follows:

(4.2) $\quad$ (The $C_1 \& C_2 \& \ldots \& C_8)$-world nearest to the actual world a_0 is a $(p(x) = w(x))$-world.

However, as Nowak rightly emphasizes, (4.2) certainly is not the whole story to be told about the counterfactual character of the law of value or any other scientific law. Thus, even if one accepts Lewis' explication of counterfactuals in the framework of possible world semantics (or some similar account) such an explication is seriously incomplete. What is missing might be called the interplay of concretization and idealization. Science is not content to state some counterfactual idealized laws about ideal objects but rather strives to get rid of these idealizations, at least partially. This may sound somewhat paradoxical from the view point of common sense, as

Nowak has pointed out, but there are good reasons to consider this procedure as the − or at least an important − method of science. Nowak describes this approach labelled as *the* Marxian or even as *the* scientific method [cf. Cartwright 1989: 204] of building theories more precisely as follows [1980: 21f]:

(1) Marx introduces some assumptions which he knows a priori to be false in empirical reality.

(2) With the above assumptions in mind he proposes the formula revealing what the phenomena in question depend on. The formula is based, then, on assumptions that do not hold in empirical conditions.

(3) These counterfactual assumptions are then removed, and the consequent of the law in question is corrected correspondingly. Thus, we obtain the "transformed forms" of the initial law which deal with conditions that are less and less abstract (i.e., satisfy less and less counterfactual assumptions). At the same time, those conditions come closer and closer to the empirical ones.

We now want to show that this Marxian procedure can be reconstructed in the structuralist framework with the aid of idealization nets introduced in the preceding section. More precisely, we reconstruct Marx's approach as building up a rather special idealization net $N(T_0)$ such that the assertion of the law of value can be identified with the empirical claim of that net.

The base element $T_0 = <K_0, I, D_0>$ of the Marxian net can be described as a theory element whose semilattice D_0 of counterfactual deformation operators is generated by operators $d_1, d_2, \ldots, d_8$ that correspond to the conditions $C_1, C_2, \ldots, C_8$ respectively. The class M_0 of models of T_0 satisfies the highly idealized law of value $(p(x) = w(x))$. The net itself is a chain

$$T_8 \leq T_7 \leq \ldots \leq T_1 \leq T_0.$$

The semilattices of counterfactual deformation operators D_i of the $T_i = <K_i, I, D_i>$ are generated by the operators $d_1, \ldots, d_i$ respectively, and the classes of models M_i are characterized by the condition that their elements satisfy concretized versions $f_i(p(x), w(x))$, the f_i being certain functions of $p(x)$ and $w(x)$. Then the "paradoxical" way of idealization and concretization of Marxian science in the case of the law of value can be diagrammatically described as follows:

$$d_0(x) \in M_0$$
$$d_1(x) \in M_1$$
$$d_2(x) \in M_2$$
$$\cdots \qquad \cdots$$
$$x \qquad \longrightarrow \qquad d_8(x) \in M_8$$

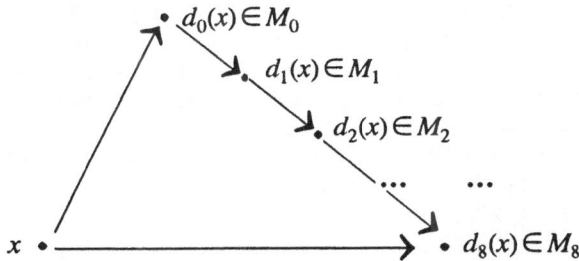

Thus, as the upshot of our sketch of *structuralism cum idealization* we can characterize a scientific law as a cascade of counterfactual propositions defined in the framework of an idealization net of theory elements.

Andoni Ibarra
Departamento de Lógica y Filosofía de la Ciencia
Univ. del País Vasco/Euskal Herriko Unibertsitatea
1249 Posta Kutxa/Apartado 1249
20008 Donostia/San Sebastian, Spain

Thomas Mormann
Institut für Philosophie, Logik und Wissenschaftstheorie
Ludwig-Maximilians-Universität München
Ludwigstrasse 31
80539 München, Germany

REFERENCES

Balzer, W., Moulines, C.-U. and Sneed, J.D. [1987]. *An Architectonic for Science*. Dordrecht: Reidel.

Cartwright, N. [1989]. *Nature's Capacities and Their Measurement*. Oxford: Clarendon Press.

Hamminga, B. [1989]. Sneed versus Nowak: an Illustration in Economics. *Erkenntnis* **30**, 247–265.

Krajewski, W. [1977]. *Correspondence Principle and Growth of Science*. Dordrecht: Reidel.

Kuokkanen, M. [1988]. The Poznań School Methodology of Idealization and Concretization from the Point of View of a Revised Structuralist Theory Conception. *Erkenntnis* **28**, 97–115.

Laymon, R. [1982]. Scientific Realism and the Counterfactual Path from Data to Theory. In: P. Asquith and T. Nickles (Eds.). *PSA 1982*, vol. 1. East Lansing, Philosophy of Science Association, 107–121.

Laymon, R. [1985]. Idealizations and the Testing of Theories by Experimentation. In: P. Achinstein and O. Hannaway (Eds.). *Observation, Experiment and Hypothesis in Modern Science*. Boston: MIT Press and Bradford Books, 147−173.

Laymon, R. [1987]. Using Scott Domains to Explicate the Notions of Approximate and Idealized Data. *Philosophy of Science* **54**, 194−221.

Lewis, D. [1986] (1973). *Counterfactuals*. Oxford: Basil Blackwell.

Miroiu, A. [1984]. A Modal Approach to Sneed's 'Theoretical Functions'. *Philosophia Naturalis* **21**, 273−286.

Niiniluoto, I. [1986]. Theories, Approximations, Idealizations. In: R. Barcan Marcus, G.J.W. Dorn and P. Weingartner (Eds.). *Logic, Methodology, and Philosophy of Science VII*. Amsterdam, North-Holland, 255−289.

Nowak, L. [1980]. *The Structure of Idealization. Towards a Systematic Interpretation of the Marxian Idea of Science*. Dordrecht: Reidel.

Nowak, L. [1989]. On the (Idealizational) Structure of Economic Theories. *Erkenntnis* **30**, 225−246.

Nowak, L. [1991]. Thoughts are Facts in Possible Worlds, Truths are Facts of a Given World. *Dialectica* **45**, 273−287.

Stegmüller, W. [1979]. *The Structuralist View of Theories − A Possible Analogue of the Bourbaki Programme in Physical Science*. Berlin etc.: Springer.

Poznań Studies in the Philosophy
of the Sciences and the Humanities
Vol. 42, pp. 95–108

Ilkka A. Kieseppä

ASSESSING THE STRUCTURALIST THEORY
OF VERISIMILITUDE

In a number of papers, including Kuipers [1982], [1987b] and [1992], Theo
A.F. Kuipers has developed a structuralist theory of truthlikeness. In his
most recent paper, he is concerned with the problem of *theoretical* truthlikeness rather than with the one of *descriptive* truthlikeness; *i.e.* with the
truthlikeness of hypotheses characterizing the subset of *empirical possibilities*, when the set of *conceptual possibilities* is given, rather than with that
of hypotheses about the actual possibility, or the possibility that has been
realized. However, the problem of theoretical truthlikeness reduces to the
descriptive one when there is just one empirically possible alternative, the
one that has been realized, and thus his formalism can be used for both purposes. In what follows, I shall first prove a theorem which shows the
limited applicability of the Kuipers's definition when dealing with the latter
problem, and then construct an example of the former problem in which the
definition leads to counterintuitive results.

I shall concentrate on a particularly important special case consisting of
cognitive problems which typically arise when predictions about physical
experiments and systems are made; namely, I shall discuss the case in
which the outcome of an experiment, or the state of a physical system about
which the two theories make predictions, can be characterized by n quantitative variables or real numbers $r_1, r_2, \ldots, r_n$. In this case, the conceptually
possible states of the system can be thought of as elements of $\mathbb{R}^n$.

Examples of systems falling under both of these restrictive assumptions
would be, *e.g.*, a gas characterized by its pressure, temperature and volume, and a set of n identifiable classical particles whose state can be
characterized by the their $3n$ position coordinates and $3n$ velocity components provided that n is known. An example of a situation not satisfying
the restrictive assumptions is the case in which two theories about a system
of this kind give differing answers, not only to the question what the

positions and velocities of the particles are, but also to the question of how many particles there are.

I
The Refined Definition

Kuipers considers a set of potential models M_p corresponding to the conceptual possibilities and a subset $T \subseteq M_p$, representing the set of empirically possible cases. Other subsets $X, Y \subseteq M_p$ can then be used to represent various hypotheses stating which models are empirically possible. Kuipers distinguishes between *a naive definition* and *a refined definition* of truthlikeness, the former being simply a definition made in terms of the symmetric difference of T and each hypothesis, *i.e.* Y is closer to T than X if and only if $Y \Delta T \subseteq X \Delta T$. It is the latter definition that I shall be concerned with in what follows.

In the refined definition, it is presupposed that a *relation s of betweenness* has been defined on M_p; *i.e.* that $s(x,y,z)$ is valid for some $x,y,z \in M_p$ if and only if y is in some sense between x and z. Kuipers considers various properies that such a relation s might or might not have. What he calls *minimal s-conditions* are the conditions $(1)-(3)$ as follows:[1]

(1) $s(x,x,x)$ for each $x \in M_p$,

(2) if $s(x,y,x)$ for some $x,y \in M_p$, then $x=y$,

(3) if $s(x,y,z)$ for some $x,y,z \in M_p$, then $s(x,x,y)$, $s(x,x,z)$, $s(y,y,z)$, $s(x,y,y)$, $s(x,z,z)$ and $s(y,z,z)$.

With the help of the three-placed relation s Kuipers defines a *two-placed relation $r(x,z)$ of relatedness*:[2]

(4) $r(x,z)$ if and only if there is a y such that $s(x,y,z)$.

The claim that $r(x,y)$ means, in a sense, that models x and y can be compared with each other.

Taking the three-placed relation s satisfying $(1)-(3)$ on M_p as given, Kuipers's definition of truthlikeness amounts to the following:[3]

---

1 Kuipers [1992], p. 316.

2 *Ibid.*, p. 315.

3 *Ibid.*, p. 320.

Definition 1. If M_p, s and r are defined as above and if $X,Y,T \subseteq M_p$, then Y is *at least as close to T* as X if and only if the following conditions 1.1. and 1.2. obtain:

    1.1.   For all $x \in X$ and $z \in T$, if $r(x,z)$, then there is a $y \in Y$ such that $s(x,y,z)$.

    1.2.   For all $y \in Y - X \cup T$ there are $x \in X - T$ and $z \in T$ such that $s(x,y,z)$.

And it sounds natural to make the following addition:

Definition 2. If M_p is as above and $X,Y,T \subseteq M_p$, then Y is *closer to T than X* if Y is at least as close to T as X and X is not at least as close to T as Y.

Thus, Kuipers's definition states essentially that *1)* for any empirical possibility z and for any possibility x allowed by the hypothesis X there is a possibility y allowed by the hypotheses Y which is between z and x, and thus at least as close to z as x, and that *2)* for any possibility y incorrectly allowed by Y there is a possibility x incorrectly allowed by X which is still worse in the sense that it is further from some actual possibility z than y; *i.e.*, it is such that y is between x and z.

The cases in which either the set representing the truth or the hypotheses are *convex* are particularly important for Kuipers, as several of his theorems are concerned with these special cases. Here convexity is defined in terms of the relation s: *i.e.*, a set X is convex if and only if it is the case that if x,y,z are such that $x,z \in X$ and $s(x,y,z)$, then $y \in X$. Kuipers justifies the assumption of convexity by noting that "[i]f, for instance, T is not convex, this has the consequence that there is an empirical *im*possibility ... between two empirical possibilities, and this is unlikely as far as nature is continuous."[4] However, I shall give in part III below a counter-example which is, in my view, by no means untypical and which demonstrates how actual scientific theories lead to situations in which hypotheses or the set of empirical possibilities are not convex.

II
Applying the Refined Definition to Problems of Quantitative Nature

Next, I shall try to form a more concrete idea of what this proposal means for the special case mentioned above with which this paper is concerned, or

---

4 *Ibid.*, p. 319.

for the case in which the conceptual possibilities correspond to elements of $\mathbb{R}^n$.

In what follows I shall take the liberty of representing the conceptual possibilities simply as n-tuples of real numbers, so that $M_p = \mathbb{R}^n$, instead of introducing the whole structuralist machinery for this special case. If definitions 1 and 2 are used when discussing the problem of descriptive truthlikeness, it has to be assumed that $T = \{(r_1, .., r_n)\}$ is a set containing just one element; when discussing theoretical truthlikeness T is, of course, normally an infinite subset of $\mathbb{R}^n$.

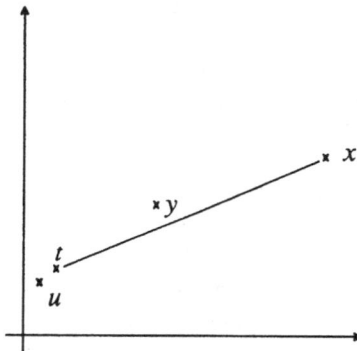

Figure 1.

The relation s of betweenness in $\mathbb{R}^n$ has to be fixed before giving any verdict on the usefulness of Kuipers's proposal for the special case considered in this paper. An idea that first comes into mind is to take the concept of betweenness literally and to choose the relation s_1 given by

(5) $s_1(x,y,z)$ iff $(\exists t \in [0,1])(y = x + t(z-x))$

where the operation of addition in $\mathbb{R}^n$ and the multiplication of elements of $\mathbb{R}^n$ by the scalar t are defined in the usual way. The meaning of $s_1(x,y,z)$ is that y is literally between x and z: i.e. that x, y, and z are all on the same straight line and y is between x and z on it.

This definition of betweenness is, however, clearly unsatisfactory, as is made clear by thinking of hypotheses X and Y in definition 1 as point estimates and the truth T as one point. Surely one would wish to say that in figure 1, y is between x and t so that $Y = \{y\}$ is closer to $T = \{t\}$ than $X = \{x\}$ in the sense of definition 2, although y is slightly displaced from the straight line leading from x to t.

Indeed, Kuipers proposes the following relation of betweenness, which I shall denote by s_2, for multi-dimensional conceptual structures:[5]

(6)   If $x=(x_1,..,x_n)$, $y=(y_1,..,y_n)$ and $z=(z_1,..,z_n)$, then $s_2(x,y,z)$ if and only if for each $i=1,..,n$, $x_i \leq y_i \leq z_i$ or $z_i \leq y_i \leq x_i$.

Thus, $s_2(x,y,z)$ holds if and only if each component of y is between the corresponding components of x and z. A unintuitive consequence of this definition is that no comparisons of hypotheses that are on different "sides" of the truth can be made with its help; thus, if we take s to be s_2 in definitions 1 and 2, it cannot be said that, in figure 1, the hypothesis $U=\{u\}$ is closer to the truth $T=\{t\}$ than $X=\{x\}$, although this can be said of $Y=\{y\}$. A third possibility would be to clearly separate the relation s of betweenness from any geometric images of being between, and to take s to be s_3 defined by

(7)   If $x=(x_1,..,x_n)$, $y=(y_1,..,y_n)$ and $z=(z_1,..,z_n)$, then $s_3(x,y,z)$ if and only if for each $i=1,..,n$, $|z_i-y_i| \leq |z_i-x_i|$.

Thus, according to the third proposal, y is between x and z if for each coordinate $i=1,..,n$, the distance from z_i to y_i is not larger than the distance from z_i to x_i, even if y_i should be on the other side of z_i from x_i. It is obvious that for each of the three relation s_1, s_2 and s_3 the minimal s-conditions $(1)-(3)$ as well as *transitivity* — i.e., $s(x,y,z)$ and $s(y,z,t)$ imply $s(x,y,t)$ and $s(x,z,t)$ — are trivially satisfied, and that $r(x,y)$ is valid for any $x,y \in \mathbb{R}^n$.

As already stated, definition 1 above has originally been intended to apply to the problem of *theoretical* verisimilitude. However, nothing prevents us from applying this definition to the descriptive case by choosing a set of empirically possible models consisting of one model only; namely, the actual one. In particular, if one thought of the initial state of the modelled system as given, believed that the system was governed by deterministic laws and was interested in its state at an exactly specified point of time, one would have to take only one model to represent a genuine empirical possibility.

In what follows, I shall make a further restrictive assumption and discuss the case in which the hypotheses are represented by *closed* (with respect to the usual topology of $\mathbb{R}^n$, of course) and *bounded* sets. This is just a special

---

5 *Ibid.*, p. 318.

case of definition 1, but it is an important one, which includes hypotheses represented by finite curves containing their end points, bounded surfaces containing their boundary, and, as a matter of fact, any bounded hypersurface that is smooth enough and contains its own boundary. By considering this special case, I shall demonstrate that only a small part of the information contained in hypotheses is relevant for judgements of relative truthlikeness in Kuipers's sense. In what follows, I shall use the relation of betweenness s_2 given by (6) which has been proposed by Kuipers, but the reader is invited to verify that the lemma and theorem below remain valid if one substitutes s_3 given by (7) for s_2 in the proofs that follow.

First, let us define,

(8) $\quad In_t(X) = \{x \in X \mid \sim (\exists y \in X)(y \neq x \wedge s_2(x,y,t))$

and

(9) $\quad Out_t(X) = \{x \in X \mid \sim (\exists y \in X)(y \neq x \wedge s_2(y,x,t))$

for each $X \subseteq \mathbb{R}^n$ and for each $t \in \mathbb{R}^n$. Intuitively, $In_t(X)$ contains the points of X that are closest to the truth and $Out_t(X)$ contains the points furthest away from the truth. The following lemma is needed in what follows:

Lemma. Suppose that $X \subseteq \mathbb{R}^n$ is closed and bounded, that $t \in \mathbb{R}^n$, and that $In_t(X)$ and $Out_t(X)$ are given by (8) and (9), respectively. Then, for every $x \in X$ there is a $y \in In_t(X)$ such that $s_2(x,y,t)$ and a $z \in Out_t(X)$ such that $s_2(z,x,t)$.

Proof. Let $x \in X$ be arbitrary. Consider the continuous function

$$f : \mathbb{R}^n \to \mathbb{R}, f(u) = \sum_{j=1}^n |u_j - t_j|,$$

and the sets X' and X'' given by

$$X' = X \cap \{u \in \mathbb{R}^n \mid s_2(x,u,t)\} =$$
$$X \cap \{u \in \mathbb{R}^n \mid \text{for all } i = 1,\ldots,n, \, t_i \leq u_i \leq x_i \vee x_i \leq u_i \leq t_i\}$$

and

$$X'' = X \cap \{u \in \mathbb{R}^n \mid s_2(u,x,t)\} =$$
$$X \cap \{u \in \mathbb{R}^n \mid \text{for all } i = 1,\ldots,n, \, t_i \leq x_i \leq u_i \vee u_i \leq x_i \leq t_i\}$$

respectively. The sets X' and X'' are compact, being the intersection of a compact and a closed set, so that their images under the continuous mapping f, $f(X')$ and $f(X'')$, are compact and therefore closed. Define the numbers d and D by

$d = \inf f(X')$ and $D = \sup f(X'')$

respectively. Obviously, if $s_2(x,u,t)$ then $f(u) \le f(x)$ and if $s_2(u,x,t)$ then $f(x) \le f(u)$ so that $0 \le d \le D$. Because X'' is bounded as a subset of the bounded set X, $D < \infty$. Because $f(X')$ and $f(X'')$ are closed, $d \in f(X')$ and $D \in f(X'')$. Thus, there are $y \in X'$ and $z \in X''$ such that $f(y)=d$ and $f(z)=D$.

Next it will be shown that these y and z satisfy the requirements of the lemma. By the definitions of X' and X'' it follows that $s_2(x,y,t)$, $s_2(z,x,t)$, and that $x,y \in X$. Now, obviously, $y \in In_t(X)$, because if there were some $u \in X$, $u \ne y$ such that $s_2(y,u,t)$, it would follow from the transitivity of s_2 that $s_2(x,u,t)$, and that $\sum_{j=1}^{n} |u_j - t_j| < d$, which together contradict the definition of d. Similarly, $z \in Out_t(X)$, because if there were some $u \in X$, $u \ne z$ such that $s_2(u,z,t)$, it would follow from the transitivity of s_2 that $s_2(u,x,t)$, and that $\sum_{j=1}^{n} |u_j - t_j| > D$, which together contradict the definition of D. $\square$

In a sense, the above lemma states that closed sets contain their boundaries defined in terms of the relation $s_2(\cdot,\cdot,t)$. The following theorem follows now easily:

Theorem. Suppose that $M_p = \mathbb{R}^n$, $X,Y \subseteq M_p$ are non-empty, closed and bounded and that $t \in \mathbb{R}^n$. Then Y is at least as close to $\{t\}$ as X in the sense of definition 1 if and only if
  i) $(\forall x \in In_t(X))(\exists y \in In_t(Y))s_2(x,y,t)$, and
  ii) $(\forall y \in Out_t(Y))(\exists x \in Out_t(X))s_2(x,y,t)$.
Proof. To see the necessity of the conditions *i)* and *ii)*, suppose that Y is at least as close to $\{t\}$ as X. To prove *i)*, choose an arbitrary $x \in In_t(X)$. By condition 1.1. in definition 1, a $y' \in Y$ can be found such that $s_2(x,y',t)$. The lemma above shows that we can choose $y \in Y$ so that $y \in In_t(Y)$ and $s_2(y', y,t)$. It follows that from the transitivity of s_2 that $s_2(x,y,t)$.

Similarly, in order to prove *ii)* suppose that $y \in Out_t(Y)$ is arbitrary. If $y=t$, then it is trivially true of any $x \in Out_t(X)$ that $s_2(x,y,t)$. If $y \in X$, there is by the lemma an $x \in Out_t(X)$ such that $s_2(x,y,t)$. Suppose then that $y \in Out_t(Y)-X \cup \{t\} \subseteq Y-X \cup \{t\}$. By condition 1.2. of definition 1, there is an $x' \in X$ such that $s_2(x',y,t)$. The lemma above implies that there is an $x \in Out_t(X)$ such that $s_2(x,x',t)$, so that, by the transitivity of s_2, $s_2(x,y,t)$.

To prove the sufficiency of the conditions, suppose that conditions *i)* and *ii)* are valid. Choose an arbitrary $x \in X$. The lemma implies that there is an $x' \in In_t(X)$ such that $s_2(x,x',t)$. It follows from the condition *i)*, there is a $y \in In_t(Y) \subseteq Y$ such that $s_2(x',y,t)$. By the transitivity of s_2, $s_2(x,y,t)$. This proves the condition 1.1. of definition 1.

Suppose then that $y \in Y-X \cup \{t\}$. The lemma implies that there is a $y' \in$ $Out_t(Y)$ such that $s_2(y',y,t)$. Because $y \neq t$, it follows that $y' \neq t$. By $ii)$, there is a $x \in Out_t(X) \subseteq X$ such that $s_2(x,y',t)$. Because $y' \neq t$, it follows that $x \neq t$. By the transitivity of s_2, $s_2(x,y,t)$. This proves condition 1.2. of definition 1, and completes the proof of this theorem. $\square$

This result means that Kuipers's definition makes use only of information concerning the parts of each hypothesis closest to or furthest from the truth. In particular, it does not differentiate between hypotheses such as A and B depicted in figure 2. In this two-dimensional case, in which the truth is represented by the point t, each of the hypotheses A and B will be judged to be as close to the truth as the other, although most of A lies nearer to the truth than most of B.

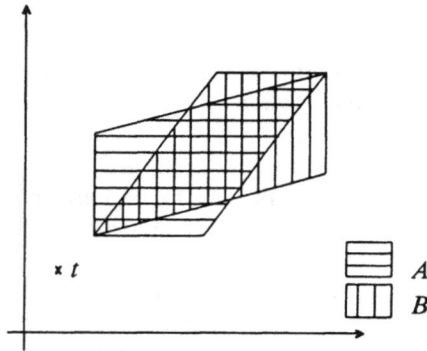

Figure 2.

Should this be considered to be a weakness of definition 1? It contradicts at least the intuitions that underlie both the average measure of truthlikeness proposed by Graham Oddie[6] and the various measures of Ilkka Niiniluoto[7]. The hypothesis A has a smaller average distance from the truth than B so that it is closer to the truth than B in Oddie's sense. Similarly, if one follows Niiniluoto whose favorite measure, the *min-sum-measure* $M_{ms}^{\gamma\gamma'}$,[8] measures distance from the truth by *1)* the sum of the distances of all false alternatives not excluded the hypothesis and by *2)* the minimum distance of

---

6 Oddie [1986], p. 49.
7 Niiniluoto [1987], pp. 222–232.
8 This measure is defined on p. 228 in *ibid.*

the hypothesis from the truth, one must conclude that A is closer to the truth if one replaces the sum in Niiniluoto's definition by a surface integral.

Of course, the intuition that A should count as being closer to the truth than B might simply be denied; indeed, discussions on conflicting intuitions of verisimilitude can easily turn into what have been referred to as "Byzantine disputations" by one participant in them.[9] Anyway, it can be concluded that if one uses the definition proposed by Kuipers, one cannot make all the judgements of relative truthlikeness that have been considered to be intuitively obvious in the earlier literature on the subject.

III
A Counter-example to the Refined Definition

Another, perhaps more serious criticism of Kuipers's proposal is that it seems to suit only the case in which the truth is represented by a convex set. It has already been observed that Kuipers seems to think that this case was typical. Making an excursion into the problem of theoretical verisimilitude, I shall give a simple physical example which serves to show that such an assumption is unwarranted. Even very simple systems that are actually discussed by special scientists, e.g. by physicists as in the case below, lead to non-convex sets of empirical possibilities, and in such cases Kuipers's definition has implausible consequences.

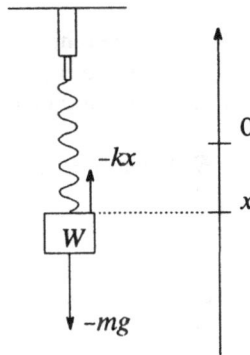

Figure 3.

---

9 Bonilla [1992], p. 352.

Suppose that there is a spring to which a body W of mass m is attached [see figure 3]. The state of this physical system can be characterized at each moment of time by the displacement x of the weight from the height at which the spring exerts no force on it, and by its velocity v. Without introducing the whole of structuralist machinery for this simple example, the models for the state of this system can be represented by the pairs (x,v). In what follows, an idealizing assumption, which is customary in elementary physics textbooks, is made. It is supposed that the force F exerted by the spring on the weight is proportional — with some constant of proportionality k — to the displacement from the height $x=0$, so that the total force acting on W, which is the sum of the force of gravity $-mg$ and the force due to the spring, is

(10) $F = -kx - mg$.

Suppose that the weight was originally at rest, which means that originally its displacement was

$$x_0 = -mg/k = constant,$$

so that the potential energy $E_{p,0}$ due to the expansion of the spring was originally

$$E_{p,0} = \tfrac{1}{2}kx_0^2 = \tfrac{1}{2}m^2g^2/k.$$

Suppose further that at time $t=0$ the weight was given the kinetic energy E by a sudden push upwards. It follows from the conservation of energy that, after $t=0$, the displacement x and the velocity v of the weight are related by

$$\tfrac{1}{2}mv^2 + \tfrac{1}{2}kx^2 + mgx = E + mgx_0 + E_{p,0},$$

which is equivalent with

$$\tfrac{1}{2}mv^2 + \tfrac{1}{2}kx^2 + mgx = E - \tfrac{1}{2}m^2g^2/k,$$

and also with

(11) $\tfrac{1}{2}mv^2 + \tfrac{1}{2}k(x + mg/k)^2 = E$.

Finally, suppose that two competing mini-theories make statements on the empirically possible states of the system after time $t=0$. Ignoring any forces other than F given by (10) which might act on W, it can be said that the set T of empirically possible models is the set of pairs (x,v) satisfying (12):

(12) $\quad T = \{(x,v) \mid \frac{1}{2}mv^2 + \frac{1}{2}k(x + mg/k)^2 = E\}$

This is the equation of an ellipse, which turns into a circle with the center $(-mg/k,0)$ if the units are so chosen that m and k have the same numerical value [cf. figure 4]. More generally, if the notation

(13) $\quad Z_{E'} = \{(x,v) \mid \frac{1}{2}mv^2 + \frac{1}{2}k(x + mg/k)^2 = E'\}$

is introduced for each non-negative value of E', then, in particular, $Z_E = T$, and each set $Z_{E'}$ is represented by a circle which has the same center with T in figure 4.

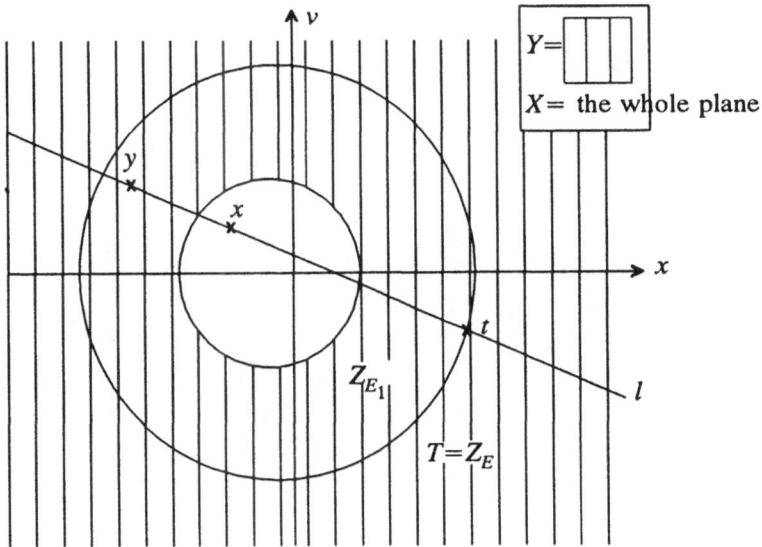

Figure 4.

Now, Kuipers's refined definition has several contra-intuitive consequences when applied to this quite simple example. First, suppose that hypotheses X and Y are given by

$$X = \bigcup \{Z_{E'} \mid E' \geq 0\} \text{ and } Y = \bigcup \{Z_{E'} \mid E' > E_1\},$$

respectively, where

$$0 < E_1 < E.$$

Thus, X is the statement that any combination of x and v is possible and Y is a true hypothesis stating that the energy of W is necessarily larger than E_1. Which one is closer to the truth, X or Y? According to Kuipers's refined definition, they are *both just as close to the truth*. First, it is obvious that Y is at least as close to the truth as X in the sense of definition 1, because it contains T (so that condition 1.1. of definition 1 is trivial) and because $Y \subseteq X$ (so that condition 1.2. of that definition is also trivial). But X is at least as close to the truth as Y just as well, since X also contains T, so that condition 1.1. of definition 1 is trivial, and if an arbitrary $x \in X-Y \cup T = X-Y$ is chosen, it is possible, by drawing a straight line l through the point x [see figure 4], to choose a $y \in Y-T$ which is on l inside the circle T. Using t to denote the point of intersection of T with l on the other side of x from y, it can be concluded that $s_2(y,x,t)$.

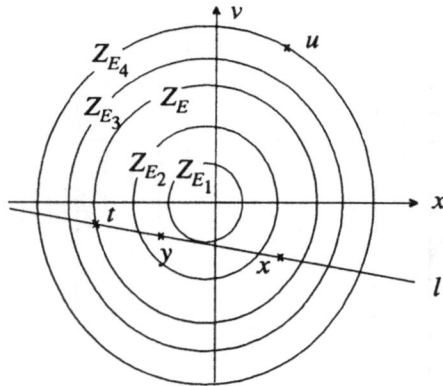

Figure 5.

This result is strange, as it seems obvious that one should say that the informative and true hypothesis Y is closer to the truth than the true but uninformative hypothesis X. One gets results stranger still if one compares two hypotheses that result from different interval estimates for the energy of W. Suppose that $0 < E_1 < E_2 < E < E_3 < E_4$ as in figure 5, and that there are two physicists A and B who do not know exactly what the value of E is although they have characterized the system correctly otherwise. A estimates that $E_1 \leq E \leq E_3$, and B estimates that $E_2 \leq E \leq E_4$. Their hypotheses on the empirically possible states of the system can be represented as X_A and X_B given by

$$X_A = \bigcup \{Z_{E'} \mid E_1 \leq E' \leq E_3\} \text{ and } X_B = \bigcup \{Z_{E'} \mid E_2 \leq E' \leq E_4\}$$

respectively.

Which hypothesis is closer to the truth, X_A or X_B? There might be differing intuitions on this matter. It might be said that they cannot be compared at all, since the shortcomings of the two physicists are different; this would correspond to the intuition behind the original Popperian definition of truthlikeness. It might also be said the results should depend on the relative sizes of the areas X_A-X_B and X_B-X_A, and thus, on the numerical values of the parameters E_1, E_2, E_3 and E_4. However, Kuipers's definition yields the result that X_A *is closer to T than X_B* whenever $E_1 < E_2 < E < E_3 < E_4$, quite independently of the numerical values of the parameters E_1, E_2, E_3 and E_4.

To see this, observe first that X_B cannot be as close to T as X_A, because if some particular $u \in Z_{E4} \subset X_B$-$T \cup X_A$ is chosen, there cannot possibly be any $x \in X_A$ and $t \in T$ such that $s_2(x,u,t)$, thus contradicting condition 1.2 of definition 1. But X_A is at least as close to T as X_B, because condition 1.1. of definition 1 is trivially valid as $T \subset X_A$, and its condition 1.2. can be proved by the same construction that we have already once used. Suppose that $y \in X_A$-$T \cup X_B = X_A$-X_B. [*Cf.* figure 5.] One draws a straight line l through the point y and chooses for $x \in X_B$-T some point belonging to X_B which is on l inside the ellipse $T = Z_E$, and for $t \in T$ the point of intersection of Z_E and l on the other side of y from x. In this way, it can be concluded that $s_2(x,y,t)$, as required. This result means that, in a sense, however small the set X_B-X_A outside T should be in comparison with the set X_A-X_B inside T, it will be judged more to be more important.

These results follow because, if T is not convex, one can choose the models $x \in X$-T and $z \in T$, needed in the second clause of definition 1, in ways which obviously cannot have been intended when the definition was chosen. Indeed, it is my position that these two counterintuitive results show the inadequacy of the second clause proposed by Kuipers in the case in which the set of empirically possible models is not convex.

IV
Conclusion

When comparing predictions of the values of a finite number of quantitative variables, the refined definition of verisimilitude proposed by Theo A.F. Kuipers can be criticized for not yielding any judgement of relative verisimilitude in some cases in which such a judgement seems intuitively obvious. Further, in some cases other than comparing predictions, it can be criticized for yielding quite counterintuitive results. However, it should be observed that, rather than the theory of verisimilitude as such, the main topic of Kuipers [1992] is the relationship between the notion of verisimilitude and the distinction between theoretical and non-theoretical terms, and some methodological rules. I have not discussed these topics here, and the question of whether his theory provides us with new insights on them lies outside the scope of my paper.

Department of Philosophy
University of Helsinki
P.O. Box 24
00014 University of Helsinki, Finland

REFERENCES

Balzer, W., Moulines, C.-U. and Sneed, J.D. [1987]. *An Architectonic for Science. The Structuralist Program*. Dordrecht: Reidel.
Bonilla, J.P.Z. [1992]. Truthlikeness Without Truth. A Methodological Approach. *Synthese* **93**, 343 – 72.
Dugundji, J. [1966]. *Topology*. Boston: Allyn and Bacon Inc.
Kuipers, T.A.F. [1982]. Approaching Descriptive and Theoretical Truth. *Erkenntnis* **18**, 343 – 78.
Kuipers, T.A.F. (Ed.) [1987a]. *What is Closer-to-the-truth? A parade of approaches to truthlikeness*. Amsterdam: Rodopi.
Kuipers, T.A.F. [1987b]. A Structuralist Approach to Truthlikeness. In: Kuipers (Ed.) [1987a], pp. 79 – 99.
Kuipers, T.A.F. [1992]. Naive and Refined Truth Approximation. *Synthese* **93**, 299 – 341.
Niiniluoto, I. [1984]. *Is Science Progressive?* Dordrecht: Reidel.
Niiniluoto, I. [1987]: *Truthlikeness*. Dordrecht: Reidel.
Oddie, G. [1986]. *Likeness to Truth*. Dordrecht: Reidel.
Oddie, G. [1987]. Truthlikeness and The Convexity of Propositions. In: Kuipers [1987a], pp. 197 – 216.

IDEALIZATION, APPROXIMATION
AND THEORY FORMATION

Poznań Studies in the Philosophy
of the Sciences and the Humanities
Vol. 42, pp. 111–126

Leszek Nowak

REMARKS ON THE NATURE OF GALILEO'S
METHODOLOGICAL REVOLUTION

The present writer has already had an opportunity to put a conjecture that the methodological breakthrough made by Galileo consisted in introducing the method of idealization and concretization to physics [Nowak 1971, pp. 35–36]. In this paper, I would like to transform this loose conjecture into a hypothesis, that is, to substantiate it more thoroughly and to analyze its explanatory power.

I

There is a consensus among philosophers of science that the methodological breakthrough in the natural sciences dates back to Galileo. According to an old stereotype, this breakthrough is taken to have consisted in Galileo's rejection of the a priori dogmas of Aristotle's physics thus allowing the observation of the world unprejudiced by any assumptions. "It is by relying on observation that Galileo became the founder of modern physics" states Ernest Mach [1976, p. 101]. The legend has it that Galileo has used the tower of Pisa to drop various heavy objects to the ground and proved — contrary to Aristotle — that they fall with an equal speed[1].

Today, it is easy to correct the legend. As the historians of science made evident, Pisa's tower had been used not by Galileo but by Coresio, an Aristotelian, in order to prove that the heavier the objects the quicker they fall[2]. What is more important, Galileo's breakthrough did not consist in returning to the observation of nature. This would have been superfluous

---

1 It happens that the legend is repeated in advanced handbooks of physics, e.g. Piekara [1964], p. 33.

2 See Butterfield [1958], pp. 80–81 and Cohen (I.B.) [1960], pp. 95–96.

because the Aristotelian theory was always very closely connected with everyday common-sense experience. The basic type of argumentation in Aristotle's *Physics* was a reference to everyday observation. It is not a systematic theory but a totality of explanations that take their origin in observations available to everybody [Cohen (I.B.) 1960, pp. 22ff, Crombie 1959, p. 101):

> Aristotle's physics originates from common sense data; it builds an extraordinarily coherent unity from them. Those data or facts which constitute its foundation are extremely simple, and so obvious that they could create the basis for physics based on naive realism in every epoch [Lesniak 1968, p. xviii].

And Galileo was fully aware of this claiming that "Aristotle. . .preferred sensory experience to any considerations" [Galileo 1962, p. 225]. Indeed, there was no reason for Galileo to bring Aristotle closer to the facts. He was quite close to them.

According to Aristotle, there are two types of motion: natural and forced. All bodies move toward the centre of the world (identified with the center of the Earth) naturally. Each movement in any other direction is unnatural and requires an explanation by revealing a force responsible for it. A body moves as long as the moving force operates; when the force ceases to act, the body falls down to the Earth or finds itself in the state of rest. Observation of moving objects teaches us that two factors influence the movement of bodies: external force and resistance of the environment. In order to make objects move, external force has to be greater than resistance. If the resistance of the environment is constant, then the velocity of a body will be directly proportional to the moving force. The question of how objects behave when there is no resistance was never answered by Aristotle, for first it should be made clear what is a vacuum. And vacuum has never been noticed within this world [Aristotle 1968, 216b]. Galileo knew very well, too, that vacuum does not exist in the physical world. This, however, did not prevent him from asking a question which was senseless in Aristotle's physics: how will "the perfectly round ball move on the plain which is smoothly leveled in order to eliminate all external and accidental obstacles" upon an assumption that the "resistance arising when the ball makes its way and all other obstacles that could arise" are not considered at all? [Galileo 1962, p. 155].

In order to raise this question several assumptions had to be adopted:
(1)(a) the rolling ball is perfectly round,
    (b) the plane is ideally smooth,
    (c) the plane is perfectly spherical,
    (d) the resistance of environment equals zero.

These are obviously idealizing assumptions — i.e., they are consciously adopted as deforming the empirical reality for simplification goals — and therefore all the considerations conducted under them lead to idealizational statements, i.e., conditionals necessarily possessing some idealizing conditions in their antecedents [more about these notions cf. my 1972, 1980, Chap. VIII][3]. Galileo believed that the four conditions (a–d) above are equivalent to the claim that no external forces affect a moving ball. Therefore, he was in a position to claim that if all external obstacles were removed, the movement would last as long as the plane extended, going neither downwards nor upwards. Therefore, if space was infinite, the movement would also be limitless and hence infinite [Galileo 1962, p. 158].

His answer to the above quoted question was then his law of inertia:

(G) "Imagine any particle projected along a horizontal plane without friction; ... this particle will move along this same plane with a motion which is uniform and perpetual, provided the plane has not limits" [Galileo 1963, p. 234]

which may be read as a conditional:

(2)   if x is a rolling ball which is perfectly round, projected along an ideally smooth and perfectly spherical unlimited plane and the resistance of the environment exerted upon x equals zero, then x moves along this plane with uniform perpetual motion.

Galileo was aware of the idealizational nature of thesis (2) as he commented further on: "Even horizontal motion which, if no impediments were present, would be uniform and constant is altered by the resistance of the air and finally ceases" [Galileo 1963, p. 242]. Why then formulate such a statement? For Galileo the reasons for formulating (2) are twofold.

The first way of using the law of inertia is based on the observation that although the movement described in (2) ceases due to the resistance of the air, "the less dense the body, the quicker the process" [ibid.]. And conversely, the more dense the body, the slower does the movement disappear. Therefore, there is a possibility to approximate (2) to the movement of sufficiently dense bodies. In particular, the movement of celestial bodies is,

---

3 Alternative conceptualizations of the method of idealization have been presented by Barr [1971], Suppe [1972], Ludwig [1981], Cartwright [1983, 1989], Niiniluoto [1986, 1990], Kuokkanen [1988], Kuokkanen and Tuomivaara [1992] and others. A (limited) comparative analysis of these approaches is given by Kupracz [1990].

as Galileo stresses, similar to the movement described by the law of inertia [*ibid.*, p. 251].

Another circumstances in which the law of inertia is approximated is the state of balance among the forces. Galileo considers the case of a body falling towards the Earth from a great height which originally moves with a uniformly accelerated motion but after some time the resistance of the air balances "the natural acceleration downwards common to all bodies". Then the body is supposed to move with an approximately uniform motion. This case does not strictly satisfy but merely approximates the law of inertia as there are still numerous forces affecting the body under consideration. For

> if we consider only the resistance which the air offers to the motions studied by us, we shall see that it disturbs them all and disturbs them in an infinite variety of ways corresponding to the infinite variety of form, weight, and velocity of the projectiles [Galileo 1963, p. 242].

That is why in the case of balance of forces one can at most expect an approximation to the law of inertia.

The second way of using the law of inertia is based on the presupposition that the formula (2) can be corrected for some types of motions that occur in the conditions that are far away from the ideal circumstances postulated in (2). As Galileo goes on to say immediately after the above quoted formulation of the law of inertia:

> But if the plane is limited and elevated, then the moving particle, which we imagine to be a very heavy one, will on passing over the edge of the plane acquire, in addition to its previous uniform and perpetual motion, a downward propensity due to its own weight; so that the resulting motion which I call projection, is compounded of one which is uniform and horizontal and of another which is vertical and naturally accelerated [Galileo 1963, p. 234].

And he proves [*ibid.*, p. 235ff] a theorem stating that such a movement may be described as semi-parabolic. The proof is based on the "superposition of two different states", namely, the inertial force "which if acting alone would carry the body at a uniform rate to infinity, and the velocity which results from a natural acceleration downwards common to all bodies" [*ibid.*, p. 207]. This theorem one may read as the statement [*ibid.*, p. 235]:

(3)   if x is a rolling ball which is perfectly round, projected along an ideally smooth and perfectly spherical but limited and elevated plane so that x is carried with a naturally accelerated vertical

motion and the resistance of the environment exerted upon x equals zero, then the path of motion of x is semi-parabola.

(3) may be claimed to be a concretization of (2): some of the idealizing assumptions are waived being replaced with their realistic negations and appropriate corrections to the consequent of the conditional are introduced [see more about this notion in my 1980, Ch. VIII]. Then, according to Galileo, the idealizational law of inertia may be applied to actual cases either by approximation or by concretization[4].

Let us add that, as is well known, Galileo's formulation of the law of inertia is not, in the light of Newton's mechanics, entirely correct [e.g. Krajewski 1974, Such 1978]. Galileo's law of inertia was still based on some elements of Aristotelian conceptual background [shared also by another scientific revolutionary, Copernicus – see Dingle 1959, pp. 24/25].

---

4 It is presupposed here that the law of inertia is a hypothesis of a special sort. As is well known, there is also a view that the principle of inertia is a mere hidden definition of the notion of force. This view was expressed in the writings of Ernest Mach. According to him, "every science aims at finding constancies of connection and of combination and interdependence of (human) reactions" [Mach 1976, p. 98]. With such a general view on science, Mach had to find the principle of inertia troublemaking: in what sense the law referring to the ideal vacuum can "combine human reactions" that always appear in the conditions of lack of vacuum? None the less, he makes several descriptive remarks on the role of idealization in physics [ibid., pp. 140, 354, 355]. Similar observations concerning the role of idealization in science are made by Ostwald [1908]. And in doing this Ostwald [1908, p. 50] is similarly inconsistent as Mach since he stresses that the method of incomplete induction (combined with deduction of new observations from empirical generalizations) is the main method of science. Also an outstanding historian of physics, Koyre [1955, p. 335], states that "The Galilean solution of the problem of trajectory of the falling body is ... extremely ingenius and elegant. Unfortunately, it is quite false. It is even so obviously false ... that one may wonder that Galileo did not see it himself" whereas the law of free fall is not "false" but idealizational. These are examples of an interesting, more general phenomenon. It happens quite often that methodologists who perfectly know science (e.g., they are at the same time outstanding scientists, like in the cited cases) miss to see the difference between what they know from science (its "spontaneous methodology" with which they are acquainted) and what they are able to methodologically conceptualize in their terms (the "official methodology" which they declare). By consequence, they fail to satisfy the basic duty of a philosopher of science which is the translation of the spontaneous methodology of science into his/her methodological doctrine. Instead, they simply repeat what they know from science not even trying to reconstruct this in their conceptual apparatuses. So, a bit of ignorance in science might be of use for a philosopher of science protecting him/her from loosing his proper perspective when crossing the border of science.

First and foremost the notion of movement is a bit burdened by the notion of natural movement: Every body − states Galileo − which finds itself in the state of rest but is left to itself and able to move − begins to move, if the Nature offers it an inclination to reach a definite place [Galileo 1962, p. 19]. Hence the movement counts not from a place on but toward a place. For this reason Galileo rejects the notion of infinite movement:

> Motion along the straight line is in its very nature infinite since the straight line is infinite. . .It is then inconceivable to claim that something moves along the straight line, that is, to the goal which is unreachable [*ibid.*, p. 18].

Galileo postulating the movement of an ideal object rolling infinitely on an ideally smooth surface presupposed at the same time that this surface is at each point equidistant from the centre of the Earth. His law of inertia states that a body moving with a constant velocity around a circle does not stop until external forces begin to operate. This is the reason why Galileo regarded the motion around a circle as natural. A paradigmatic example was for him the case of a ship that could move around the Earth without stopping, if no external obstacles appear. Briefly, for Galileo the movement around the circle could be an inertial motion, for Newton, it could be not; for Newton the movement around the circle is instead an accelerated motion which is preserved due to a certain force.

These are, however, substantive limitations of Galileo's discovery. As far as their methodological background is concerned one may say that the very method with which he invented his law of inertia was a real novelty in comparison to the then widespread methods of investigation. He gave up Aristotelian inductivism to adopt a method of making assumptions "even if they [are] not strictly true" [Galileo 1963, p. 241].

The Galilean breakthrough consisted then in systematically imagining what a given phenomenon would be like if the factors considered to be secondary did not act upon this phenomenon at all. And that is what was typical for the innovation Galileo brought into the body of methods applied in the natural sciences[5]. Galileo systematically applied the method of ideali-

---

5 The author is not enough competent in the history of science to put forward the claim of the historical priority of Galileo as a literally historical thesis. What I mean is rather that Galileo applied the method of idealization historically successfully, that is, he had not only invented it but also succeeded in making impact on other physicists so that they were following him. I should say that this correction of my earlier formulation [cf. Nowak 1971] is due to Such [1978] who claims that it was Archimedes who applied the method of idealization for the first time. It is worth of adding that Galileo had made

zation. That was the real sense of the revolution in the natural sciences which bears his name[6]. Let us look at his method a little more carefully.

II

This is how Galileo presents his theoretical goal in the analysis of the phenomenon of free fall:

> we have decided to consider the phenomena of bodies falling with an acceleration such as actually occurs in nature and to make this definition of accelerated motion exhibit the essential features of observed accelerated motion [Galileo 1963, p. 154].

The point is which features pertaining to the "observed falling bodies" are claimed to be essential and which secondary. "[T]he intimate relationship between time and motion" belongs to the former and due to it "we may picture to our mind a motion as uniformly and continuously accelerated when, during any equal intervals of time whatever, equal increments of speed are given to it" [*ibid.*, p. 155]. More strictly,

---

reference to Archimedes himself: "In his Mechanics and in his firt quadrature of the parabola he takes for granted that the beam of a balance or steelyard is a straight line, every point of which is equidistant from the common center of all heavy bodies, and that the cords by which heavy bodies are suspended are parallel to each other" [Galileo 1963, p. 241].

6 Another attempt to get out of the alternative of the Platonistic and hypothetico-deductive interpretations is that presented by Wallace [1974]. He claims that Galileo's point was to employ the "demonstration *ex suppositione*" which, on Wallace's reconstruction, "can be expressed in the form, 'if p then q' ... p stands for a result that is attained in nature regularly or for the most part, whereas q states an antecedent cause or condition necessary to produce that result" [Wallace 1974, p. 95]. Interpreting Wallace's idea in terms of Lukasiewicz's classification of reasonings into deductive and reductive [see Kwiatkowski 1992], one could say that the reasoning *ex suppositione* is not deductive but reductive. In this respect, Wallace's interpretation goes in the same direction as one presented above because making idealization in the style of Galileo is a reductive reasoning as well. However, that is reduction of a special sort as the results obtained − i.e. Wallace's q's − apply to the ideal conditions different from those stated in his p's. For this reason a special operation of returning from ideal worlds to the real one, viz. concretization, is necessary. On Wallace's reading of Galileo, there is no operation of the kind. Analyzing the same example of derivation the equation of semi-parabola [cf. (4)−(6)], Wallace reconstructs it as falling under the modus ponendo ponens scheme (if p, then (if p then q), then q) [Wallace 1974, p. 97] which disregards the status of idealizational premisses and the role of concretization at all.

The spaces described by a body falling from rest with a uniformly accelerated motion are to each other as the squares of the time-intervals employed in traversing these distances [*ibid.*, p. 167].

One of the secondary factors is the air resistance. It is a negligeable influence because

it disturbs all [the movements] and disturbs them in an infinite variety of ways, corresponding to the infinite variety in the form, weight and velocity of the projectiles. . .Of these properties of weight, of velocity, and also of form, infinite in number, it is not possible to give any exact description; hence, in order to handle this matter in a scientific way, it is necessary to cut loose from these difficulties; and having discovered and demonstrated the theorems, in the case of no resistance, to use them and apply them with such limitations as experience will teach [Galileo 1963, p. 242].

Thus, Galileo adopts the idealizing assumption that the body falls in perfect vacuum, i.e., in the conditions in which the forces of friction equal zero.

Apart from that Galileo adopted an assumption to the effect that the Earth's gravity g is constant. For him, it was simply a constatation of fact, that is a realistic condition.

Galileo's formulation of the law of free fall may be reconstructed as follows:

(4)   if $fb(x,e)$ & $v_0(x) = 0$ & $r(x) = 0$ & $g(e) = const$,
    then $s(x) = \frac{1}{2}gt^2(x)$,

where
'$fb(x,e)$' means: x is a body falling in the direction of the Earth (e),
v_0 stands for the initial velocity,
r stands for the forces of resistance of the medium,
s is a distance covered by the falling body,
g is the constant of the Earth's gravitation,
t is a time of fall.

Of these assumptions, two are realistic. One delineates the universe of discourse $(fb(x,e))$ and the other postulates that the Earth gravitation is constant. The remaining two postulate that the initial velocity equals zero $(v_0(x)=0)$ and there is the lack of any resistance of the medium, i.e., the perfect vacuum $(r(x)=0)$. They may be taken to be idealizing assumptions.

Taking into account the falling bodies whose initial velocity differs from zero one must waive the condition $v_0(x)=0$ and correct the Galilean equation[7]:

(5)   if $fb(x,e)$ & $v_0(x) > 0$ & $r(x) = 0$ & $g(e) = const$,

then $s(x) = v_0(x)t(x) + \frac{1}{2}gt^2(x)$.

The procedure of concretization of the law (3) with respect to the assumption postulating the free fall in the perfect vacuum is marked by Galileo only intuitively. He refers to the law of Archimedes (a body submerged in fluid loses as much weight as is the weight of fluid pushed aside by this body) and takes it that a body falling in air loses as much weight as is the weight of the air pushed aside by this body. As he claims,

> Assuming. . .that all falling bodies acquire equal speeds in a medium which. . . offers no resistance to the speed of the motion, we shall be able accordingly to determine the ratios of the speeds of. . .bodies moving. . .through different space-filling, and therefore resistant media. This result we may obtain by observing how much the weight of the medium detracts from the weight of the moving body, which weight is the means employed by the falling body to open a path for itself and to push aside the parts of the medium. . .And since it is known that the effect of the medium is to diminish the weight of the body by the weight of the medium displaced, we may accomplish our purpose by diminishing in just this proportion the speeds of the falling bodies, which in a non-resisting medium we have assumed to be equal [Galileo 1963, p. 72].

It is quite visible that the operation performed on the law of free fall (3) is one of concretization. It is, however, equally visible that it was done on an intuitive level.

Later Boyle succeeded to create the "pneumatical engine" as he called his air pump and to empirically determine that in the physical vacuum bodies with different shapes fall down with the same velocity, exactly as Galileo's law predicted [Newton 1962, vol. 2, p. 543]. This was a(n) approximate) confirmation of Galileo's law [Conant 1953, pp. 52ff][8].

---

7 For this reason it is incorrect to say that for Galileo assumption "$v_0(x)=0$" is a realistic condition which is, surprisingly, "concretized" [Such 1978, p. 71, note 16].

8 Even if it were true that Galileo invented his law of free fall due to experiments, it does not follow that the law is of an inductive nature [as is claimed, e.g., by Harré 1981, p. 90]. An experiment, or even a crude observation, may namely play the same role as a thought experiment, namely as a source of invention leading to the formula

III

The interpretation outlined above shares some traits with the two main interpretations of Galileo's method, viz. the Platonist and hypothetico-deductive. Both interpretations stand against the inductivist stereotype mentioned at the beginning of this paper but in quite opposite ways. According to the Platonist interpretation, the breakthrough made by Galileo consisted in the fact that he opened the "book of nature written in the language of mathematics" for the first time. This presupposes

> that "the world of thoughts and the world of phenomena correspond to one another[,] . . .that the laws. . .which as such embrace freely general notions and connections between notions, still have reality and validity in the Nature; in other words, that the reasonable is also real" [Snell 1858, p. 41].

Today, this view is held for example by Koyre [1968] who claims that for Galileo the empirical reality is merely manifestation of eternal Platonic ideas and his alleged experiments were thought experiments making his abstract reasonings easier to get along. According to the hypothetico-deductive interpretation [Drake 1973, Shapere 1972], the role of actual experiments was absolutely crucial for Galileo's way of making science − putting forward hypotheses, deducing the observational consequences and testing the hypotheses with reference to the data: "At each stage of inquiry,

---

(consequent) of an idealizational law.

I would not like to insist that the present interpretation of Galileo's method as idealization is the only admissible one. As every interpretation, also this one has its weak points, that is, some fragments of Galileo's text that can scarcely, if at all, be understood on the assumption that what he had in mind was the method of idealization. Here is one of them: "SALVIATI. The request (to illustrate experimentally the theoretical conclusions) which you, as a man of science, make, is a very reasonable one this is the custom. . .in those sciences where mathematical demonstrations are applied to natural phenomena. . .where the principles, once established by well-chosen experiments, become the foundations of the entire superstructure" [Galileo 1963, p. 171]. This passage can be, one must admit, easily read under the empiricist (and the more so under the hypothetico-deductive) interpretation of Galileo's method. However, it can also be read under the idealizational interpretation, if we remember that experiments are methods of approximating idealizational laws. The passage would then claim that first basic idealizational laws should be confirmed experimentally and then all the remaining statements are to be derived from them. Such understood, this passage would express the attitude characteristic of the empirical-idealizational stage of development of science [cf. Magala and Nowak 1985].

sense experience must be combined with reasoning and with mathematics to afford a sound basis of deduction" [Drake 1972, p. 266].

The interpretation outlined above attempts to give justice to both the idea of discovering abstract traits of reality in the language of mathematics – this is how the basic idealizational laws are found, and the idea of empirically testing the correctness of these discoveries – this is why the concretization of idealizational laws is indispensable. Still, it differs from both of them. On the one hand, idealizational laws are hypotheses about ideal worlds which may prove to be easily false in them. If magnitudes H, m and n influence F, then an idealizational law which states, say actual, dependence of F on H but abstracts from n and fails to abstract from m would be false in a world characterized by F-facts and H-facts but lacking both m- and n-facts. Already this fact that idealizational statements can be false in the ideal domains determined by their idealizing antecedents suffices to distinguish them (e.g. Galileo's law of free fall) from mathematical theorems which are supposed to hold in the domains determined by their assumptions. On the other hand, the hypothetico-deductive method does not embrace the method of idealization. I have already had an opportunity to support this opinion [cf. Nowak 1980, Chap. V], so I shall restrict myself here to one new argument. As is well known, Popper's scheme of testing a theory is that of modus tollens [Popper 1959, para. 18]. If law t implies basic (observational) statement b and if *non-b* holds then t cannot hold, i.e., *non-t* holds. Assume, however, that t is an idealizational statement of the form $[n=0 \rightarrow F=f(H)]$. What it implies then is not b but an *idealizational conditional* $[n=0 \rightarrow b]$ which cannot be tested by direct observation; hence b cannot be possibly rejected on this basis. That is why the operation of concretization is necessary. So, from $[n=0 \rightarrow F=f(H)]$ its concretization $[n \neq 0 \rightarrow F=f'(H,n)]$ is derived. If it implies $[n \neq 0 \rightarrow b]$, then, provided that $n \neq 0$ is *realistic*, b can be obtained. A possible rejection of b can therefore *ceteris paribus* testify to the falsity of the initial premiss $[n=0 \rightarrow F=f(H)]$.

IV

Obviously, the interpretation of Galileo's method outlined above is an interpretation, nothing more. How is it possible to prove its superiority over the two rival interpretations? Obviously there is no such way. There are, however, some ways in which such a superiority could be argued for. One of them is the following. Let all the three readings of Galileo establish some methodological facts in Galileo. It is likely that all of them would agree that the three methods are characteristic of Galileo's way of doing research: (a) mathematical language, (b) experiments, (c) idealization. And they do not differ in stating these historical facts but in the significance they attach to them. The question is now which of these three ways of reading Galileo's methodology has a greater explanatory power. It seems to me that (c) could be perhaps derived from (a).[9] As I tried to argue a moment ago, (b) is unable to explain (c) and I do not see how (b) could explain (a).

In contrast to the two alternatives, the idealizational interpretation can, I conjecture, explain both (a) and (b) in terms of (c). It will be seen that the idealizational statement $[n=0 \rightarrow F=f(H)]$ presupposes in its consequent a substantive instantiation of the mathematical formula $Y=f(X)$; in other words, the formula $F=f(H)$ of this statement obtains by a substantive interpretation of the variables Y (as F) and X (as H) in $Y=f(X)$. Similarly, the consequent of its concretization obtains by the substantive interpretation of the mathematical formula $Y=f'(X,Z)$ which in a special case transforms into $Y=f(X)$. In a way, idealizing conditions are necessary to deform the phenomena to the effect that the idealized counterparts of those phenomena fall under the mathematical functions. Our world is not written in the language of mathematics but its idealized deformations are. That is why idealization has a logical priority before the application of mathematics in a given domain.

The significance of the experimental method can also be explained in terms of idealization. Consider the same idealizational law as above. If it is not known how n influences F, or if the mathematical formalism allowing to build f-expressions of two variables is not yet known at all, the concretization of this law is excluded. The best way of testing the law is then a rough approximation. However, it may happen, and often does, that in the actually appearing conditions the influence of n upon F is not negligeable. But if conditions where it is negligeable can be technically created, then this statement could be approximated: $[E \ \& \ n \approx 0 \rightarrow F \approx f(H)]$, where E is a condition limiting application of this statement to the experimentally created circumstances. One of significant functions of the experiment is then to (approximately) test idealizational laws.

---

9 According to Pitt [1991, pp. 90−91], there were two main points in Galileo's *Dialogue*: the use of geometric demonstration and the use of terrestrial phenomena as a basis for explaining physical processes both on the Earth and in the heaven. Snell [1858, p. 39f], the author of the Platonist interpretation of Galileo, makes several interesting remarks on the nature of experiment as applied by Galileo, but from the Platonist standpoint the crucial question is why the laws which are mathematicals embodied in our world are to be empirically tested at all.

In sum, if we hypothetically assume that it is idealization which is the core of the Galilean method, we become able to explain the other important components of his methodology, viz. the mathematization of his theory and the experimentalization of his research practice. And if I am correct that it would be difficult to explain the other two elements of Galileo's method from the point of view of either of the remaining alternatives, then it appears that the proposed interpretation has some advantage in comparison to them.

Conclusion

The Galilean revolution consisted in making evident the misleading nature of the world image which senses produce. We only see phenomena which are the joint effect of all the relevant influences. As a result, senses do not contribute in the slightest to the understanding of the facts. In order to understand phenomena the work of reason is necessary which selects some features of the objects through idealization and in their idealized models recognizes some other features of the empirical originals. These models differ a great deal from their sensory prototypes, what is more, they present images of hidden relationships which could not be grasped with the aid of experience at all. Science idealizing phenomena opposes common sense:

> experiences which clearly state against the annual movement — says Galileo — are seemingly so contrary to the theory that. . .I cannot find the words to express my admiration for Aristarches and Copernicus who managed to put reason into a frame which forced the sense to withdraw their trust in the apparent meaning of sensory data. . .[This proves how great is] the elevation of these minds which accepted these views and took them as true ones overcoming the testimony of their own senses with the quickness of minds and preferring that which reason dictated to what senses and experiments seemed to offer [Galileo 1962, pp. 353−54]. [That is why] The philosopher-geometrician who wants to investigate in reality what has abstractly been proved whould exclude the interfering influences of matter from the calculations [Galileo 1962, p. 225].

This gap between the abstract world of laws and the world of senses can be filled with the aid of concretization which takes into account what has been previously abstracted from. Due to this, abstract laws become more and more realistic and the distance between them and the actual facts diminishes. Idealization and concretization constitute the essence of the method whose adoption in physics Galileo had initiated. This method had been

124

systematically applied by Newton. Also the understanding of it had been deepened in Newton's *Principia*. But that is a separate story.

Department of Philosophy
University of Poznań
Szamarzewskiego 89C
60-586 Poznań, Poland

REFERENCES

Aristotle [1968]. *Fizyka* (Physics). Warszawa: PWN.

Barr, W.O. [1971]. A Syntactic and Semantic Analysis of Idealizations in Science. *Philosophy of Science* 38, 258−72.

Brzezinski, J. (Ed.) [1985]. *Consciousness: Methodological and Psychological Approaches* (*Poznań Studies in the Philosophy of the Sciences and the Humanities* 8). Amsterdam: Rodopi.

Brzezinski, J., Fr. Coniglione, T.A.F. Kuipers and L. Nowak (Eds.) [1990]. *Idealization I: General Problems* (*Poznań Studies in the Philosophy of the Sciences and the Humanities* 16). Amsterdam/Atlanta: Rodopi.

Brzezinski, J. and L. Nowak (Eds.) [1992]. *Idealization III: Approximation and Truth* (*Poznań Studies in the Philosophy of the Sciences and the Humanities* 25). Amsterdam/Atlanta: Rodopi.

Butterfield, H. [1958]. *The Origins of Modern Science 1300−1800*. London: Bell & Sons (quoted after Polish translation, Warszawa 1968).

Cartwright, N. [1983]. *How the Laws of Physics Lie?* Oxford: Oxford University Press.

Cartwright, N. [1989]. *Nature's Capacities and Their Measurement*. Oxford: Oxford University Press.

Cohen, I.B. [1960]. *The Birth of a New Physics*. London/Melbourne/Toronto: Heinemann.

Cohen, R.S. and M. Wartofsky (Eds.) [1974]. *Methodological and Historical Essays in the Natural and Social Sciences* (*Boston Studies in the Philosophy of Science* XIV). Dordrecht/Boston: Reidel.

Conant, J.B. [1953]. *On Understanding Science*. New York: The New American Library.

Coniglione, Fr., R. Poli and J. Wolenski (Eds.) [1993]. *Polish Scientific Philosophy: The Lvov−Warsaw School* (*Poznań Studies in the Philosophy of the Sciences and the Humanities* 28). Amsterdam/Atlanta: Rodopi.

Crombie, A.C. [1959]. *Medieval and Early Modern Science*. Vol. I. Cambridge: Harvard University Press (quoted after Polish translation, Warszawa 1960).

Dingle, H. [1959]. Copernicus and the Planets. In: Lindsay [1959], pp. 18−27.

Drake, S. [1972]. Galileo Galilei. In: Edwards [1972], vol. 3, pp. 262−66.

Drake, S. [1973]. Galileo's Discovery of the Law of Free Fall. *Scientific American* 228, 84−93.

Edwards, P. (Ed.) [1972]. *The Encyclopedia of Philosophy*. London/New York: Macmillan & the Free Press.

Galileo, G. [1962]. *Dialog o dwoch wielkich systemach swiata, Ptolemeuszowym i Kopernikanskim* (Dialogue on Two Principal Systems of the World, Ptolemeic and Copernican). Warszawa: PWN.

Galileo, G. [1963]. *Dialogues Concerning Two New Sciences* (trans. H. Crew and A. de Salvio). New York/Toronto/London: McGraw-Hill.

Harré, R. [1981]. *Great Scientific Experiments. 20 Experiments that Changed Our View of the World*. London: Phaidon Press.

Hartkaemper, A. and H.-J. Schmidt (Eds.) [1981]. *Structure and Approximation in Physical Theories*. New York/London: Plenum Press.

Koyre, A. [1955]. *A Documentary History of the Problem of Fall from Kepler to Newton* (*Transactions of the American Philosophical Society*, vol. 45, part 4). Philadelphia: The American Philosophical Society.

Koyre, A. [1968]. *Metaphysics and Measurement*. Cambridge, Mass.: Harvard University Press.

Kuokkanen, M. [1988]. The Poznań School Methodology of Idealization and Concretization from the Point of View of a Revised Structuralist Theory Conception. *Erkenntnis* 28, 97—115.

Kuokkanen, M. and T. Tuomivaara [1992]. On the Structure of Idealizations. In: Brzezinski and Nowak [1992], pp. 67—101.

Kupracz, A. [1990]. *O dwoch ujeciach idealizacji w naukach empirycznych. Proba analizy porownawczej* (On Two Approaches to Idealization: A Comparative Analysis). Poznań: Polish Academy of Science Press.

Kwiatkowski, T. [1993]. Classification of Reasonings in Contemporary Polish Philosophy. In: Coniglione *et al.* [1993], pp. 117—67.

Lesniak, K. [1968]. Aristotle's Philosophy of Nature (Filozofia przyrody Arystotelesa). In: Aristotle [1968], pp. IX—XXXI.

Lindsay, J. (Ed.) [1959]. *A Short History of Science: Origins and Results of the Scientific Revolution. A Symposium*. Garden City, N.Y.: Doubleday.

Ludwig, L. [1981]. Imprecision in Physics. In: Hartkaemper A. and H.-J. Schmidt [1981], pp. 7—19.

Mach, E. [1976]. *Knowledge and Error*. Dordrecht/Boston: Reidel.

Magala, S. and L. Nowak [1985]. The Problem of Historicity of Cognition in the Idealizational Concept of Science. In: Brzezinski [1985], pp. 18—35.

Newton, I. [1962]. *Principia* (translated A. Motte, revised by F. Cajori). Berkeley: University of California Press.

Niiniluoto, I. [1986]. Theories, Approximations and Idealizations. In: R.B. Marcus, G.J.W. Dorn, and P. Weingartner (Eds.). *Logic, Methodology and Philosophy of Science VII*. Amsterdam: North-Holland, pp. 255—89.

Niiniluoto, I. [1990]. Theories, Approximations and Idealizations. In: J. Brzezinski, F. Coniglione *et al.* [1990], pp. 9—57.

Nowak, L. [1970]. O zasadzie abstrakcji i stopniowej konkretyzacji (On the Principle of Abstraction and Gradual Concretization). In: *Metodologiczne zalozenia 'Kapitalu' Karola Marksa* (Methodological Assumptions of Karl Marx's 'Capital'). Warszawa: KiW, pp. 117—213.

126

Nowak, L. [1971]. Galileusz nauk spolecznych (Galileo of the Social Sciences). *Nurt* 1, 39–41.

Nowak, L. [1972]. Theories, Idealization and Measurement. *Philosophy of Science* 39, 533–47.

Nowak, L. [1980]. *The Structure of Idealization. Towards a Systematic Interpretation of the Marxian Idea of Science*. Dordrecht/Boston/London: Reidel.

Piekara, A. [1964]. *Mechanika ogolna* (The General Mechanics). Warszawa: PWN.

Pitt, J.C. [1991]. Galileo, Copernicus and the Tides. *Theoria et Historia Scientiarum* 1, 83–94.

Popper, K.R. [1959]. *The Logic of Scientific Discovery*. London: Hutchinson.

Ostwald, W. [1908]. *Grundriss der Naturphilosophie*. Leipzig.

Shapere, D. [1972]. Newtonian Mechanics and Mechanical Explanation. In: Edwards [1972], vol. 5, pp. 491–96.

Snell, K. [1858]. *Newton und die mechanische Naturwissenschaft*. Leipzig: Arnold.

Such, J. [1978]. Idealization and Concretization in the Natural Sciences (*Poznań Studies in the Philosophy of the Sciences and the Humanities* 4). Amsterdam: Gruener, 49–73. (Translation of the Polish paper published in *Studia Filozoficzne*, 11/1972.)

Suppe, F. [1972]. What's Wrong with the Received View on the Structure of Scientific Theories? *Philosophy of Science* 39, 1–19.

Wallace, W.A. [1974]. Galileo and Reasoning *ex suppositione*. The Methodology of the *Two New Sciences*. In: Cohen and Wartofsky [1974], pp. 79–104.

Poznań Studies in the Philosophy
of the Sciences and the Humanities
Vol. 42, pp. 127–139

Ilkka Niiniluoto

APPROXIMATION IN APPLIED SCIENCE*

Applied science exists in two forms: predictive and design science. The former tries to establish dynamic regularities that help to predict the future state of a natural or social system; the latter attempts to establish technical norms or conditional rules of action. It is typical of both cases that idealized theoretical descriptive models are combined with empirical information. When the idealized model is concretized in Nowak's sense, the derived predictions and technical norms can be likewise improved, so that their degree of approximate truth or truthlikeness increases. These methodological ideas can be illustrated by the history of exterior ballistics.

1. Applied Research: Predictive and Design Science

In an earlier paper [Niiniluoto 1993], I have argued that there are two types of applied research. One of them is predictive science, the other design science.[1]

Predictive science resembles basic research, as they both aim at descriptive knowledge about the world. Such knowledge can be formulated by singular sentences (expressing the past, present or future state of some natural or social system) or by lawlike generalizations (expressing nomic or causal regularies in the behavior of a deterministic or probabilistic system). In the deterministic case, these generalizations typically have the form

(1) X causes A in situation B

---

* Prof. Henry Fullenwider has kindly revised the language of the paper.

1 The idea was mentioned already in Niiniluoto [1984], p. 238. Similar views have been developed by T. Kotarbiński, H. Simon, and M. Bunge.

128

or

$$(2) \quad s(t) = f(s(0), t),$$

where $s(t)$ is the state of a system at time t and f is a function which describes dynamically the trajectory of the system starting from the initial state $s(0)$.[2]

While basic research (e.g., in physics, chemistry, biology, history, psychology and sociology) employs laws of the form (1) and (2) for the purpose of explanation, predictive science (e.g., practical astronomy, meteorology, economics) uses them as "predictive models" in order to foretell the future. This presupposes that we are able to observe, with sufficient accuracy, the occurrence of the causal factors X or initial states $s(0)$.[3]

Assume that we are able to manipulate the cause X by making it happen at our will. Then (at least with some additional conditions) the generalization (1) can be turned into a *technical norm*, i.e., a conditional statement about a means—end relation:

(3)   If you want A, and (you believe) you are in situation B, then you ought to do X.

A similar conversion of (2) gives

(4)   If you want $s(t)$ at time t, prepare the system now in $s(0)$.

A technical norm (3) is true if and only if the corresponding causal statement (1) is true.[4]

Design science can now be defined as research aiming at true technical norms. This conception seems to fit well practical disciplines (e.g., aeroplane engineering, forestry, pharmacy, nursing science, didactics) which are interested in controlling natural and artificial systems. They do not describe reality, but rather tell what we ought to do in order to realize our goals.

Bunge [1966] argues that technology in the proper sense (as distin-

---

2 The law (2) is typically obtained as a solution of differential equations.

3 Deterministic chaotic systems are extremely sensitive to small variations in the initial state. Therefore, they are not predictable. See Earman [1986].

4 More precisely, (3) is true if and only if X is a necessary cause of A in situation B. Cf. Niiniluoto [1993]. Bunge [1966] asserts that "rules can be only more or less effective", but I think they derive a truth-value from the associated causal regularities.

guished from pseudotechnology and prototechnology) should be science-based: its rules should be "grounded" by scientific theories. However, technical norms cannot always be directly derived "from above", i.e., from descriptive regularities provided by basic research. In many situations, there is no basic theory from which such a derivation could be made. Then the researcher typically employs background theoretical information and constructs a mathematical model, with manipulable and dependent variables, and tries to obtain relevant empirical information by experimentation and computer simulation. For example, in the optimization models of Operations Research, the aim A is construed as the maximization of a goal function $g(F_1,...,F_n,G_1,...,G_m)$, the situation B includes the estimated values of the functions $G_1,...,G_m$ and the boundaries for the possible values of $F_1,...,F_n$, and the desired action X is found by choosing the best values for $F_1,...,F_n$.

These divisions permit many scientific fields (e.g., medicine) to include both basic and applied research. As we shall see, there are also disciplines (e.g., ballistics) which serve both the predictive and design purposes.

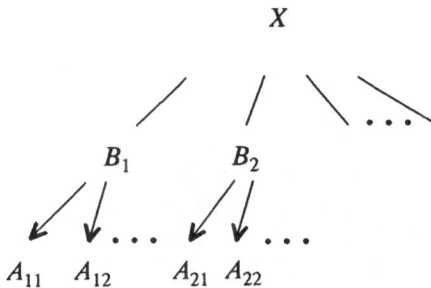

Fig. 1. A theory in basic science

To the extent that both basic and applied research rely on regularities of the type (1), one might claim that there is not much difference between them. However, I think the distinction helps us to understand important structural differences among sciences.[5] Thus, a theory in basic science typically exhibits the application of one causal factor X (such as gravity in the Newtonian mechanics) in a variety of situations $B_1,B_2,...$ with the resulting effects $A_{11},A_{12},...$ This structure, which is illustrated in Fig. 1, is familiar from the presentation of "theory-evolution" by the structuralist

---

5 The distinction has also important consequences for the ethics of science. See Niiniluoto [1993].

school.[6] Applied research is instead interested in various means $X_{11}, X_{12}, \ldots$ for obtaining, in different situations $B_1, B_2, \ldots$, the given single goal A [see Fig. 2]. For this reason, disciplines in this area are usually identified or named by the most general aim A of the relevant technical norms (e.g., health in medicine, peace in peace research).

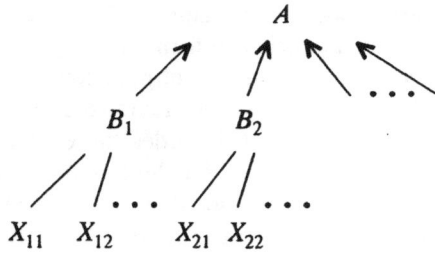

Fig. 2. A theory in applied science

2. Idealization and Approximation

It is now well established, especially through the work of Leszek Nowak [1980] and the Poznań school, that scientific laws and theories normally are idealized, i.e., they presuppose counterfactual idealizing assumptions (I). Sometimes such assumptions are made on purpose, sometimes they have to be detected. Then the original theory T can be replaced by a counterfactual conditional $I \Rightarrow T$ (If it were the case that I, then it would hold that T). The next step is the gradual removal of I by concretization. It leads to a new factual theory T' which takes into account the factors eliminated by I. If the Principle of Correspondence holds, T' entails $I \Rightarrow T$, and T turns out to be a counterfactual limiting special case of T'.

As the original theory T contains false presuppositions, it is false. However, if T can be transformed by concretization to a true theory T', then T is *truthlike* in the sense explicated by Niiniluoto [1987]. This allows us to defend critical scientific realism also in the case of idealized theories [cf. Niiniluoto 1990].

Another way of looking at the relation between the idealized theory T and reality is the following [see Fig. 3]: T is true in idealized systems which

---

6 See Balzer, Moulines, and Sneed [1987], pp. 229–243.

satisfy the counterfactual assumptions *I*, but such systems are nevertheless similar to the real system [cf. Giere 1988]. This relation, truth + similarity, induces again a concept of verisimilitude. More precisely, *T* is *approximately true*, if *T* is true in some system close to the real system.[7]

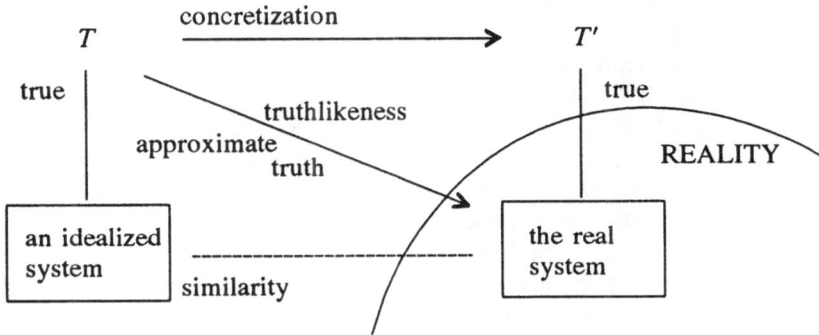

Fig. 3. Idealized theory and reality

Suppose now that we try to derive a given empirical statement *E* from an idealized theory *T*. Often such a derivation requires simplifying assumptions, where some quantity or function is replaced by an approximately equal quantity or function (e.g., $\pi \approx 3.14$; $e^x \approx 1 + x + x^2/2$ for small x). Then we may succeed to show that *T* logically entails a statement *E'* that is approximately the same as *E* [see Fig. 4]. In this case, *E* is said to be *approximately deducible* from *T*.

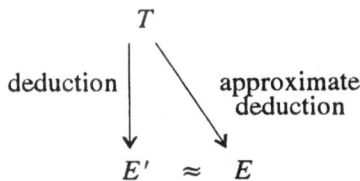

Fig. 4. *T* approximately entails *E*

Nancy Cartwright [1983] has argued that generally "approximations take us away from theory and each step away from theory moves closer towards the truth" [p. 107]. This is illustrated by an example where inaccurate

---

7 See Niiniluoto [1991].

predictions are improved by adding an empirical "correction factor" that is "not dictated by fundamental law" [p. 111]. In terms of Fig. 4, Cartwright thinks that E (e.g., the "phenomenological laws" of applied physics and engineering) are "highly accurate" [p. 127] or true, while the strict predictions E' from fundamental theories are less accurate. But then the fact that approximations lead us closer to the truth argues for the falsehood of theories [p. 15].

As we have already seen, a critical scientific realist is not surprised by the observation that fundamental theories are strictly speaking false. This does not preclude the possibility that they are nevertheless highly truthlike [cf. Fig. 3]. But, against Cartwright's instrumentalist view, a realist will also urge that in many cases our theoretical predictions are closer to the truth than empirically established "phenomenological" laws [cf. McMullin 1985]. This means that, in Fig. 4, E' is more accurate than E, i.e., E' corrects E. Then also the derivation of E' from T gives an approximate and corrective explanation of the initial statement E.

For example, Newton's theory entails corrected versions of the original empirical laws of Kepler and Galileo. If the earth is a sphere with radius R and mass M, and G is the gravitational constant, Newton's Law of Gravitation entails that the position $s_y(t)$ of a freely falling body at the height h above the surface of the earth is

$$(5) \quad s_y(t) = -\frac{GM}{2(R+h)^2} t^2$$

[see Niiniluoto 1990, p. 38]. By making the approximation

$$(6) \quad h = 0,$$

equation (5) reduces to Galileo's law of free fall

$$(7) \quad s_y(t) = -\frac{GM}{2R^2} t^2 = -gt^2/2$$

where $g = GM/R^2$ is a constant independent of h.

The important point here is that (5) is more accurate than (7). What is more, the approximation (6) is a typical idealizing assumption, so that (5) can be obtained from Galileo's (7) by concretization in Nowak's sense.

Fig. 4 thus applies to cases where theory T helps us to improve an empirical statement E by entailing another statement E' which is more concrete and more accurate than E. The existence of such situations, where corrections come from "the top down" rather than from "the ground up", has been doubted by Cartwright [*op. cit.*, p. 111]. For the comfort of critical scientific realists, further examples of such situations can be easily found in applied science, as İ shall illustrate in the next section by the history of ballistics.

3. Ballistics as Applied Science

What is the role of idealization and approximation in applied science? Can we relate the account of applied science, given in Section 1, to the Poznanian method of idealization and concretization? Can we apply the concepts of approximate truth and truthlikeness to typical results of applied research?

In Ch. 12 of his book Nowak himself has treated "practical sciences" which aim at "programming of future facts" (instead of explanation). He discusses optimization models with a goal function $g(F_1,...,F_n, G_1,...,G_m)$ and shows that by concretizing the G-functions it is possible to gain improved F-functions as solutions. I think this idea is correct and important, but it covers only a special case of applied research.

I shall use ballistics as a case study illustrating the characteristic ways in which applied research combines theoretical and empirical information in its attempt to find improved predictions and rules of action.

The history of ballistics follows the method of idealization and concretization. In the case of *interior ballistics*, this can be seen in the textbook of J. Corner [1950]. The basic problem is to find ways of determining the desired velocity for a shot at the peak of a gun − but not to greater accuracy than 5 ft/sec. [p. 6]. Given the characterics of shot, charge and gun, the task is to calculate the muzzle velocity and peak pressure [p. 14]. Until 1930 gun behaviour was studied under "extreme conditions": the shot was assumed to be infinitely heavy. But then the "abnormalities" were "traced in modified form right into the region of normal conditions" [p. vii]. The goal of the theory is thereby reduced to finding exact and numerical "solutions of approximations to the true equations" [p. 20].

Exterior ballistics studies the behaviour of the shot after it leaves the peak of the gun. According to the medieval impetus theory [Jean Buridan], the path of a projectile is a straight line until it suddenly drops down because the impetus or "motive force" is continually diminished by the resistance of air. It was known to Niccolo Tartaglia in 1546 that the path is a nonlinear curve. He also asserted that the maximum distance is reached by making the initial angle equal to 45°. Galileo Galilei argued in 1638 that the curve of a projectile is parabolic when there is no air resistance. Isaac Newton in 1687 tried to account for the resistance of air with a force proportional to the square of velocity. Jean Bernoulli formulated in the early 18th century the classical differential equations for the ballistic problem. Their approximate solutions were soon sought by series [de Borda] and linear interpolation [Euler], but more effective numerical methods were introduced only in the 1920s [Runge-Gutta]. This theoretical approach was complemented with experimental work by military engineers, who tried to determine shooting tables for the air resistance relative to different types of shots.

The theoretical and practical wisdom[8] in exterior ballistics is nicely summarized in the textbook *Äussere Ballistik* [1972] by Günter Hauck. He starts by making a clear distinction between the true path of a projectile, a mathematical model of the path, and the numerical approximation of the model. The simplest mathematical model is the *Parabolic Theory*, based upon the following assumptions:

1° The shot is a mass point with constant mass m.
2° Motion is described in an inertial system in R^3; rotation of the earth is ignored.
3° The only force affecting the shot is gravity ($F_y = -mg$, where g is a constant).
4° The medium is vacuum.
5° The initial position is (0,0,0).
6° Newton's laws are valid.

When the initial velocity is v_0 and the angle with the x-axis is α, the assumptions $1° - 6°$ imply that the position $(s_x(t), s_y(t))$ at time t is given by

$$(8) \quad \begin{aligned} s_x(t) &= t v_0 \cos\alpha \\ s_y(t) &= t v_0 \sin\alpha - g t^2/2. \end{aligned}$$

If $v_0 = 0$ and $\alpha = -90°$, equation (8) gives Galileo's law (7) as a special case.

The error theory of ballistics studies the effects of small variations in v_0 and α to the paths described by (8).

The *Classical Theory* of exterior ballistics is obtained by allowing,

---

8 From an ethical point of view, one might well argue that applied military research is an instance of abuse of human reason. I leave that out of the discussion in this paper. It is worth pointing out that the theory of ballistics of course has also useful nonmilitary applications.

instead of assumption 3°, that the shot is also influenced by a force W due to the resistance of air. To determine W the following assumptions are introduced:

I Air is an ideal gas.
II Temperature is a known function of the height above the earth surface.
III Atmosphere is evenly distributed relative to the earth surface.

Given this idealized model of atmosphere, the application of the principles of aerodynamics leads (with some further idealizations) to the equation

$$W = \frac{1}{2} \rho v^2 A c_w(M),$$

where ρ is the density of air, v is the velocity of the shot, A is its cross section, and M is the velocity of the shot in Mach-numbers. This equation can be rewritten in the form

$$(9) \quad W = c_0 m \frac{p(y)}{p_N} \Phi(M),$$

where $p(y)$ is the air pressure at height y, p_N the normal air pressure, c_0 is a ballistic coefficient

$$c_0 = \kappa p_N A / 2m,$$

κ is the adiabitic exponent of air, and $\Phi(M) = M^2 c_w(M)$.

The form of the function $\Phi(M)$ is determined experimentally for each type of shot. The value of $c_w(M)$ is roughly constant when $0 \leq M \leq 0.8$ and $M \geq 5$, between $0.8 \leq M \leq 1.2$ it increases rapidly, and then it starts to slowly decrease. Thus, $\Phi = M^2 c_w$ can be taken to be linear for intermediate values of M $(1.2 \leq M \leq 5)$ and quadratic for small and large M. If $p(y)$ is assumed to be a constant in (9), two important special cases are obtained: (a) $W = -\beta v$, and (b) $W = -\beta v^2$ for theoretically and experimentally determined constants $\beta > 0$.

(a) For the linear case $W = -\beta v$ the equations (8) are replaced by

$$s_x(t) = \frac{m v_0 \cos\alpha}{\beta} \left(1 - e^{-\beta t/m}\right)$$

(10)

$$s_y(t) = -\frac{mg}{\beta} t + \left[\frac{m^2 g}{\beta^2} + \frac{m v_0 \sin\alpha}{\beta}\right] \left(1 - e^{-\beta t/m}\right)$$

If $\beta t/m$ is small, (10) gives approximately

(11)
$$s_x(t) \approx t v_0 \cos\alpha (1 - \beta t/2m)$$
$$s_y(t) \approx -gt^2/2 + t v_0 \sin\alpha (1 - \beta t/2m).$$

The limiting correspondence principle is satisfied: if $\beta \to 0$, the equations (11) approach (8).

(b) For the quadratic case $W = -\beta v^2$ we have

$$v_x(t) = \frac{v_0 \cos\alpha}{1 + \dfrac{\beta v_0 \cos\alpha}{m} t}$$

and

$$s_x(t) = \int v_x(t)\, dt = \frac{m}{\beta} \log\left[1 + \frac{\beta v_0 \cos\alpha}{m} t\right].$$

Hence, approximately

(12) $\quad s_x(t) \approx t v_0 \cos\alpha - \dfrac{\beta v_0^2 \cos^2\alpha}{2m} t^2.$

Again, if $\beta \to 0$, (12) reduces to the value given in (8).

These results illustrate how ballistics, both as a predictive science and as a design science, can improve its results by successful concretization.

First, consider *singular predictions*. The parabolic equations (8) imply that the shot hits the ground (x-axis) after time $t_1 = (2v_0 \sin\alpha)/g$ at the distance

(13) $\quad s_x(t_1) = \dfrac{2v_0^2}{g} \sin\alpha \cos\alpha.$

The equations (10), with a linear function for air resistance, imply that

$$t_1 = \dfrac{2v_0 \sin\alpha}{g + \dfrac{\beta v_0 \sin\alpha}{m}}$$

and

(14) $\quad s_x(t_1) = \dfrac{2v_0^2 m^2 g \sin\alpha \cos\alpha}{\left(mg + \beta v_0 \sin\alpha\right)^2},$

which is smaller than (13). Equation (14) allows us to make a more accurate prediction than (13).

Secondly, the *curve* defined by the corrected equations (11) is itself closer to the true path than the parabolic curve given by (8).[9] Concretization may thus increase the truthlikeness of predictive models.

Thirdly, also *technical norms* can be improved by concretization. For example, as the distance (13) is maximized by choosing $\alpha = 45°$, the parabolic equations (8) entail *Tartaglia's rule*:

(15) $\quad$ To obtain maximum distance, use the initial angle $\alpha = 45°$.

However, the corrected equation (14) entails:

(16) $\quad$ To obtain maximum distance, use the initial angle α that satisfies

$$\sin\alpha = \dfrac{\sqrt{\beta^2 v_0^2 + 8m^2 g^2} - \beta v_0}{4mg}.$$

Since the value given in (16) is always smaller than $\sin 45° = \sqrt{2}/2$, (16) recommends the choice of $\alpha < 45°$. Again the Correspondence principle holds for (15) and (16): if $\beta = 0$, (16) reduces to Tartaglia's rule (15).

For another illustration, suppose that a ball is thrown upwards (i.e.,

---

9 For distances between curves (functions), see Niiniluoto [1987, 1990].

$\alpha = 90°$ and $s_x(t) = 0$). By the parabolic theory (8), the maximum value of $s_y(t)$ is $v_0^2/2g$. Hence,

(17)   If you want the ball to reach the height h, choose v_0 so that

$$v_0 = \sqrt{2gh} \ .$$

With linear air resistance, (11) implies that the maximum of $s_y(t)$ is $v_0^2/2(g + \beta v_0/m)$. Hence, the conclusion of (17) is replaced by

(18)   Choose v_0 so that

$$v_0 = \frac{\beta h}{m} + \sqrt{\left[\frac{\beta h}{m}\right]^2 + 2gh} > \sqrt{2gh} \ .$$

Again, (18) reduces to (17), when $\beta \to 0$.

Let us say that a technical norm is truthlike if and only if the underlying causal regularity is truthlike (cf. (3) and (1)). Then our results show that concretization may increase the degree of truthlikeness of technical norms.

4. Concluding Remark: Theory and Practice

Suppose that a theory entails a technical norm which turns out to be effective. Bunge [1966] argues that such practical success is "not a truth criterion for the underlying hypothesis", since "the roads from success to truth are infinitely many and consequently theoretically useless or nearly so".

Ballistics as an applied science certainly is practically successful. The classical theory (with the shooting tables employed in the Second World War) is more successful than the parabolic theory. This means, I have argued, that the latter theory in itself is more truthlike than the former. But in order to evaluate Bunge's thesis, we have to note that Newton's theory plays an essential role in achieving this successful concretization. Does this provide a "truth criterion" for Newton's theory?

In the light of Einstein and quantum theory we have good reason to think that Newton's theory is false. Bunge is therefore right if his claim is simply that practical success does not prove the *truth* of a theory. As we know the falsity of this theory, not even the *probability* of its truth will increase. But if Newton's theory were completely mistaken, it would be difficult to understand how it can achieve successful concretization.[10] For this reason, the practical success of a theory is an indicator of its truth-likeness.

Department of Philosophy
University of Helsinki
P.O. Box 24
00014 University of Helsinki, Finland

REFERENCES

Balzer, W., Moulines, C.U. and Sneed, J. [1987]. *An Architectonic for Science: The Structuralist Program*. Dordrecht: D. Reidel.

Bunge, M. [1966]. Technology as Applied Science. *Technology and Culture* 7, 329–347. Reprinted in: F. Rapp (Ed.). *Contribution to a Philosophy of Technology*. Dordrecht: D. Reidel, 1974, pp. 19–39.

Cartwright, N. [1983]. *How the Laws of Physics Lie*. Oxford: Oxford University Press.

Corner, J. [1950]. *Theory of the Interior Ballistics of Guns*. New York: J. Wiley.

Earman, J. [1986]. *A Primer on Determinism*. Dordrecht: D. Reidel.

Giere, R. [1988]. *Explaining Science*. Chicago: The University of Chicago Press.

Hauck, G. [1972]. *Äussere Ballistik: Eine Einführung in die Theorie der Geschossbewegung*. Berlin: Militärverlag der Deutschen Demokratischen Republik.

Laymon, R. [1982]. Scientific Realism and the Hierarchical Counterfactual Path from Data to Theory. In: P.D. Asquith and T. Nickles (Eds.). *PSA 1982*. Vol. 1, East Lansing: Philosophy of Science Association, pp. 107–121.

McMullin, E. [1985]. Galilean Idealization. *Studies in the History of Philosophy of Science* 16, 247–273.

Niiniluoto, I. [1984]. *Is Science Progressive?*. Dordrecht: D. Reidel.

Niiniluoto, I. [1987]. *Truthlikeness*. Dordrecht: D. Reidel.

Niiniluoto, I. [1990]. Theories, Approximations, and Idealizations. In: J. Brzezinski et al. (Eds.). *Idealization I: General Problems*. (*Poznań Studies in the Philosophy of the Sciences and the Humanities* 16), Amsterdam: Rodopi, pp. 9–57.

Niiniluoto, I. [1991]. Realism, Relativism, and Constructivism. *Synthese* 89, 135–162.

Niiniluoto, I. [1993]. The Aim and Structure of Applied Research. *Erkenntnis* 38, 1–21.

Nowak, L. [1980]. *The Structure of Idealization*. Dordrecht: D. Reidel.

---

10 This is one way of interpreting Laymon's [1982] claim that a theory is "confirmed", if the use of more realistic initial conditions leads to more accurate predictions.

Poznań Studies in the Philosophy
of the Sciences and the Humanities
Vol. 42, pp. 141–158

Elke Heise, Peter Gerjets and Rainer Westermann

IDEALIZED ACTION PHASES: A CONCISE RUBICON THEORY

Introduction

If human goal-directed behavior is to be analyzed from an action theoretical perspective, a theory is required comprising all relevant processes between the deliberation of potential action goals and the final evaluation of action outcomes. The Rubicon theory of action phases [Gollwitzer, 1990, 1991; Heckhausen, 1989; Heckhausen & Gollwitzer, 1987] offers an adequate framework for a sequential description of goal-directed activities. It distinguishes four action phases:

(1)   In the *predecisional phase* potential action goals are deliberated until the psychological Rubicon of intention formation is crossed, i.e. one of them is chosen as a goal intention.
(2)   In the *preactional (postdecisional) phase*, action planning takes place.
(3)   The *actional phase* is signified by the initiation of goal-relevant overt activities.
(4)   When these activities are completed, the *postactional phase* begins where action outcomes and their consequences are evaluated.

This sequence of action phases is an idealization in several respects, which will become evident in the course of our present argument.

    In the literature, the Rubicon theory is presented in two main versions: Heckhausen's original textbook version [Heckhausen, 1989], and a more elaborated and detailed version developed by Gollwitzer [1991]. Both presentations, however, lack precision. Conceptual ambiguities make it difficult to identify the fundamental assumptions of the theory and to distinguish these from special laws that are applicable only in special domains. A reconstruction of the theory within the structuralist view [Balzer, Moulines & Sneed, 1987] has led to considerable conceptual clarifications and to a detailed comparison between the two theory versions [Gerjets, Westermann & Heise, 1992; Westermann, Gerjets & Heise, 1993].

In this paper, a more concise version of the Rubicon theory will be outlined in structuralist terms, comprising only those assumptions that are essential for either theoretical or empirical reasons. An assumption is classified as essential if it uses fundamental explanatory concepts of the Rubicon theory, if it has been empirically corroborated, or if it is at least expected to underlie future applications. All other assumptions, especially those which refer only to different operationalizations of higher-order concepts will be omitted. On the basis of the concise structuralist version of the theory, it is possible to illustrate the idealizing assumptions of the theory in more detail.

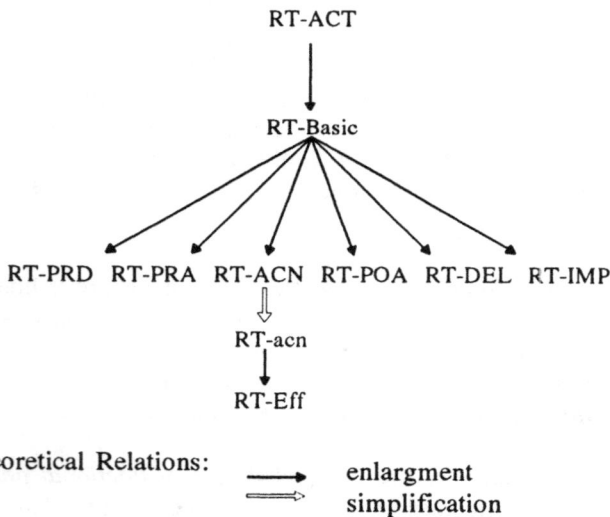

Figure 1: Part of the Rubicon Theory Net

In structuralist terminology, the Rubicon theory can be described as a net of interconnected theory elements, each theory element T being defined as an ordered pair consisting of a formal core K and a set I of intended applications: $T = <K, I>$. The formal core comprises among others the set of models. For some important theory elements, these models will be defined by semi-formal set-theoretical axiomatizations and explained in the following, whereas several other aspects of a comprehensive structuralist theory presentation, such as theoreticity or constraints, are not included in the present context. Figure 1 outlines the net of theory elements to be discussed.

I General Theory Elements

Actions as Goal-Directed Activities (RT-ACT)

$x \in M(RT\text{-}ACT)$ iff

(1) $x = \langle A,G,T,GRA,GI,AS,ACT \rangle$

(2) A is a non-empty set (activities)
         $T \subseteq \mathbb{N}$ is a non-empty set (points of time)
         G is a non-empty set (represented goals)

(3) $GRA \subseteq G \times A \times T$ (goal-relevant activities)
         $GI \subseteq G \times T$ (goal intentions)
         $AS \subseteq Po(A \times T)$ (activity sequences)
         $ACT \subseteq AS \times G$ (actions)

(4) $act_i = (as_i,g_i) \in ACT$ iff $(g_i,a_i,t_i) \in GRA$ for all $(a_i,t_i) \in as_i$ and $(g_i,t_j) \in GI$ for at least one t_j with $(a_j,t_j) \in as_i$.

According to this definition, an entity x must satisfy certain conditions in order to be a model of the theory element RT-ACT. It must be a structure of a certain conceptual form (1), consisting of the three base sets and the four relations introduced in conditions (2) and (3), and it must satisfy one substantial assumption (4).

Before an action can be divided into phases according to the Rubicon theory, a fundamental concept of an action must be introduced. The theory element RT-ACT thus defines an action as a sequence of goal-relevant activities with the goal being a goal intention for at least one point of time within the sequence. The concept of time is interpreted psychologically, i.e., points of time can be of variable length and are described by natural numbers with the convention that t_i+1 immediately follows t_i.

Consider axiom (4) as an example of a conceptual idealization. According to this law, an action is characterized by one and only one element from the set G of goals for which all the activities in the sequence are relevant. As a consequence of this idealization, we get different actions if the same sequence of activities is combined with more than one goal or if different sequences of activities are used to reach the same goal.

The Concept of Action Phases (RT-Basic)

$x \in$ M(RT-Basic) iff

(1) $x = $ <A,G,T,GRA,GI,AS,ACT,PRD,PRA,ACN,POA,IFO,INI, TER> and
        <A,G,T,GRA,GI,AS,ACT> $\in$ M(RT-ACT)

(2)   PRD $\subseteq$ A x T (predecisional activities)
        PRA $\subseteq$ A x T (preactional activities)
        ACN $\subseteq$ A x T (actional activities)
        POA $\subseteq$ A x T (postactional activities)
        IFO $\subseteq$ A x T (intention formations)
        INI $\subseteq$ A x T (initiations of activities)
        TER $\subseteq$ A x T (terminations of activities)

(3)   For all $act_i = (as_i,g_i) \in$ ACT: $as_i = PRD_i \cup \{ifo_i\} \cup PRA_i \cup \{ini_i\} \cup ACN_i \cup \{ter_i\} \cup POA_i$ and $PRD_i \subseteq$ PRD, $ifo_i \in$ IFO, $PRA_i \subseteq$ PRA, $ini_i \in$ INI, $ACN_i \subseteq$ ACN, $ter_i \in$ TER, $POA_i \subseteq$ POA with all sets being mutually exclusive

(4)   For all $act_i = (as_i,g_i) \in$ ACT with $(a_i,t_{ifo}) = ifo_i$, $(a_j,t_{ini}) = ini_i$, $(a_k,t_{ter}) = ter_i$ and $t_{ifo} < t_{ini} < t_{ter}$ and for all $(a_m,t_m) \in as_i - \{ifo_i, ini_i, ter_i\}$:
        (a) If $t_m < t_{ifo}$, then $(a_m,t_m) \in PRD_i$
        (b) If $t_{ifo} < t_m < t_{ini}$, then $(a_m,t_m) \in PRA_i$
        (c) If $t_{ini} < t_m < t_{ter}$, then $(a_m,t_m) \in ACN_i$
        (d) If $t_{ter} < t_m$, then $(a_m,t_m) \in POA_i$

(5)   For all $act_i = (as_i,g_i) \in$ ACT and for all $(a,t) \in as_i$:
        If $(a,t) \in PRA_i \cup ACN_i$, then $(g_i,t) \in$ GI

Axiom (2) introduces different classes of activities which are used in axiom (3) to segment the activity sequence. Each activity can be classified as belonging to one and only one class, i.e. the sets of activities are mutually exclusive. In axiom (4) the activities are temporarily ordered. In our opinion, the transition points between action phases are correctly interpreted as pretheoretically given. By means of these fixed points of time, the action is segmented into successive phases. This segmentation is supposed to underlie every empirical application of the Rubicon theory and can thus be interpreted as its fundamental law.

Note that the classes of activities are not explicitly characterized with respect to their contents. The set *PRA* of preactional activities, for example, may include some deliberating activities besides the activities of action planning. This counteracts the original idealized characterization of action phases by homogeneous types of activities but seems to correspond more closely to reality.

Axiom (5) postulates that at least during the preactional and actional action phases, the goal of an action becomes a goal intention; that is, in these phases the actor has commited himself or herself to pursue the chosen goal, so that planning and execution of relevant actional activities may begin.

The basic element of the Rubicon theory net describes an idealized action insofar as the four phases of the sequence are passed consecutively in only one direction. Surely it is possible to subsume under the theory special types of actions in which an action phase is skipped or omitted by postulating that one of the subsets, for example PRD_i, is empty. But a possible action in which the actor suddenly steps back into the phase of action planning after overt actional activities have already been initiated in the actional phase, cannot be described from the Rubicon perspective, because it violates the presupposed ideal sequence of activities.

Starting with the element RT-Basic, special theory elements for the four action phases are developed. In each action phase element, additional concepts are introduced and new substantial axioms are formulated, so that these theory elements can be regarded as enlargements of the basic element.

II Action Phase Elements

The Predecisional Action Phase (RT-PRD)

$x \in M(\text{RT-PRD})$ iff

(1) $x = \ <A,G,T,GRA,GI,AS,ACT,PRD,PRA,ACN,POA,IFO,INI,$
        $TER,com,mot,fc,fc_0,sac,cig>$ and
        $<A,G,T,GRA,GI,AS,ACT,PRD,PRA,ACN,POA,IFO,INI,$
        $TER> \ \in M(\text{RT-Basic})$

(2) $LG \subseteq G$ (long-term goals)
$\quad$ $IG \subseteq G$ (impulse goals)

(3)   com: $A \times T \longrightarrow \mathbb{R}$ (completeness of deliberating)
$\quad$ mot: $G \times T \longrightarrow \mathbb{R}$ (resulting motivational tendency)
$\quad$ fc: $\quad A \times T \longrightarrow \mathbb{R}$ (facit-tendency)
$\quad$ fc_0: $\quad ACT \longrightarrow \mathbb{R}$ (threshold for facit-tendency)
$\quad$ sac: $A \times T \longrightarrow \mathbb{R}$ (significance of action consequences)
$\quad$ cig: $A \times T \longrightarrow \mathbb{R}$ (costs of information gathering)

(4)   For all act_i, $act_j \in ACT$ with $act_i = (as_i, g_i)$, $act_j = (as_j, g_j)$ and for all $(a_i, t_i) \in PRD_i$, $(a_j, t_j) \in PRD_j$ with (g_i, t_i), $(g_j, t_j) \notin GI$ and for all $g_k \in G-\{g_i\}$:
$\quad$ (a) If $com(a_i, t_i) > com(a_j, t_j)$, then $fc(a_i, t_i) > fc(a_j, t_j)$
$\quad$ (b) If $fc(a_i, t_i) > fc_0(act_i)$ and $mot(g_i, t_i) > mot(g_k, t_i)$,
$\qquad$ then $(a_k, t_i + 1) = ifo_i$
$\qquad$ and for all $t_m > t_i + 1$ with $(a_m, t_m) \in as_i-POA_i$: $(g_i, t_m) \in GI$

(5)   (a) For all $act_i = (as_i, g_i)$, $act_j = (as_j, g_j) \in ACT$ with $g_i \in LG$:
$\qquad$ $PRD_i = PRD_j$ and $ifo_i = ifo_j$
$\quad$ (b) For all $act_i = (as_i, g_i) \in ACT$ with $g_i \in IG$: $PRD_i = \emptyset$

(6)   For all act_i, $act_j \in ACT$ with $(a_i, t_i) \in PRD_i$, $(a_j, t_j) \in PRD_j$:
$\quad$ If $sac(a_i, t_i) > sac(a_j, t_j)$ or $cig(a_i, t_i) < cig(a_j, t_j)$,
$\quad$ then $fc_0(act_i) > fc_0(act_j)$

Axiom (4) describes the processes that lead from the deliberation of goals to the formation of a goal intention. Intention formation primarily depends on two determinants: the resulting motivational strength of a goal, and a so-called "facit-tendency". The facit-tendency increases when the subject thinks that his or her deliberating of goals has reached completeness.

The threshold of the facit-tendency is an action-specific value. The higher the significance of action consequences and the lower the costs of gathering more information, the more will deliberation take place before an intention is formed (6).

In axiom (5), the special cases of actions with long-term goals and impulse goals are described. If an action is directed by a long-term goal, such as an identity goal [Gollwitzer, 1987], predecisional deliberation and intention formation will take place only once, even if planning and overt goal-relevant behavior extend over long periods of time. Impulse goals are situationally and emotionally determined and lead to the immediate generation of goal intentions without any preceding deliberation of alternatives.

The Preactional Action Phase (RT-PRA)

$x \in M(RT\text{-}PRA)$ iff

(1) $x = <A,G,T,GRA,GI,AS,ACT,PRD,PRA,ACN,POA,IFO,INI,$
      $TER,PLA,mot,vol,mis,fail,urg,fav,fi,fi_0,din,dre>$ and
      $<A,G,T,GRA,GI,AS,ACT,PRD,PRA,ACN,POA,IFO,INI,$
      $TER> \in M(RT\text{-}Basic)$

(2) $PLA \subseteq T \times AS$ (plans)
      $mot: G \times T \longrightarrow \mathbb{R}$ (resulting motivational tendency)
      $vol: GI \longrightarrow \mathbb{R}$ (volitional strength)
      $mis: GI \longrightarrow \mathbb{R}$ (number of missed opportunities)
      $fail: GI \longrightarrow \mathbb{R}$ (number of failed attempts)
      $urg: GI \longrightarrow \mathbb{R}$ (urgency of goal intention)
      $fav: GI \longrightarrow \mathbb{R}$ (favorability of opportunity)
      $fi: GI \longrightarrow \mathbb{R}$ (fiat-tendency)
      $fi_0: ACT \longrightarrow \mathbb{R}$ (threshold for fiat-tendency)
      $din: GI \longrightarrow \mathbb{R}$ (difficulty of initiation)
      $dre: GI \longrightarrow \mathbb{R}$ (difficulty of realization)

(3) For all $act_i = (as_i,g_i)$, $act_j = (as_j,g_j) \in ACT$ with $ib_i = (a_i,t_i) \in act_i$
      and $ib_j = (a_j,t_j) \in act_j$:
      If $mot(g_i,t_i) > mot(g_j,t_j)$, then $vol(g_i,t_i+1) > vol(g_j,t_j+1)$

(4) For all $gi_i = (g_i,t_i)$, $gi_j = (g_j,t_j) \in GI$ with $(a_i,t_i), (a_j,t_j) \in PRA$:
      If (a) $mis(gi_i) < mis(gi_j)$ or (b) $fail(gi_i) > fail(gi_j)$,
      then $vol(gi_i) > vol(gi_j)$

(5) For all $gi_i = (g_i,t_i)$, $gi_j = (g_j,t_j) \in GI$ with $(a_i,t_i), (a_j,t_j) \in PRA$:
      If $urg(gi_i) > urg(gi_j)$, then $fav(gi_i) > fav(gi_j)$

(6) For all $gi_i = (g_i,t_i)$, $gi_j = (g_j,t_j) \in GI$ with $(a_i,t_i), (a_j,t_j) \in PRA$:
      If (a) $vol(gi_i) > vol(gi_j)$ or (b) $fav(gi_i) > fav(gi_j)$,
      then $fi(gi_i) > fi(gi_j)$

(7) For all $act_i = (as_i, g_i) \in ACT$ with $(g_i, t_i) \in GI$:
If $fi(g_i, t_i) > fi_0(act_i)$ and $fi(g_i, t_i) > fi(g_j, t_i)$ for all (g_j, t_i),
then $(a_i, t_i + 1) = ini_i \in as_i$

(8) For all $act_i = (as_i, g_i) \in ACT$ with $(g_i, t_i) \in GI$:
If $din(g_i, t_i) > 0$ or $dre(g_i, t_i) > 0$, then there is $(a_j, t_j) = ifo_i$ with
$(t_j, as_j) \in PLA$ and $(g_i, a_l, t_l) \in GRA$ for all $(a_l, t_l) \in as_j$ or there is
$(a_k, t_k) \in PRA_i$ with $(t_k, as_k) \in PLA$ and $(g_i, a_m, t_m) \in GRA$ for all
$(a_m, t_m) \in as_k$

Following the preactional action phase, overt activities relevant to a chosen goal intention are initiated. According to axiom (7), this happens as soon as the fiat-tendency of a goal intention exceeds the action specific threshold and at the same time there is no competing goal intention with a higher fiat-tendency. The fiat-tendency itself is determined by the volitional strength of the goal intention and the favorability of the opportunity to initiate it (6).

The volitional strength of a goal intention is positively related to the resulting motivational tendency of the corresponding goal (3) and to the number of failed attempts in the past to implement it (4). It is a negative-monotonous function of the number of missed opportunities to implement the goal intention.

The more urgent the goal intention is, the more favorable will be the subjectively perceived opportunity for the initiation of relevant activities (5). In case of urgent intentions, the favorability of the opportunity might be overestimated and thus lead to a hasty initiation of activities.

If difficulties concerning a certain goal intention are anticipated with respect to the initiation or realization of corresponding activities, plans are formed to help the actor overcome these difficulties. The formation of plans may either accompany the formation of the goal intention or follow it in the preactional phase (8).

The Actional Action Phase (RT-ACN)

$x \in M(RT\text{-}ACN)$ iff

(1) $x = <A, G, T, GRA, GI, AS, ACT, PRD, PRA, ACN, POA, IFO, INI,$
$TER, SGR, AG, RG, CG, mot, dac, vol, vol_{op}, int, eff, fi, gr>$ and
$<A, G, T, GRA, GI, AS, ACT, PRD, PRA, ACN, POA, IFO, INI,$
$TER> \in M(RT\text{-}Basic)$

(2) $\quad$ SGR $\quad \subseteq$ G x G $\qquad$ (subgoal relation)

$\quad$ AG $\quad \subseteq$ G $\qquad$ (goals on the level of activities)

$\quad$ RG $\quad \subseteq$ G $\qquad$ (goals on the level of results)

$\quad$ CG $\quad \subseteq$ G $\qquad$ (goals on the level of consequences)

$\qquad$ with $(a,b) \in$ SGR only if $(a \in$ AG and $b \in$ RG$)$ or $(a \in$ RG and $b \in$ CG$)$

$\quad$ mot: $\quad$ G x T $\longrightarrow$ IR $\qquad$ (resulting motivational tendency)

$\quad$ dac: $\quad$ A x T $\longrightarrow$ IR $\qquad$ (difficulty of activity)

$\quad$ vol: $\quad$ GI $\quad \longrightarrow$ IR $\qquad$ (volitional strength)

$\quad$ vol$_{op}$: ACT $\quad \longrightarrow$ IR $\qquad$ (optimal volitional strength)

$\quad$ int: $\quad$ A x T $\longrightarrow$ IR $\qquad$ (intensity of activity)

$\quad$ eff: $\quad$ A x T $\longrightarrow$ IR $\qquad$ (efficiency of activity)

$\quad$ fi: $\quad$ GI $\quad \longrightarrow$ IR $\qquad$ (fiat-tendency)

$\quad$ gr: $\quad$ GI $\quad \longrightarrow$ G $\qquad$ (goal representation)

(3) $\quad$ For all $act_i = (as_i, g_i)$, $act_j = (as_j, g_j) \in$ ACT and for all $(a_i, t_i) \in ACN_i$, $(a_j, t_j) \in ACN_j$:

$\quad$ (a) $\enspace$ If $mot(g_i, t_i) > mot(g_j, t_j)$ or $dac(a_i, t_i) > dac(a_j, t_j)$, then $vol(g_i, t_i) > vol(g_j, t_j)$

$\quad$ (b) $\enspace$ If $vol(g_i, t_i) > vol(g_j, t_j)$, then $int(a_i, t_i) > int(a_j, t_j)$

$\quad$ (c) $\enspace$ If $vol(g_i, t_i) < vol(g_j, t_j)$ and $vol(g_i, t_i) < vol_{op}(act_i)$ and $vol(g_j, t_j) < vol_{op}(act_j)$, then $eff(a_i, t_i) < eff(a_j, t_j)$

$\quad$ (d) $\enspace$ If $vol(g_i, t_i) < vol(g_j, t_j)$ and $vol_{op}(act_i) < vol(g_i, t_i)$ and $vol_{op}(act_j) < vol(g_j, t_j)$, then $eff(a_i, t_i) > eff(a_j, t_j)$

(4) $\quad$ For all $act_i = (as_i, g_i) \in$ ACT and for all $(a_i, t_i) \in ACN_i$: If there is $act_j = (as_j, g_j) \in$ ACT and there is $(a_j, t_j) \in PRA_j$ and $fi(g_j, t_j) > fi(g_i, t_i)$, then there is $(a_k, t_k) \in ACN_i$ with $t_k > t_i$ so that $eff(a_k, t_k) < eff(a_i, t_i)$

(5) $\quad$ For all $(g_i, t_i) \in$ GI: $vol(g_i, t_i) \leq mot(g_i, t_i)$

(6) $\quad$ For all $act_i = (as_i, g_i) \in$ ACT with $(a_i, t_i) \in ACN_i$ and $gi_i = (g_i, t_i) \in$ GI: $int(a_i, t_i) \leq vol(gi_i)$

(7) $\quad$ For all $act_i = (as_i, g_i) \in$ ACT and for all $(a_i, t_i) \in ACN_i$:

$\quad$ (a) $\enspace$ If $dac(a_i, t_i) \leq 0$, then $gr(g_i, t_i) \in$ RG $\cup$ CG

$\quad$ (b) $\enspace$ If $dac(a_i, t_i) > 0$, then $gr(g_i, t_i) \in$ AG

The volitional strength of a goal intention in the actional phase depends on two variables (3a): The higher the resulting motivational tendency of the corresponding goal or the higher the difficulty of the current activity is, the more volitional strength will be provided for the goal intention. Volitional strength, however, cannot exceed motivational strength (5). With higher volitional strength, the intensity of the activity increases (3b), i.e. the actor increases his or her efforts to carry out goal-relevant activities with volitional strength being the upper limit of the intensity (6). This mechanism facilitates the persistence of an intention in case of difficulties.

Volitional strength also determines the efficiency of activities (3c,d). It is postulated that for every action there exists an optimal value of volitional strength, and that for suboptimal values the efficiency increases with increasing volitional strength, whereas it decreases for superoptimal levels of volitional strength. The efficiency of actional activities also decreases if at the same time there is a competing action in its preactional phase that has a goal intention with a fiat-tendency higher than that of the current intention (4). This axiom reconstructs Heckhausen's idea that in such a situation the process of deciding whether to give up the current activity in favor of a competing one draws off attentional resources from the current activity, making it less efficient [Heckhausen, 1989, p. 214].

According to axiom (7), there are three different levels of goal representation. Which level is actually chosen depends on the difficulty of the current activity. In case of no difficulty, the goal will be represented on more abstract levels such as possible results or consequences of results. If difficulties occur, however, the goal intention is represented on the level of concrete activities.

The Postactional Action Phase (RT-POA)

$x \in M(RT\text{-}POA)$ iff

(1) $x = <A,G,T,GRA,GI,AS,ACT,PRD,PRA,ACN,POA,IFO,INI,$
TER,acc,att,fi,end$>$ and
$<A,G,T,GRA,GI,AS,ACT,PRD,PRA,ACN,POA,IFO,INI,$
TER$> \in M(RT\text{-}Basic)$

(2) acc: GI $\longrightarrow \{0,1\}$ (goal accomplishment)
att: ACT $\longrightarrow$ IN (number of attributions)
fi: GI $\longrightarrow$ IR (fiat-tendency)
end: ACT $\longrightarrow$ T (end of postactional phase)
with end(act_i) $= t_i$ iff there is $(a_i,t_i) \in POA_i$
and there is no $(a_i,t_i+1) \in POA_i$

(3)　For all $act_i = (as_i, g_i) \in$ ACT and all $(a_i, t_i) \in POA_i$:
　　　If $acc(g_i, t_i) = 1$, then $(g_i, t_i + 1) \notin$ GI

(4)　For all $act_i = (as_i, g_i)$, $act_j = (as_j, g_j) \in$ ACT and for all $(a_i, t_i) \in$
　　　POA_i, $(a_j, t_j) \in POA_j$ and (g_i, t_i), $(g_j, t_j) \in$ GI:
　　　If $acc(g_i, t_i) = 0$ and $acc(g_j, t_j) = 1$, then $att(act_i) > att(act_j)$

(5)　For all $act_i = (as_i, g_i)$, $act_j = (as_j, g_j) \in$ ACT:
　　　with $(a_i, t_i) = ter_i$, $(a_j, t_j) = ter_j$, $(a_k, t_k) \in POA_i$ and $(a_m, t_m) \in POA_j$:
　　　If there are (g_k, t_k), $(g_m, t_m) \in$ GI with $g_k \neq g_i$, $g_m \neq g_j$ and $fi(g_k, t_k)$
　　　$> fi(g_m, t_m)$, then $(end(act_i) - t_i) < (end(act_j) - t_j)$

According to axiom (3), a goal intention is deactivated as soon as it has been successfully accomplished. Especially in the case of failure to achieve a pursued goal, the actor will tend to look for an explanation. Therefore, after failure more causal attributions will occur than after success (4).

　　Axiom (5) postulates that the postactional phase will be shortened if another goal intention is present. The higher the fiat-tendency of the new intention is, the sooner the postactional phase of the current action will end. This formalizes Heckhausen's hypothesis that announcing a new task (inducing a new goal intention) will shorten postactional evaluation processes.

III Mind-Set Elements

In Gollwitzer's version of the theory, each action phase is accompanied by a distinct mind-set. Each mind-set (deliberative, implemental, actional, or evaluative) is described as a cognitive orientation leading to a characteristic way of information processing. In our concise Rubicon theory, however, we make use of only two mind-set elements, because only for the deliberative and implemental mind-sets are elaborated assumptions and empirical studies reported.

The Deliberative Mind-Set (RT-DEL)

$x \in$ M(RT-DEL) iff
(1)　$x = <$A,G,T,GRA,GI,AS,ACT,PRD,PRA,ACN,POA,IFO,INI,
　　　　TER,del,sac,$proc_{df}$,rec,$real_{df}$,met,fc,$fc_0 >$ and
　　　　$<$A,G,T,GRA,GI,AS,ACT,PRD,PRA,ACN,POA,IFO,INI,
　　　　TER$> \in$ M(RT-Basic)

(2) del: $A \times T \longrightarrow IR$ (degree of deliberative cognitive orientation)

sac: $A \times T \longrightarrow IR$ (significance of action consequences)

$proc_{df}$: $A \times T \longrightarrow IR$ (degree of processing of information concerning desirability and feasibility)

rec: $A \times T \longrightarrow IR$ (receptivity for information)

$real_{df}$: $A \times T \longrightarrow IR$ (degree of realism in processing of information concerning desirability and feasibility)

met: $A \times T \longrightarrow IR$ (degree of meta-motivation)

fc: $A \times T \longrightarrow IR$ (facit-tendency)

fc_0: $ACT \longrightarrow IR$ (threshold for facit-tendency)

(3) For all $act_i = (as_i, g_i) \in ACT$ and for all (a_i, t_i), $(a_j, t_j) \in as_i$:
If $(a_i, t_i) \in PRD_i$ and $(a_j, t_j) \notin PRD_i$, then $del(a_i, t_i) > del(a_j, t_j)$

(4) For all $act_i = (as_i, g_i) \in ACT$ and for all (a_i, t_i), $(a_j, t_j) \in PRD_i$:
If $sac(a_i, t_i) > sac(a_j, t_j)$, then $del(a_i, t_i) > del(a_j, t_j)$

(5) For all (a_i, t_i), $(a_j, t_j) \in A \times T$:
If $del(a_i, t_i) > del(a_j, t_j)$, then a) $proc_{df}(a_i, t_i) > proc_{df}(a_j, t_j)$, and b) $rec(a_i, t_i) > rec(a_j, t_j)$ and c) $real_{df}(a_i, t_i) > real_{df}(a_j, t_j)$ and d) $met(a_i, t_i) > met(a_j, t_j)$

(6) For all $act_i = (as_i, g_i)$, $act_j = (as_j, g_j) \in ACT$ with $(a_i, t_i) \in PRD_i$, $(a_j, t_j) \in PRD_j$:
If $del(a_i, t_i) = del(a_j, t_j)$ and $fc(a_i, t_i) < fc_0(act_i)$ and $fc(a_j, t_j) > fc_0(act_j)$ and $(a_i, t_i+1) \notin PRD_i$ and $(a_j, t_j+1) \notin PRD_j$, then $del(a_i, t_i+1) > del(a_j, t_j+1)$

A deliberative mind-set is generated during the predecisional action phase. As axiom (3) indicates, predecisional activities are accompanied by a higher degree of deliberative cognitive orientation than are other activities. The degree of this cognitive orientation increases with higher significance of (anticipated) action consequences (4).

As a consequence of a deliberative cognitive orientation, the actor becomes more receptive to information in general (5b) and the processing of special information about desirability and feasibility of possible goals is intensified (5a). This information is processed in an objective and realistic way (5c). According to Heckhausen [1989, p. 204], a deliberative cognitive orientation also leads to the occurrence of meta-motivation (5d). Meta-motivation refers to information about the processing of motivational information and its improvement.

Gollwitzer [1990, 1991] postulates that the cognitive tuning effects of a deliberative mind-set will continue for some time even if the predecisional activities are interrupted before the facit-tendency reaches threshold. This resembles the assumption of an inherent inertia tendency for deliberative mind-sets and is reformulated as axiom (6). This axiom is required in those empirical studies where the predecisional action phase is externally terminated and the actors are expected to process information within the same mind set during the following test phase.

The Implemental Mind-Set (RT-IMP)

$x \in M(RT-IMP)$ iff

(1) $x = <A,G,T,GRA,GI,AS,ACT,PRD,PRA,ACN,POA,IFO,INI,$
       $TER,imp,vol,din,dre,proc_{imp},rec,real_{df},real_{imp}>$ and
       $<A,G,T,GRA,GI,AS,ACT,PRD,PRA,ACN,POA,IFO,INI,$
       $TER> \in M(RT-Basic)$

(2)   imp: $A \times T \longrightarrow IR$ (degree of implemental cognitive orientation)
       vol: $GI \longrightarrow IR$ (volitional strength)
       din: $GI \longrightarrow IR$ (difficulty of initiation)
       dre: $GI \longrightarrow IR$ (difficulty of realization)
       $proc_{imp}$: $A \times T \longrightarrow IR$ (degree of processing of information concerning implementation)
       rec: $A \times T \longrightarrow IR$ (receptivity for information)
       $real_{df}$: $A \times T \longrightarrow IR$ (degree of realism in processing of information concerning desirability and feasibility)
       $real_{imp}$: $A \times T \longrightarrow IR$ (degree of realism in processing of information concerning implementation)

(3)   For all $act_i = (as_i,g_i) \in ACT$ and for all (a_i,t_i), $(a_j,t_j) \in as_i$:
       If $(a_i,t_i) \in PRA_i$ and $(a_j,t_j) \notin PRA_i$, then $imp(a_i,t_i) > imp(a_j,t_j)$

(4)   For all $act_i = (as_i,g_i)$, $act_j = (as_j,g_j) \in ACT$ with $(a_i,t_i) \in PRA_i$,
       $(a_j,t_j) \in PRA_j$:
       If a) $vol(g_i,t_i) > vol(g_j,t_j)$ or b) $din(g_i,t_i) > din(g_j,t_j)$ or c) $dre(g_i,t_i) > dre(g_j,t_j)$, then $imp(a_i,t_i) > imp(a_j,t_j)$

(5)   For all (a_i,t_i), $(a_j,t_j) \in A \times T$:
     If $\mathrm{imp}(a_i,t_i) > \mathrm{imp}(a_j,t_j)$, then a) $\mathrm{proc}_{\mathrm{imp}}(a_i,t_i) > \mathrm{proc}_{\mathrm{imp}}(a_j,t_j)$, and b)
     $\mathrm{rec}(a_i,t_i) < \mathrm{rec}(a_j,t_j)$, and c) $\mathrm{real}_{\mathrm{imp}}(a_i,t_i) > \mathrm{real}_{\mathrm{imp}}(a_j,t_j)$, and d)
     $\mathrm{real}_{\mathrm{df}}(a_i,t_i) < \mathrm{real}_{\mathrm{df}}(a_j,t_j)$

The implemental cognitive orientation is associated with preactional activities of action planning (3). It is facilitated by high volitional strength of the goal intention (4a) and by (anticipated) difficulties of initiation or realization (4b, c). As a consequence of an implemental orientation, the general receptivity for information decreases (5b), whereas the processing of information concerning the implementation of the current goal intention is intensified (5a). This information is compatible with the mind-set and is thus processed more realistically than is information about desirability and feasibility of the goal (5c, d).

IV Action Phases and Mind-Sets in Empirical Studies

In our structuralist Rubicon theory, assumptions concerning action phases on the one hand and mind-sets on the other are formulated in different theory elements. Although both Heckhausen and Gollwitzer postulate a close relationship between these concepts, the definition of separate theory elements can be justified by an analysis of the empirical applications of the theory. Each empirical study refers to either special laws of distinct action phases or to special laws characterizing mind-sets, but never to both simultaneously. Thus, mind-sets and action phases are represented by different parts of the Rubicon theory net, which are connected to a common basic element RT-Basic. The correspondence between action phase and mind-set is formulated in each mind-set element as a substantial axiom.

    The theory elements in both parts of the net have to be further specialized before direct reference points of empirical studies are reached. An example of an "empirical theory element" within the partial net of action phase elements is given in the following section.

Efficiency of Activities (RT-Eff)

The theory element RT-Eff pertains to an empirical study by Heckhausen and Strang [1988; see also Heckhausen, 1989, pp. 215−216], who investigated the effect of overoptimal exertion during the actional phase on the efficiency of activities. In the experiment, basketball players had to run circuits under either normal or record performance demands. Heart rate and lactate concentration were measured as indicators of intensity, and hit rate was taken as an index of efficiency of the activities. Results show that under record conditions intensity increases whereas efficiency of performance decreases.

Under the Rubicon perspective, the record condition may have lead to an overoptimal level of volitional strength, which according to axiom (3) of RT-ACN lowers the efficiency of an actional activity. On the basis of RT-ACN, the empirical theory element RT-Eff is developed by a simplification followed by an enlargement. On the one hand, not all assumptions of RT-ACN are needed to describe the empirical application from a Rubicon point of view; on the other hand, new special assumptions are required.

$x \in M(RT\text{-}Eff)$ iff

(1) $x = <A,G,T,GRA,GI,AS,ACT,PRD,PRA,ACN,POA,IFO,INI,$
        $TER,dac,vol,vol_{op},int,eff,lac,hrt,hit>$ and
        $<A,G,T,GRA,GI,AS,ACT,PRD,PRA,ACN,POA,IFO,INI,$
        $TER> \in M(RT\text{-}Basic)$

(2) $dac:$ $A \times T \longrightarrow \mathbb{R}$ (difficulty of activity)
        $vol:$ $GI \longrightarrow \mathbb{R}$ (volitional strength)
        $vol_{op}:$ $ACT \longrightarrow \mathbb{R}$ (optimal volitional strength)
        $int:$ $A \times T \longrightarrow \mathbb{R}$ (intensity of activity)
        $eff:$ $A \times T \longrightarrow \mathbb{R}$ (efficiency of activity)
        $lac:$ $A \times T \longrightarrow \mathbb{R}$ (lactate concentration)
        $hrt:$ $A \times T \longrightarrow \mathbb{R}$ (heart rate)
        $hit:$ $A \times T \longrightarrow \mathbb{R}$ (hit rate)

(3)   For all $act_i = (as_i,g_i)$, $act_j = (as_j,g_j) \in ACT$ and for all $(a_i,t_i) \in$ ACN_i, $(a_j,t_j) \in ACN_j$:
        (a) If $dac(a_i,t_i) > dac(a_j,t_j)$, then $vol(g_i,t_i) > vol(g_j,t_j)$
        (b) If $vol(g_i,t_i) > vol(g_j,t_j)$, then $int(a_i,t_i) > int(a_j,t_j)$
        (c) If $vol(g_i,t_i) < vol(g_j,t_j)$, and $vol(g_i,t_i) < vol_{op}(act_i)$, and $vol(g_j,t_j)$ $< vol_{op}(act_j)$, then $eff(a_i,t_i) < eff(a_j,t_j)$
        (d) If $vol_{op}(act_i) < vol(g_i,t_i)$, and $vol_{op}(act_j) < vol(g_j,t_j)$, and $vol(g_i,$ $t_i) < vol(g_j,t_j)$, then $eff(a_i,t_i) > eff(a_j,t_j)$

(4)   For all $act_i = (as_i,g_i)$, $act_j = (as_j,g_j) \in ACT$ and for all $(a_i,t_i) \in$ ACN_i, $(a_j,t_j) \in ACN_j$:
        (a) If $int(a_i,t_i) > int(a_j,t_j)$, then $lac(a_i,t_i) > lac(a_j,t_j)$

(b) If $int(a_i,t_i) > int(a_j,t_j)$, then $hrt(a_i,t_i) > hrt(a_j,t_j)$

(c) If $eff(a_i,t_i) > eff(a_j,t_j)$, then $hit(a_i,t_i) > hit(a_j,t_j)$

A comparison of RT-Eff with RT-ACN shows that axiom (3) of RT-Eff is taken from the theory element for the actional action phase, whereas axiom (4) formulates new special assumptions for the distinct experimental situation created by Heckhausen and Strang (1988). By simplification of RT-ACN, a theory element RT-acn is generated with only one substantial assumption (3). In a second step, this element is enlarged by axiom (4) above. Since the concept of mind-sets is not needed for a description of the empirical situation under discussion, RT-Eff can be considered an "empirical action phase element" [for more empirical theory elements see Gerjets, Westermann & Heise, 1992; Westermann, Gerjets & Heise, 1993].

Conclusions

From the proceeding discussion it has become obvious, that the original Rubicon theory makes use of idealizations and simplifications in many respects. It therefore seems to be applicable only to sequences of activities that pass through all four phases in predetermined order with homogeneous activities taking place in each phase. The concise structuralist version de-idealizes the original theory with respect to the homogeneity of activities during action phases, thus bringing it closer to reality. A number of idealized assumptions, however, are maintained:

- According to RT-ACT, an action comprises one and only one goal guiding the sequence of activities.
- The Rubicon basic element precludes steps back from an action phase into a preceeding one. In its present formulation, the theory does not comprise any feedback loops within the course of an action.
- In RT-PRA action plans are interpreted as planned sequences of activities; whereas in most real actions, plans will have the form of conditional plans with their execution depending on the presence of distinct situational conditions.
- Although the concise Rubicon theory leaves the quality of the activity classes unspecified, the assumption of ideal homogeneous activities within each action phase is implicitly preserved. First, intended interpretations of the concepts may characterize predecisional activities as deliberating and preactional activities as planning. Second, these concepts are used within different substantial assumptions. Those assump-

tions which describe the preactional planning activities, for example, are supposed not to hold in the predecisional phase and vice versa.
— The theory ignores the influence of objective situational determinants, including the activities of others, on the action under consideration. Situational factors occur only in internally interpreted form, such as the (subjective) favorability of the opportunity to implement a goal intention. In the structuralist formalization, no set of actors needs to be introduced, because interactions between actors are not within the scope of the Rubicon theory.

If certain non-ideal actions are to be analyzed from a Rubicon point of view, new special theory elements have to be incorporated into the theory net. Consider, for example, clinically relevant activities such as compulsive actions. Compulsive goals may be regarded as emotional impulse goals (see RT-PRD above), so that the intention to perform a compulsive ritual can be formed without predecisional deliberations. During the actional phase, the degree of subjective resistance against the compulsive activity could be introduced as a determinant of the difficulty of the activity. Special laws could connect the compulsive activities to typical emotional experiences such as anxiety and discomfort. But even many non-pathological actions may require enlargements and specializations of the Rubicon theory before they can be successively described. For example, unexpected or unintended action outcomes could be incorporated into the theory, or the Rubicon perspective could be extended to the analysis of interactive behavior.

Despite several idealizations and simplifications, the Rubicon theory of action phases can, in our opinion, be used as an integrative theoretical approach towards a comprehensive description and explanation of human goal-directed activities. As indicated above, specializations in different directions will make the theory applicable to a wider and less idealized class of actions.

Institut für Psychologie
Georg-August-Universität
Gosslerstrasse 14
D-37073 Göttingen, Germany

REFERENCES

Balzer, W., Moulines, C.U. and Sneed, J.D. [1987]. *An Architectonic for Science: The Structuralist Program*. Dordrecht: Reidel.

Gerjets, P., Westermann, R. and Heise, E. [1992]. Strukturalistische Analysen zur Rubikontheorie der Handlungsphasen. (Arbeitsbericht) Göttingen: Institut für Psychologie der Georg-August-Universität.

Gollwitzer, P.M. [1987]. The Implementation of Identity Intentions: A Motivational-Volitional Perspective on Symbolic Self-Completion. In: F. Halisch and J. Kuhl (Eds.). *Motivation, Intention, and Volition.* Berlin: Springer, pp. 349–369.

Gollwitzer, P.M. [1990]. Action Phases and Mind-Sets. In: E.T. Higgins, R.M. Sorrentino (Eds.). *Handbook of Motivation and Cognition: Vol. 2. Foundations of Social Behavior.* New York: Guilford Press, pp. 53–92.

Gollwitzer, P.M. [1991]. *Abwägen und Planen. Bewußtseinslagen in verschiedenen Handlungsphasen.* Göttingen: Hogrefe.

Heckhausen, H. [1989]. *Motivation und Handeln* (2. Aufl.). Berlin: Springer.

Heckhausen, H. and Gollwitzer, P.M. [1987]. Thought Contents and Cognitive Functioning in Motivational versus Volitional States of Mind. *Motivation and Emotion* 11, 101–120.

Heckhausen, H. and Strang, H. [1988]. Efficiency under Record Performance Demands: Exertion Control – an Individual Difference Variable? *Journal of Personality and Social Psychology* 55, 489–498.

Westermann, R., Gerjets, P. and Heise, E. [1993]. Präzisierung und Strukturierung von Handlungstheorien. In: L. Montada (Hrsg.). *Bericht über den 38. Kongreß der Deutschen Gesellschaft für Psychologie in Trier 1992* (Band 2). Göttingen: Hogrefe, s. 791–797.

Poznań Studies in the Philosophy
of the Sciences and the Humanities
Vol. 42, pp. 159—177

Klaus G. Troitzsch

MODELLING, SIMULATION, AND STRUCTURALISM

Introduction

In contrast to earlier publications on the same topic [cf. Troitzsch 1990a, pp. 124—177, 1990b, 1992], this paper aims at a logical reconstruction of some different approaches to modelling processes of self-organization in the social sciences, and, moreover, toward showing that computer simulation supports structuralist reconstruction of theories if it is done in a certain way. In contrast especially to Troitzsch [1992] we do not concentrate on a specific theory of attitude change, although we shall take the axiomatizations carried out there as but one example of a theory of self-organization.

First, we have to discuss the term "self-organization". Following Haken [1988, p. 11] we call a system self-organized

> "if it acquires a spatial, temporal or functional structure without specific inter-
> ference from the outside. By 'specific' we mean that the structure or functioning is
> not impressed on the system, but that the system is acted upon from the outside in
> a nonspecific fashion."

Furthermore we refer to Hayek [1942, p. 288], who wrote even fifty years ago about the social sciences that

> "the problems which they try to answer arise only in so far as the conscious
> action of many men produce undesigned results, in so far as regularities are ob-
> served which are not the result of anybody's design. If social phenomena showed no
> order except in so far as they were consciously designed, there could be no room for
> theoretical sciences of society... It is only in so far as some sort of order arises as
> a result of individual action but without being designed by any individual that a
> problem is raised which demands a theoretical explanation."

During the last twenty years, the interdisciplinary research efforts of synergetics have invented methods that make new and powerful tools available for the social scientist as well. It should be interesting to try a structuralist reconstruction of these theoretical tools.

Modern approaches to modelling processes of self-organization have in common that they

- are carried out with methods of mathematics and computer science,
- model complex stochastic systems,
- represent individuals and populations (or groups or other collectives of people) at the same time, as well as
- interactions between these two (or even more) levels, and, in many cases,
- interactions among the objects of each level.

We have to discriminate between two groups of approaches: In a first group we have models in which individuals of one or more populations do not interact directly, but change the state of the whole system (or of their respective population) by their behaviour and react on the change of this collective with individual changes of their behavioural state. This is the case in typical synergetic models, both in the social science and in physics, chemistry, biology etc. where the most famous example of the latter seems to be the slime mould case [Wurster 1988]. The point is here that in the beginning we have an "aggregate" of initially unconnected individuals, which by itself turns into a system due to the effect that the individuals are endowed with the ability to move in a potential which is built up as a result of the sheer existence of these individuals. For a comparison of the slime mould case and a theory of attitude change in human populations see Troitzsch [1991].

On the other hand we have models in which individuals interact directly, mostly in a network or, as it were, in a cellular automaton. Here too, the population is modelled homogeneously in the beginning, then stochastically influenced interactions change the state of the interacting individuals and the relations between each pair of interacting individuals, and in the end we find stable clusters, groups, subnets, or strata in the whole population. For references, see Hegselmann [1993], Helbing [1992a,1992b] and Drogoul and Ferber [1994].

I
Models with Indirect Interaction

Collective
"Meso level"

Individuals
"Micro level"

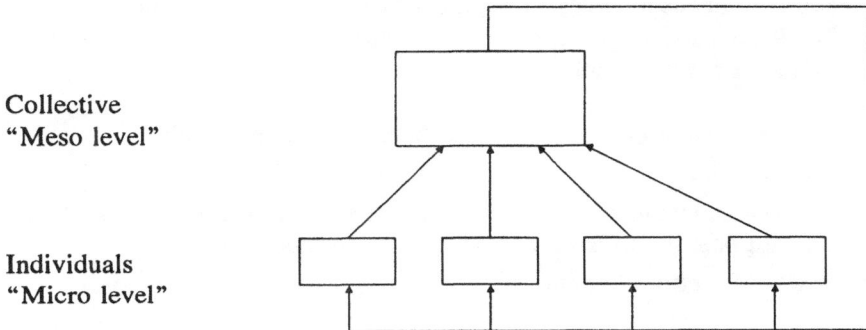

Figure 1: Models with indirect interaction between individuals and the collective

Elements on the lowest level are endowed with the capability to react on a potential (or, more generally: on the state of a higher level) and at the same time to change this potential or higher level state (see figure 1, where arrows point from causes to effects: changes at the arrowhead are effected by the state of the arrow's origin). In the following example, the individual have individual attitudes which change the majority or the "public opinion" on the collective level. On the other hand, they change their individual attitudes depending on "public opinion", trying to adapt to the major opinion trends.

We shall now take this example to show what computer simulation and structuralist theory reconstruction have in common. Thus we take one version of the miniature theory of attitude change considered in Troitzsch [1992] and quote the major definitions of this paper (where we may easily drop the empirical applications of this miniature theory which is outside the scope of this paper which is devoted to a comparison between simulation techniques and theory reconstruction).

We first define the theory element of **AC**:

Def TE (AC): TE(AC) := $\langle K(AC), I(AC) \rangle$ where

$K(AC) := \langle M_p(AC), M(AC), M_{pp}(AC), Po[M_p(AC)], M_p(AC) \rangle^1$ and $I(AC) \subseteq M_{pp}(AC)$, intended applications of AC, is such that members of I(AC) are data sets describing homogeneous populations whose individual members have attitudes which may be observed; there are exactly two continuous attitudes — say on a "left"-"right" continuum and on a "political satisfaction" continuum — which must (and can) be asked for in one or several consecutive surveys or even panels.

By this description of intended applications of AC we include

- singular surveys,
- repeated anonymous surveys encompassing the same population such that it is not possible to build individual attitude time series,
- panels of arbitrary length.

We continue with a definition of a potential model of AC.

Def $M_p(AC)$: η is a potential model of AC, i.e. $\eta \in M_p(AC)$, iff there exist $I, C, A, T, \ell, \Omega, a, \phi, \delta$ and ρ such that

1. $\eta = \langle I, C, A, T, \mathbb{R}, \ell, \Omega, a, \phi, \delta, \rho \rangle$.

2. I is a non-empty, finite set [of persons].

3. C is a non-empty finite set of subsets of I [i.e. of populations or collectives], $C \subseteq Po(I)$.

4. $\langle A, \mathcal{A} \rangle$ is a measurable space [of attitudes] with $A = \mathbb{R}^2$.

5. T is a set [of instants].

6. $\ell : T \to \mathbb{R}$ is a bijective function [labeling (or coordinatizing) the instants with real numbers].

7. $\langle \Omega, \mathcal{F}, P \rangle$ is a probability space with
    (a) Ω is a sample space,
    (b) $\omega : I \to \Omega$ is bijective,
    (c) $\mathcal{F} \subseteq \Omega$ is a family of events,
    (d) P is a probability measure defined on $\mathcal{F}$.

8. a is a function with $Dom(a) = I \times T$ and $Rge(a) = A$, such that (with $i \in I$) $a(i,t)$ is the attitude of i at time t.

---

1 Since we do not introduce any constraints here, every set of potential models, i.e. every member of $Po[M_p(AC)]$ satisfies all conceivable constraints; and since we do not introduce any links, all potential models satisfy all the theory's links with other theories.

9. ϕ is a function with $\text{Dom}(\phi) = C \times T$ and $\text{Rge}(\phi) = \{f \mid f : A \to \mathbb{R}^+;$ $\int_A f \, da = 1\}$, such that (with $C \in \mathcal{C}$) $\phi(C,t)$ is the function which describes the probability density of each $i \in C$ having attitude $\alpha \in A$ at time $t \in T$, i.e. $\phi(C,t)(\alpha)$ is the probability density of finding a $i \in C$ with attitude $\alpha \in A$ at time $t \in T$.

10. δ is a function with $\text{Dom}(\delta) = \mathcal{C} \times T \times T$ and $\text{Rge}(\delta) = \{d \mid d : A \to A\}$ such that (with $C \in \mathcal{C}$) $\delta(C,s,t)$ is the integral of the trend coefficient of the diffusion process from s to t, i.e. $\delta(C,s,t)(a) = \int_s^t \mu[a(.,\tau),\tau] d\tau$. This means that $a(i,t)$ would be equal to $a(i,s) + \int_s^t \mu[a(.,\tau),\tau] d\tau$ if there were no noise.

11. $(\rho_t)_{t \in T}$ is a stochastic process on the probability space $\langle \Omega, \mathcal{F}, P \rangle$ with values in the measurable space $\langle A, \mathcal{A} \rangle$ such that ρ is bivariate Gaussian white noise without cross-correlation which may be written $\rho(i,t)$ further on.

It should be clear that in empirical applications of our miniature theory, all terms mentioned up to item 9 of **Def M_p(AC)** are measurable without any theory concerning attitude *changes*. The measurement of a, of course, necessitates theories of attitude measurement, as the measurement of ϕ necessitates probability theory, but in both cases we do not have to know anything about the process of attitude change.

Only δ and ρ are AC-theoretical terms — as will be discussed below.

We are now in a position to compare the definition of the potential model of **AC** to a skeleton of a MIMOSE program which is complete but for some function applications:

```
pop := { params    : list of real
                   := ... persons.att ... ;
         kparams   : list of real
                   := ... ;
         persons   : list of person
       };
person := { r              : real
                          := ... ;
            att           : list of real
                          := ... population.params_1 ... ;
            population : pop
          };
```

From table 1 we see that MIMOSE models a function from $I \times T$ to an arbitrary set S as an attribute of the object type corresponding to I whose type is the type corresponding to S. If S is a set of functions, then the corresponding MIMOSE attribute is of type `list of real`, i.e. the space of real valued parameters of the functions $\in S$.

AC	MIMOSE
K	object type person
C	object type pop
A	type list of real of the attribute att of object type person
T	the set of simulation steps which is always implicit in MIMOSE
ℓ	no match, since time is always discrete in MIMOSE (and on any digital computer)
Ω	the set of seeds of MIMOSE's random number generators, each random attribute of each instantiation of each object type having its own seed
a	the attribute att of object type person
ϕ	the attribute params of object type pop — the range of params being the space of ϕ's function parameters
δ	the attribute kparams of object type pop — the range of kparams being the space of δ's function parameters
ρ	the attribute r of object type person

Table 1: Match between terms of **AC** and identifiers of the related MIMOSE simulation program

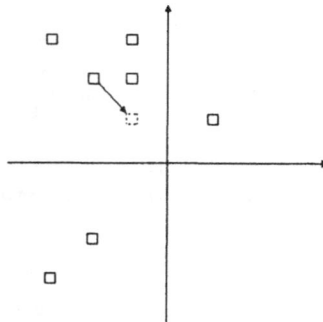

Figure 2: Attitude space of **AC** with some individuals plotted. One individual is changing its attitude position to achieve a more densely populated position, thus changing the whole distribution on the attitude space.

In our example we suppose a two-dimensional continuous attitude space (for a graphical representation see figure 2). We restrict ourselves to a function ϕ whose domain consists of probability density functions of the type

$$
\begin{aligned}
(1) \quad f(a;\theta) = \exp\{ \quad & \theta_{00} \quad + \theta_{10}a_1 \quad + \theta_{20}a_1^2 + \cdots + \theta_{n0}a_1^n \\
& + \theta_{01}a_2 + \theta_{11}a_1a_2 + \cdots \quad + \theta_{n-1,1}a_1^{n-1}a_2 \\
& + \cdots \\
& + \cdots \quad + \theta_{ij}a_1^i a_2^j + \cdots \\
& + \theta_{0n}a_2^n\} \\
(2) \quad = \exp\{ & -V(a)\}
\end{aligned}
$$

that is, to probability density functions of the exponential family whose exponent is a polynomial up to fourth degree in a. This is not an important restriction, however, because supposing a polynomial $V(a)$ up to fourth degree in the exponent of the probability density function only means taking a Taylor expansion of an arbitrary function $\tilde{V}(a)$ including only the lower powers up to the biquadratic terms. The last term of the Taylor expansion of $\tilde{V}(a)$ included into $V(a)$ must, of course, always be of even degree, since otherwise $\exp\{-V(a)\}$ will be no probability density function (and $V(a)$ will be no potential).

The θ's may be estimated by one of the two procedures (named calc-pdfparams in the MIMOSE program text below) described in Troitzsch [1987, pp. 160–170] and Troitzsch [1990b, pp. 360–361], Herlitzius [1990], respectively.

In this case, the attitude change function δ describes a diffusion process with

$$
(3) \quad \mu(a,t) = \frac{-\partial \gamma V(a)}{\partial a}
$$

$$
(4) \quad \Sigma(a,t) = \sigma_\rho^2 I
$$

with V the (negative) exponent of the probability density function of equations 1 and 2, and γ is a constant in which the dependence of δ on C becomes manifest. Perhaps we should have inserted an additional AC-theoretical term $\gamma : C \rightarrow I\!R$ into our definition of potential models of AC.

In this continuous case, ρ is a bivariate Gaussian white noise process with no cross correlation (i.e. with $D^2(\rho) = \sigma_\rho^2 I$ where σ is a constant in which a dependence of ρ on C unmentioned so far becomes manifest.

Perhaps an additional **AC**-theoretical term $\sigma : \mathcal{C} \to I\!R^+$ would have been appropriate.

Here, the definition of a model of **AC** reads:

Def M(AC): ζ is a model of **AC**, i.e. $\zeta \in$ **M(AC)**, iff there exist I, $\mathcal{C}$, A, T, ℓ, Ω, a, ϕ, δ and ρ such that

1. $\zeta = \langle I, \mathcal{C}, A, T, I\!R, \ell, \Omega, a, \phi, \delta, \rho \rangle$.

2. $\zeta \in$ **M$_p$(AC)**

3. V is an (auxiliary) function with $\mathrm{Dom}(V) = A \times T$ and $\mathrm{Rge}(V) = I\!R - \{-\infty\}$ with

$$V(a,t) = -\ln \phi(C,t)(a)$$

and

$$\lim_{\Delta \to 0} \frac{\delta(C,t,t+\Delta t)(a)}{\Delta t} = -\gamma_c \frac{\partial V(a,t)}{\partial a}$$

such that (informally)

$$a(i,t) = a(i,s) + \gamma_c \int_s^t -\frac{\partial V(a(i,\tau),\tau)}{\partial a} d\tau + \int_s^t \rho(i,\tau) d\tau$$

where γ_c is a constant function.

This leads to a complete MIMOSE program which reads

```
pop := { gamma    : real;
         sigma    : real;
         params   : list of real
                    := calcpdfparams(persons.att);
         kparams  : list of list of real
                    := derivative(params);
         persons  : list of person
       };
person := { r            : list of real
                           := normal;
            att          : list of real
                           := langevin(att_1,
                                       pop.gamma,
                                       pop.kparams,
                                       pop.sigma,
                                       r);

            population : pop
          };
```

where the function named `langevin` performs the operations required by Definition **M(AC)**.2.

Here, too, the parameters of the attitude change function δ can be measured by means of **AC** as was shown in Troitzsch [1990b].

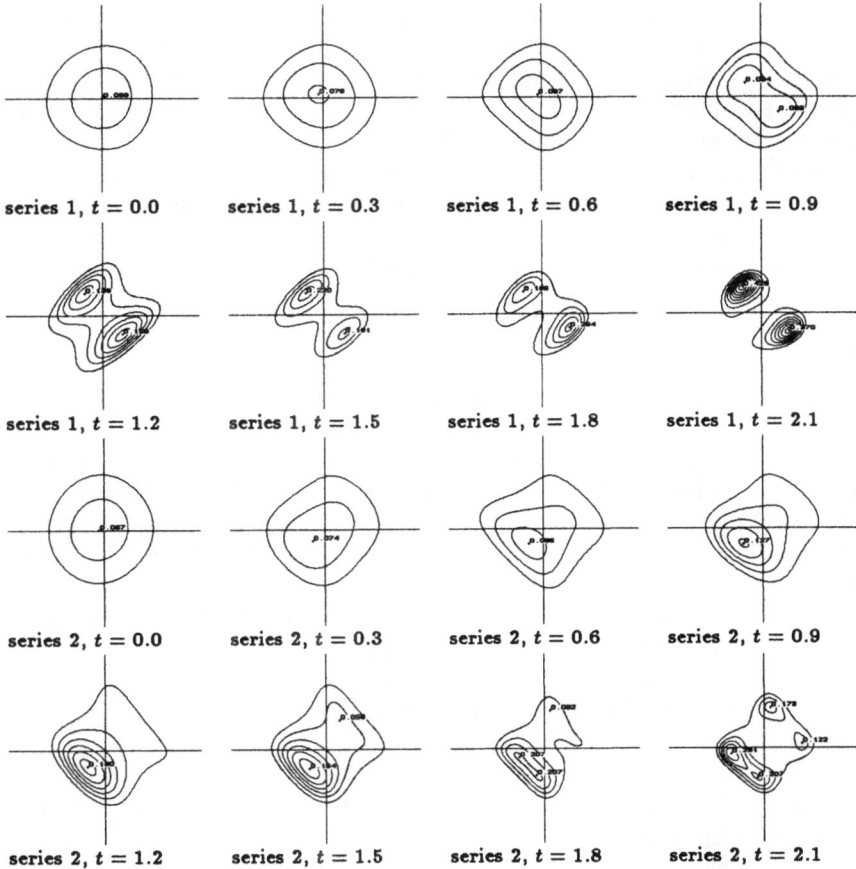

Figure 3: Two realizations of the stochastical process of the continuous case over eight time steps; distance between contour lines is 0.025, where the maxima are lower than 0.200, and 0.050 else.

Figure 3 shows two realizations of the stochastic process on the collective level. Both realizations start from the same initial conditions on the individual level; they differ only in the seed of the random number number generator. Both realizations are approximately normally (at least: unimodally) distributed in the beginning. Their distributions develop into multimodality after a while, apparently becoming stationary in the end. There are, of course, also realizations that seem to remain unimodal forever.

It should be noted here that the user interfaces provided by MIMOSE (which we shall not describe in this paper) have some facilities to hide attributes corresponding to theoretical terms from the user. A MIMOSE user using GEMM (the graphical editor for the MIMOSE modeling language, Klee and Troitzsch [1993]) will even be guided through the modeling process in a similar manner as the one adopted by structuralism: In step 1 of GEMM the user will have to define object types (base sets), while in step 2 attribute types (function ranges) are defined; step 3 allows the definition of function domains, while step 4 defines the function bodies which correspond to the axioms of the definition of the actual model.

II
The Partial Potential Models of Theory AC

We have defined the gradient function δ and the stochastic process ρ as AC-theoretical terms of our miniature theories. It is obvious that in every conceivable empirical application of our theory as described in the previous subsection the influences of δ and ρ cannot be separated. Of course, it is possible to measure the attitudes of all $i \in C$ at two different points of time and to arrive at estimates of the gradient function of the diffusion process of the continuous case. It is, however, first necessary to *know* (or *assume*) that the observed process is the diffusion process of equations 3 and 4.

This is why we have to define the partial potential models of **AC** in the following manner:

Def $M_{pp}(AC)$: ξ is a partial potential model of AC, i.e. $\xi \in M_{pp}(AC)$, iff there exists η such that

1. $\eta = \langle I, C, A, T, I\!\!R, \ell, \Omega, a, \phi, \delta, \rho \rangle \in M_p(AC)$ and
2. $\xi = \langle I, C, A, T, I\!\!R, \ell, \Omega, a, \phi \rangle$

III
Models with Direct Interaction between Individuals

A completely different approach to modelling processes of self-organization must be adopted when individuals are allowed to interact directly. Here, the individual is no longer isolated from the other individuals (and not only knows about their opinions by observation of the population as a whole, e.g. by listening to the radio or by reading newspapers) but changes its attitude due to the locally observable attitude of those individuals with whom it has personal contacts. Here, for example, we may suppose that the strength of *alter*'s influence on *ego* depends on how often *alter* and *ego* interacted in the past, and that the frequency of interactions depends on the similarity of attitudes. Thus again we have two coupled attributes, the frequency of pairwise interactions, and the individual attitude.

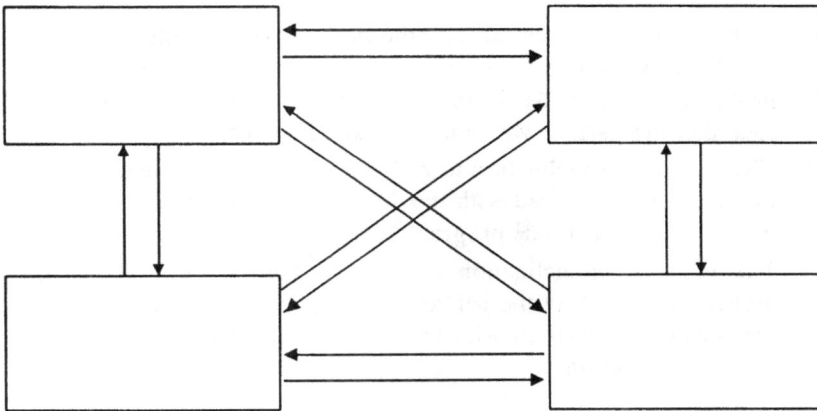

Figure 4. Models with direct interaction between the individuals

The classical example of this type of model seems to be Kirk's and Coleman's [1967] model of pairwise interaction in a three person's group which goes back to Simon's [1957, pp. 99–114] famous formalization of Homans's [1950] theory of interaction in social groups. In this model, each of three individuals has a certain inclination to interact with one of the two others. This inclination is corroborated by any realized interaction. Moreover, sympathy towards all possible partners is decreased in every time step. Whether an interaction occurs in a certain time step is decided by a randomly selected individual (the dominant individual) which then opens the interaction with respect to the individual it likes most. The third partner is left empty-handed in this round. Kirk and Coleman describe several variants of this model: first according to the probability that a certain individual becomes dominant in a certain round: this probability may

• be constant and equal for all three individuals, or

- depend on the frequency of interactions a certain individual has taken part in up to this round – either: the more frequent the participation up to now, the higher the probability to be dominant in the next round; or: the less frequent the participation up to now, the higher the probability to be dominant in the next round;

secondly according to the amount by which the inclination to interact with a certain partner is increased by an interaction:

- If the interaction occured with the desired partner, the inclination to re-interact with him or her will be increased by more (or with higher probability, RA > RB in the MIMOSE program below) than in the case that the partner was not the desired partner.
- The increase in inclination to re-interact is not dependent on whether the interaction occured with the desired or with the third partner (RA = RB in the MIMOSE program below).
- The increase in inclination to re-interact with the same partner is higher (RA < RB in the MIMOSE program below) in the case when the interaction occured with the third partner than in the case of the desired interaction.

The possible outcomes of this model are the following:

- In the long run, all individuals participate equally in the interactions.
- In the long run, one individual is left out of the interactions while the two others form a stable pair.
- In the long run, one individual participates in nearly every interaction while the two others nearly never interact with each other.

Thus, in Kirk's and Coleman's [1967] model, too, we have the case of emergent structure in an initially homogeneous collective, and hence the case of self-organization in the sense of Haken [1988, p. 11] since the triad "acquires a ... functional structure without specific interference from the outside." The only inference from outside is the random effect which transforms a person's inclination to interact into a real interaction.

Returning to Hayek [1944, p. 288] we find that "regularities are observed which are not the result of anybody's design" since the consciousness of our model individual does not even refer to the concept of a "stable pair" or a "stable leader".

The simulation gives quite clear results as to which of the three possible outcomes may be expected in the nine (three times three) variants of the model (see figure 5). The variation of the inclination increase alternatives plays only a minor role, whereas the dominance regulations are clearly correlated with the outcomes: Constant and equal dominance probability leads to an equal participation, increased dominance as a consequence of high participation leads to a stable pair, and increased dominance as a consequence of low participation leads to a stable leadership.

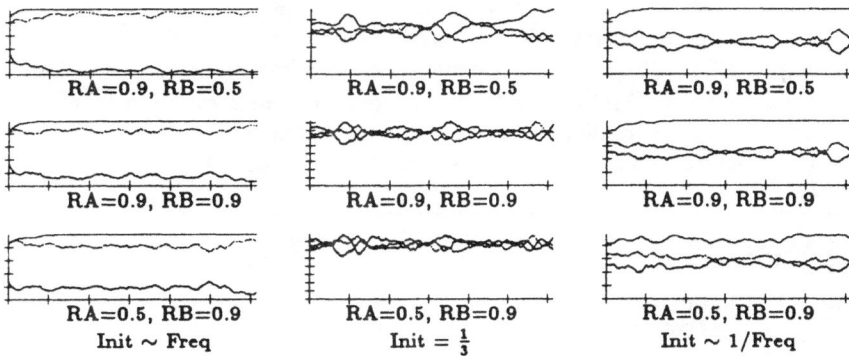

RA=0.9, RB=0.5	RA=0.9, RB=0.5	RA=0.9, RB=0.5
RA=0.9, RB=0.9	RA=0.9, RB=0.9	RA=0.9, RB=0.9
RA=0.5, RB=0.9	RA=0.5, RB=0.9	RA=0.5, RB=0.9
Init ~ Freq	Init = $\frac{1}{3}$	Init ~ 1/Freq

Figure 5: Simulation results for various versions of the model

We now come to the presentation of a MIMOSE program of Kirk's and Coleman's [1967] model (the variants are hidden in the functions dominance and incsymp which are of no interest for the definition of a potential model of the theory).

```
system :=
{ dominant : int := dominance(tripel.symp);
  tripel : list of person;
}
person :=
{ number : int;
  sys     : list of system;
  partners : list of person;
  symp    : list of real
          := updsymp(symp_1, corr, decrement, realpartner);
  desiredpartner : int
                 := dp(symp_1, uniform(1, 0.0, pluslist(symp_1)), 1);
  realpartner : int
              := sys.dominant_1
                 if partners.desiredpartner_1[sys.dominant_1] =
                 number
               else desiredpartner_1
                 if sys.dominant_1 = number
               else 0;
```

```
corr : real
    := 1.0 if ((realpartner > 0) &&
                (((desiredpartner_1 = realpartner) &&
                 (uniform(2, 0.0, 1.0) < RA)
                ) ||
                 ((desiredpartner_1 != realpartner) &&
                 (uniform(2, 0.0, 1.0) < RB)
                )))
         else 0.0;
}
```

From this simulation program a structuralist reconstruction of Kirk's and Coleman's [1967] theory should be possible. To achieve this goal we set up table 2 analogous to table 1. We remember that a variable MIMOSE attribute x of an object type y corresponds to a function from $Y \times T$ to X while a constant MIMOSE attribute z corresponds to a function from Y to Z. Furthermore we remember that we do not yet need the MIMOSE state change functions (not even their signatures, because a potential model

MIMOSE	KC
the set of simulation steps which is always implicit in MIMOSE	T
the set of seeds of MIMOSE's random number generators, each random attribute of each instantiation of each type having its own seed	Ω
the type list of real of the attribute symp	$A = \mathbb{R}^{|\mathcal{P}|}$
object type system	S
attribute dominant	$d: S \times T \rightarrow \{1,...,|\mathcal{P}|\}$
attribute tripel	—
object type person	$\mathcal{P}$
attribute number	$n: \mathcal{P} \rightarrow \{1,...,|\mathcal{P}|\}$
attribute sys	—
attribute partners	—
attribute symp	$\sigma: \mathcal{P} \times T \rightarrow A$
attribute desiredpartner	$\delta: \mathcal{P} \times T \rightarrow \{1,...,|\mathcal{P}|\}$
attribute realpartner	$\rho: \mathcal{P} \times T \rightarrow \{0,...,|\mathcal{P}|\}$
attribute corr	$\gamma: \mathcal{P} \times T \rightarrow \{0,1\}$

Table 2: Match between identifiers of the Kirk–Coleman [1967] MIMOSE simulation program and the terms of **KC**

corresponds to a MIMOSE program with everything cancelled between each
" : =" and the next semicolon). We must only observe implicit applications
of random functions like uniform which are allowed in MIMOSE.

Hence, we are ready to write down the definition of a potential model of
KC − here we do not yet need the MIMOSE state change functions (not
even their signatures, because a potential model corresponds to a MIMOSE
program with everything cancelled between each " : =" and the next semicolon).

Def M_p(KC): η is a potential model of **KC**, i.e. $\eta \in M_p$(KC) iff there exist
$\mathcal{P}, \mathcal{S}, A, T, \Omega, d, \sigma, \delta, \rho$ such that

1. $\eta = \langle \mathcal{P}, \mathcal{S}, A, T, \Omega, d, \sigma, \delta, \rho \rangle$.

2. $\mathcal{P}$ is a non-empty, finite set [of persons].

3. $\mathcal{S}$ is a non-empty finite set of subsets of $\mathcal{P}$ [i.e. of triads or, for the
    sake of generality, of m-ads, or systems], $\mathcal{S} \subseteq \text{Po}(\mathcal{P})$.

4. $\langle A, \mathcal{A} \rangle$ is a measurable space with $A = \mathbb{R}^{|\mathcal{P}|}$ of vectors containing as
    elements the amounts of sympathy a person feels towards each
    person.

5. T is a set [of instants].

6. $\langle \Omega, \mathcal{F}, P \rangle$ is a probability space with
    (a) Ω is a sample space,
    (b) $\omega : I \to \Omega$ is bijective,
    (c) $\mathcal{F} \subseteq \Omega$ is a family of events,
    (d) P is a probability measure defined on $\mathcal{F}$.

7. d is a function with $\text{Dom}(d) = \mathcal{S} \times T$ and $\text{Rge}(d) = \{1,...,|\mathcal{P}|\}$, such
    that $(s \in \mathcal{S}$ and $t \in T)$ $d(s,t)$ is the number of the dominant person in
    system s at time t.

8. σ is a function with $\text{Dom}(\sigma) = \mathcal{P} \times T$ and $\text{Rge}(\sigma) = A$, such that
    (with $p \in \mathcal{P}$ and $t \in T$) $\sigma(p,t)$ is the sympathy vector of p at time t.

9. δ is a function with $\text{Dom}(\delta) = \mathcal{P} \times T \times T$ and $\text{Rge}(\delta) = \{1,...,|\mathcal{P}|\}$,
    such that (with $p \in \mathcal{P}$ and $t \in T$) $\delta(p,t)$ is the number of the desired
    partner of person p at time t.

10. ρ is a function with $\text{Dom}(\rho) = \mathcal{P} \times T \times T$ and $\text{Rge}(\rho) = \{0,...,|\mathcal{P}|\}$,
    such that (with $p \in \mathcal{P}$ and $t \in T$) $\rho(p,t) > 0$ is the number of the real
    partner of person p at time t, or $\rho(p,t) = 0$ means that person p has no
    partner at time t.

The definition of a model of **KC** is derived from the complete MIMOSE
program in a straightforward manner, too:

Def M(KC): ζ is a model of **KC**, i.e. $\zeta \in$ **M(KC)**, iff there exist $\mathcal{P}$, S, A, T, Ω, d, σ, δ, ρ such that

1. $\zeta = \langle \mathcal{P}, S, A, T, \Omega, d, \sigma, \delta, \rho \rangle$.
2. $\zeta \in \mathbf{M}_p(\mathbf{KC})$.
3. $d(s,t) = \ldots$
4. $\sigma(p,t) = \ldots$
5. $\delta(p,t) = \ldots$
6. $\rho(p,t) = \ldots$

where the "..." in items 3 to 6 stand for the MIMOSE state change functions of the program above which need not necessarily be written down here since the example serves only to make clear which steps must be taken to transform a MIMOSE simulation program into a structuralist reconstruction of the underlying theory.

Conclusion

We have seen in this paper that the formalization of a theory in a MIMOSE simulation program is completely analogous to a structuralist reconstruction of this theory − which should not come as a surprise since the design of MIMOSE was partly guided by the thoughts of the founders of the structuralist program.

To perform the transformation from a simulation program into a structuralist reconstruction the following steps are necessary:

- Take the object types of the MIMOSE model as base sets of the definition of the potential model.
- Take T as a set of instants and − if any random functions are applied in the MIMOSE program − $\langle \Omega, \mathcal{F}, P \rangle$ as a probability space to be additional base sets of the definition of a potential model.
- Take any constant attribute of any MIMOSE object type as a function from this object type to the attribute type (for the sake of simplicity, we identify the "type" with the "set of its instances").
- Take any variable attribute of any MIMOSE object type as a function from the cross product of object type and the set of instants to the attribute type.
- After these first four steps the definition of the potential model is complete.

- Take any function application (state change function, between ":=" and the next semicolon) as an axiom of the definition of the model. This completes the definition of the model of the theory.

The transformation of a structuralist theory reconstruction into an executable MIMOSE program is also straightforward:

- Take any base set from the definition of the potential model (except a set of instants and a probability space) and transform it into a MIMOSE object type. This is done by inserting

```
<base_set_name> := {
}
```

anywhere into the MIMOSE program (but of course not between braces).
If there is no set of instants the theory will not be about a dynamical process, and a simulation program is of no use. If the set of instants is continuous, a discretization will always be necessary for any kind of (digital) computer simulation. So first a redesign of the theory will be necessary.
- Take any function from the definition of the potential model and transform it into an attribute of the respective object type, considering the domain of the respective function. This is done by inserting

```
<function_name> : <function_range> ;
```

anywhere into the corresponding type definition (but of course only just after a semicolon).
- Take any axiom from the definition of the model of the theory and transform it into a MIMOSE state change function. This is most easily done by inserting

```
:= <axiom_name>(<terms in the right hand side of
the axiom>)
```

between the corresponding <function_range> and the semicolon.
- Use MIMOSE's user interface to initialize all constants and variable attribute which must have taken values at simulation start time, and to fix the simulation parameters (time step size, break and stop condition), and run the program.

Institute of Social Science Informatics
Koblenz–Landau University
Rheinau 1
D-56075 Koblenz, Germany

176

ACKNOWLEDGEMENTS

I am indebted to my colleagues and students at the Institute of Social Science Informatics at Koblenz—Landau University for helpful criticism, and to Tom King for giving my English a final polish. This paper originates in the research project "Micro and multilevel modelling and simulation software development (MIMOSE)" directed by Michael Möhring and the author and funded by the Deutsche Forschungsgemeinschaft under grant no. Tr 225/3—1 and —2.

REFERENCES

Drogoul, A. and Ferber, J. [1994]. Multi-agent Simulation as a Tool for Studying Emergent Processes in Societies. In: J. Doran and N. Gilbert (Eds.). *Simulating Societies: the Computer Simulation of Social Phenomena.* London: University of London College Press, pp. 127—142.

Haken, H. [1988]. *Information and Self-Organization. A Macroscopic Approach to Complex Systems.* Springer, Berlin, Heidelberg, New York: Springer Series in Synergetics. Vol. 40.

Hayek, F.A. [1942—1944]. Scientism and the Study of Society. *Economica* 9, 10, 11: 267—291, 34—63, 27—39.

Hegselmann, R. [1993]. Experimentelle Moralphilosophie. Computersimulationen zu Klassen, Cliquen und Solidarität. In: R. Hegselmann and H.-O. Peitgen (Eds.). *Ordnung und Chaos in Natur und Gesellschaft.* Wien: Hölder—Pichler—Tempski. In prep.

Helbing, D. [1992a]. A Mathematical Model for Attitude Formation by Pair Interactions. *Behavioral Science* 37, 190—214.

Helbing, D. [1991/92b]. A Mathematical Model for Behavioral Changes by Pair Interactions and its Relation to Game Theory. *Angewandte Sozialforschung* 17, 179—194.

Herlitzius, L. [1990]. Schätzung nicht-normaler Wahrscheinlichkeitsdichtefunktionen. In: J. Gladitz and K.G. Troitzsch (Eds.). *Computer Aided Sociological Research. Proceedings of the Workshop "Computer Aided Sociological Research" (CASOR '89), Holzhau/DDR, October 2nd—6th, 1989,* Berlin: Akademie-Verlag, pp. 379—396.

Homans, G.C. [1950]. *The Human Group.* New York: Harpers.

Kirk, J. and Coleman, J.S [1967]. Formalisierung und Simulation von Interaktionen in einer Drei-Personen-Gruppe. In: R. Mayntz (Ed.). *Formalisierte Modelle in der Soziologie.* Volume 39 of *Soziologische Texte.* Neuwied: Luchterhand.

Klee, A. and Troitzsch, K.G. [1993]. Chaotic Behaviour in Social Systems: Modeling with GEMM. In: K.G. Troitzsch (Ed.). *Catastrophe, Chaos, and Self-Organization Invited Papers of a Seminar Series on Catastrophic Phenomena in Soviet Society and Self-Organized Behaviour of Social Systems. Held at the Institute of Sociology of the Academy of Sciences of the Ukrainian Republic, Kiev, September 4 to 11, 1992.* Koblenz 1993.

Simon, H.A. [1957]. *Models of Man, Social and Rational. Mathematical Essays on Rational Human Behavior in a Social Setting.* New York: Wiley.

Troitzsch, K.G. [1987]. *Bürgerperzeptionen und Legitimierung. Anwendung eines formalen Modells des Legitimations-/Legitimierungsprozesses auf Wählereinstellungen und Wählerverhalten im Kontext der Bundestagswahl 1980.* Frankfurt, Bern, New York: Lang.

Troitzsch, K.G. [1990a]. *Modellbildung und Simulation in den Sozialwissenschaften.* Opladen: Westdeutscher Verlag.

Troitzsch, K.G. [1990b]. Self-organisation in Social Systems. In: J. Gladitz and K.G. Troitzsch (Eds.). *Computer Aided Sociological Research. Proceedings of the Workshop "Computer Aided Sociological Research" (CASOR'89), Holzhau/DDR, October 2nd—6th, 1989,* Berlin: Akademie-Verlag, pp. 353—377.

Troitzsch, K.G. [1991]. A Comparison of Some Models of Processes of Self-Organization. In: W. Ebeling, M. Peschel and W. Weidlich (Eds.). *Models of Self-Organization in Complex Systems.* Berlin: Akademie-Verlag, pp. 106—116.

Troitzsch, K.G. [1992]. Structuralist Theory Reconstruction and Specification of Simulation Models in the Social Sciences. In: H. Westmeyer (Ed.). *The Structuralist Program in Psychology: Foundations and Applications.* Seattle, Toronto, Bern, Göttingen: Hogrefe & Huber.

Wurster, B. [1988]. Periodic Cell Communication. In: M. Markus, S.C. Müller and G. Nicolis (Eds.). *From Chemical to Biological Organization.* Berlin, Heidelberg, New York, Paris: Springer, pp. 255—260.

*Poznań Studies in the Philosophy
of the Sciences and the Humanities*
Vol. 42, pp. 179–200

Veikko Rantala and Tere Vadén

IDEALIZATION IN COGNITIVE SCIENCE
A Study in Counterfactual Correspondence

Introduction

It is suggested by Smolensky [1988] that in cognitive science the symbolic is an idealization of the subsymbolic. Both approaches are important, however, in that they are complementary ways to understand and explain cognition, and therefore we should not try to eliminate one of them in favour of the other but rather think of their relationship as providing a cognitive correspondence principle, a principle analogous to the correspondence principle much discussed in the philosophy of physics. But Smolensky's proposals are vague, and there seem to be no attempts in the literature to give them a more definite form, wherefore it is somewhat difficult to see their real significance. In this paper, we shall evaluate the proposals by relating them to the work recently done in the philosophy of science and offer a case study of the relationship between symbolic and subsymbolic representations. This relationship will be analyzed by using a method developed by Pearce and Rantala [1984a], Pearce [1987], and Rantala [1989]. It turns out that a good sense can be given to the suggestion that a particular kind of symbolic representation (or a theory of symbolic representation) is a limiting case of a particular kind of subsymbolic representation (or a theory). Due to the method used, the case study also sheds some light on the notion of idealization in (this part of) cognitive science.

I
The Symbolic Idealizing the Subsymbolic

It is often said that the behaviour of neural networks is rule-described but not rule-governed. The rules used to describe the so-called upper level behaviour of a network do not really operate at the processing level of the network but are approximations and idealizations of lower level laws leaving certain aspects unnoticed. This and similar ideas are essentially the basis of the subsymbolic approach [see Smolensky, 1988] and they suggest that the symbolic features of cognition, as defined, for example, by Fodor and Pylyshyn [1988], could be explained as competence idealizations and approximations of the performance description given by the subsymbolic account. The key idea is that the computational, classical notion of the mind is only an idealization, or a limiting case, of a more realistic, connectionist characterization of the mind.

The Subsymbolic Paradigm

The terms 'subsymbol' and 'subsymbolic' will be used in the present article in Smolensky's [1988, p. 3] sense. While symbols are something that are manipulated by syntactic operations, subsymbols do not participate in symbol manipulations but are operated upon by numerical computations, and they are, so to speak, context sensitive parts or microfeatures of symbols rather than context free symbols. Subsymbolic cognitive explanations, for instance, are explanations which use subsymbols as their basic building blocks.

It is suggested by Smolensky that the purpose of the subsymbolic paradigm is to offer a theory of the microlevel of cognition, where the microlevel is a semantic level which lies, so to speak, below the conceptual macrolevel and is called the subconceptual level. According to Smolensky, the relation of the symbolic and subsymbolic theories is much like that of classical mechanics and quantum mechanics. The microlevel is not an implementation of the macrolevel, nor does the microtheory eliminate the macrothery, but the microlevel account provides a refinement of the macrolevel account. Part of Smolensky's view of the relation of the microtheory and macrotheory is expressed in the Principle of Approximate Explanation:

(AE) Suppose it is a logical consequence of a microtheory that, within a certain range of circumstances, C, laws of a macrotheory are valid *to a certain degree of approximation*. Then the microtheory licences *approximate* explanations via the macrotheory for phenomena occurring in C. In very special cases, these phenomena may admit more exact explanations that rest directly on the laws of the microtheory (without invoking the macrotheory), but this is not to be expected generically: For most phenomena in C, *the only available explanation will be the (approximate) one provided by the macrotheory*. [Smolensky, 1988, p. 60.]

Smolensky continues to point out that the fact that the macroaccount approximately describes and explains the higher level phenomena is not an accident but a result of a systematic and explanatorically relevant relationship between the formalisms of the microtheory and macrotheory. That it is systematic manifests itself in the Principle of Approximate Explanation through a logical argument showing that if the laws of the microtheory hold at the microlevel, then the laws of the macrotheory hold at the macrolevel [Smolensky, 1988, p. 61]. This seems to be close to the idea that there is a kind of approximate reduction relation of the former theory to the latter in the standard Nagelian sense [Nagel, 1962], though Smolensky calls it 'refinement': the microtheory refines the macrotheory. That reductions provide refinements is an old idea [see, e.g., Adams, 1955] and it can again be seen in the definition of the correspondence relation in the next section.

Smolensky calls his position concerning the relationship between the microtheory and macrotheory limitivism, which means the assumption that the idealizations involved in the macrotheory become transparent when certain limits are taken. Such a limit relation obtains, for instance, between quantum mechanics and Newtonian mechanics, since the latter is a limiting case of the former when certain conditions hold. This is, as Smolensky admits, a loose way of speaking, and a better way could be to say counterfactually that Newtonian mechanics would be a limit of quantum mechanics if the conditions held. A similar relationship obtains between subsymbolic and symbolic models, and a task of the former is to delineate the situations in which symbolic approximations are valid and to explain why [Smolensky, 1988, p. 12].

There exist similar views in the philosophical literature concerning relations of physical theories, but Smolensky does not seem to be aware of them. Physicists and philosophers of science have argued that in radical scientific changes (mainly in physics) there often obtains a limiting case correspondence between two theories, or that their relation is counterfactual or exemplifies idealization. For instance, classical particle mechanics is claimed to be a limiting case of both relativistic particle mechanics and (non-relativistic) quantum mechanics. These and similar claims have appeared somewhat controversial, but there also exist quite a number of more or less exact definitions and case studies [e.g., Glymour, 1970; Lahti, 1984; Pearce and Rantala, 1984ab; Pearce, 1987; Rantala, 1989]. Now, if Smolensky is correct, there obtains a similar limiting case correspondence in cognitive science, that is, the relation of the subsymbolic and symbolic theories. The limit in question means a shift from the microlevel performance description to the macrolevel competence idealization. The microlevel formalism describes the performance of a neural network, and if certain idealizing conditions hold (e.g., that the number of units or strength of connection in a network approaches infinity), the competence idealization becomes valid (e.g., the limitations of memory capacities are idealized away in the competence theory). The microtheory describes the performance of the network as a system that parallelly and simultaneously satisfies a multitude of soft constraints, and as soon as appropriate idealizing assumptions are made, it corresponds to a macrotheoretical competence description of a system that is hard and rule-governed.

Consequently, the subsymbolic paradigm brings about a reconsideration of certain notions used in the symbolic paradigm. Such notions as rationality, logic, and rule must be given a new (their "proper") role in the Connectionist framework [see Smolensky 1988, p. 2]. The subsymbolic paradigm treats symbolic rules, for instance, as higher level idealizations and approximations, which are useful but not the whole picture. Since Kuhn [1962], this kind of meaning change between two paradigms has often been associated with the notion of incommensurability (see also Part II, below); and there are other senses, as, for instance, ontological, in which such paradigms can be considered incompatible.

The Symbolic Conditions

Another way of seeing the idealizing nature of the symbolic is to look at the very definitions and characterizations of the symbolic model, among which the concise definition by Fodor & Pylyshyn [1988] is evidently the best known. They maintain that the classical, symbolic models of cognitive architecture have the following characteristics and argue that the characteristics make a crucial difference between the classical and connectionist accounts.

Condition 1. Mental representations have combinatorial syntax and semantics.

More precisely, (a) representations are structurally atomic or molecular, (b) molecular representations have constituents which are atomic or molecular, and (c) the semantic content of a representation is a function of the semantic contents of its syntactic parts, together with its constituent structure. Condition 1 means that mental representations are symbol structures (or symbol systems). The other condition, which is dependent on the assumption that representations are structured, is the following. It is the cornerstone of symbol manipulation:

Condition 2. Mental processes are structure sensitive. Mental operations, which transform representations into other representations, apply to representations by reference to their form.

Fodor and Pylyshyn's criticism of connectionism is essentially based on the assumption that there is no distinction in connectionist networks between atomic and molecular representations; according to them all connectionist representations are atomic and hence the nets fail to exhibit compositionality. As several commentators have pointed out, Fodor and Pylyshyn have a blind spot here, since they think that the representational vehicles in connectionist nets are activities of single units and thus fail to appreciate the import of distributed representations. Once it is realized that activity patterns formed by sets of units, rather than individual units, are representational, the molecularity is restored in a natural way. It is hard to understand how this elementary fact has escaped Fodor and Pylyshyn's notice. Connectionists' answer is, accordingly, that in an approximate and contextual sense connectionist nets can have molecular representations and satisfy Conditions 1−2 displaying computational competence, as demanded by Fodor and Pylyshyn.[1]

Smolensky [1987] argues here that connectionism can approximately and contextually accept the conditions without just being an implementation of the classical view. Distributed complex representations may involve representations of constituents, but this constituency relation is not part of the causal mechanism of the network's performance and, therefore, it is not precise and uniquely defined. The notion of constituency includes some approximation, no matter what. It follows that even though connectionist nets describe states that are compositional, they do not implement a syntactic language of thought, and, moreover, the parts left out in the approximative process of the construction of the description of the constituency relation are cognitively relevant. The subsymbolic view admits Conditions 1−2 but only approximately, and, according to Smolensky [1987], by means of distributed connectionist models one gets new instantations of compositionality principles.

In summary, the classical and connectionist models do not, in Smolensky's view, differ in whether they accept the compositionality principles but in how they implement them. In the connectionist models they

---

1 There is reason to believe that Conditions 1−2 in the exact and noncontextual sense are too strict (when human cognition is concerned) and therefore the demand that a model must implement the conditions in order to be a cognitive model can be discredited [see Rantala & Vadén 1995].

are implemented by using distributed vector-representations of cognitive states and associative structure sensitive mental processes, instead of the syntactical structures and symbol manipulation, as in the symbolic models. Thus Smolensky sees himself as pursuing a search for a middle ground between implementing symbolic computation and ignoring structure altogether. This is in line with the overall program of the subsymbolic paradigm where what he calls the *cognitive correspondence principle* is held crucial:

> When connectionist computational systems are analysed at higher levels, elements of symbolic computation appear as emergent properties [Smolensky, 1987; p. 152].

This is called the correspondence principle by Smolensky because he believes that its role in developing a microtheory of cognition is analogous to the role which the quantum correspondence principle played in the development of the microtheory in physics.[2] The idea in cognitive science is that there are certain cognitive principles that arch over the symbolic and subsymbolic formalisms. The principles can be instantiated in both formalisms, but the symbolic one can be seen as yielding higher level descriptions of the principles operative at the microlevel and the classical instantations as approximations of the lower level subsymbolic formalisms. Smolensky argues that neither quantum mechanics nor the subsymbolic paradigm in cognitive science implement the respective classical accounts, and thus he does not deny the worth of the classical accounts because they, in the spirit of (AE), provide explanations of an enormous range of classical phenomena for which lower level accounts are infeasible and because they have historically guided the discovery of the lower-level principles.

II
Counterfactual Correspondence and Idealization

We have at some length studied Smolensky's views concerning the relation of the symbolic and subsymbolic accounts of cognition since he applies in an interesting way to cognitive science the notion of correspondence principle which has been so far discussed mainly in physics and its philosophy. However, the notion of correspondence, whether employed in physics or cognitive science, is hopelessly hazy, and therefore we in this section try to

---

2 Needless to say, however, that the correspondence principle and its logical relations are under some dispute in quantum mechanics and in the philosophy of science.

make it a little more exact by considering it in a logical framework developed recently in the philosophical literature.

Niels Bohr is to be considered as a main advocate of the correspondence principle [*Korrespondenzprinzip;* Bohr, 1920; see also Lahti, 1984, for correspondence and idealization in physics], which he thought of as being a useful heuristic principle in the development of quantum mechanics. It says that a future theory is to be a generalization of a classical theory in the sense that the former yields the latter as a special or limiting case. It is well known that the principle was extensively used in the early mathematical developments of quantum mechanics in the period from 1918 to 1925 [cf. van der Waerden, 1967]. Somewhat later, the principle was generalized by Heisenberg [1958] for the whole of physics; according to him the principle plays an important normative and methodological role in physics. The principle is further generalized by philosophers of science to concern all 'mature' sciences,[3] but, on the other hand, it is often considered as a special case of a still more general principle concerning scientific growth, the one called by Suppe [1977] the theory of development by reduction. The principle is often associated with the so-called Received View of science which was, in some form or other, sustained by philosophers of science up to the 70's, but which is now considered by many as having positivist undertones.

Now the import of the correspondence principle and its generalization was already challenged in the early 60's by Kuhn [1962] and other critics of the Received View, and ever since the nature of scientific change has been one of the most controversial issues in analytic theory of knowledge and philosophy of science. The gist of his objection was that even though an equation which syntactically looks like a law of the classical theory is mathematically obtainable from the laws of the new theory, for instance, by means of a limit procedure, as in the case of classical particle mechanics and relativistic particle mechanics, the equation obtained is not really a law of the classical theory since the terms occurring in it have the same meanings as they have in the new theory and those meanings are different from their meanings in the classical theory. The classical law is not obtained before the terms are reinterpreted − and that can only be done with "the explicit guidance of the more recent theory" [Kuhn, 1962, pp. 102−103]. On the other hand, since for Kuhn scientific change is not just a matter of theory change but a far more sweeping process in which one paradigm is replaced by another, the reinterpretation of scientific terms, to the extent to

---

3 See Krajewski [1977] for an overview and a general explication of the principle.

which it can be done, is a hermeneutic enterprise, something similar to what is involved in attempts to understand other cultures [see, e.g., Kuhn, 1983].

If Kuhn is right here, it clearly implies that there is no way to explain the nature of intertheory relations, the correspondence principle included, by means of purely extensional means; such means must be amended by embedding them, so to say, in appropriate hermeneutic and pragmatic environments. It is equally obvious, on the other hand, that purely hermeneutic or similar means are not sufficient either, at least in so far as exact sciences are concerned, and such means must be amended by considering the logic of intertheory relations. As pointed out in Rantala [1989, 1995], only in an amended sense can the treatment of scientific change help one understand why there obtains so little agreement about the matter.

In what follows, a quick overview of an amended method, gradually developed in Rantala [1978, 1979, 1989, 1995], Pearce & Rantala [1983, 1984ab], and Pearce [1987] will be given, and in the next section applied to symbolic and subsymbolic representations. We shall first outline the notion of correspondence relation and its special cases, counterfactual and limiting case correspondence; and as we hope, the latter notions and their applications help us see what is involved in idealization.

Let θ and θ' be (sentences which formalize) axioms (laws) of two theories T and T', respectively, and let L and L' be logics in which θ and θ' are formulated. We assume, for simplicity, that θ and θ' are the only axioms of T and T'. Then a *correspondence* of T to T' is a pair of mappings $<F,I>$ such that F assigns models of θ to specified models of θ' and I is a translation of the language of L into the language of L'. It is clear that the mappings have to satisfy appropriate logical and pragmatic conditions in order to establish sufficiently informative conceptual relationship between the two theories. In the first place, they must satisfy the following condition for all sentences φ of L and for all models $\mathbf{M}$ of θ' in the domain of F:

$$(2.1) \qquad F(\mathbf{M}) \models_L \varphi \Leftrightarrow \mathbf{M} \models_{L'} I(\varphi),$$

where $I(\varphi)$ is the translation of φ under I, thus a sentence of L', and $\models_L$ and $\models_{L'}$ are the truth (satisfaction) relations of L and L', respectively. Then, on some general assumptions concerning F, L and L', an argument of the following form can be inferred:

$$(2.2) \qquad \theta',\psi_0,\ldots,\psi_n \models_{L'} I(\theta).$$

Here $\models_{L'}$ indicates the relation of logical (semantic) consequence in L', and $\psi_0,\ldots,\psi_n$ are sentences of L' formalizing some special assumptions about T'.

If L' is a complete and sound logic, the relation is equivalent to the relation of (syntactic) deduction in L'.

If some of the special assumptions (formalized by the sentences) $\psi_0,\ldots,$ ψ_n are counterfactual (contrary-to-fact) in relation to appropriate (i.e., actual or standard, in some sense) conditions or situations (formally, if some of the sentences $\psi_0,\ldots,\psi_n$ are false in appropriate models which are thought to correspond to such situations), the correspondence $<F,I>$ is called a *counterfactual correspondence*. Let us assume, for simplicity, that there is just one counterfactual assumption among the special assumptions. Let it be ψ_0, which will be written briefly as ψ, and let us write (2.2) into a more suggestive form:

(2.3) $\theta',\psi_1,\ldots,\psi_n,\psi \models_{L'} I(\theta)$.

If, in particular, the mappings F and I involve, in a specified way explained in Pearce & Rantala [1984a], concepts of nonstandard analysis, and ψ is an appropriate limit condition, such as 'the velocity of light is infinite', $<F,I>$ is called a *limiting case correspondence*. Then I is usually of the kind that $I(\theta)$ can be (informally) read as follows:

(2.4) θ almost holds,

where *almost* means infinitesimal accuracy, whence θ holds for all practical purposes. One may still consider the special assumption ψ possible in a relevant sense even though it is not realizable in any actual situations. It follows from (2.3) that θ almost holds in the situations where θ' plus the special conditions $\psi_1,\ldots,\psi_n,\psi$ hold.

Now it is not very clear whether the relation of counterfactual correspondence is of any explanatory importance. Since (2.3) looks like the formal scheme of a deductive-nomological explanation of $I(\theta)$, we might perhaps say that it provides such an explanation under appropriate formal and pragmatic conditions of adequacy. As an explanation it should provide an answer to a relevant question. However, one cannot ask, for instance: '*Why* does θ almost hold?' since $I(\theta)$ is contrary-to-fact (in the above sense) almost as much as θ; and even if one could, (2.3) would not provide any adequate answer because one of the special assumptions is also contrary-to-fact; and similarly for other similar why-questions. Therefore, the only kind of explanatory import argument (2.3) may have is the following: by studying the special conditions and the translation $I(\theta)$, one learns something about the conceptual relationship of the two laws. As argued in Rantala [1989, 1995], however, case studies of actual scientific theories clearly indicate that if one studies a correspondence relation $<F,I>$ itself, instead of just studying its consequence (2.3), one in fact learns more about the conceptual relationship.

It is suggested by Glymour [1970] that we have to ask different questions in cases where counterfactual special assumptions are needed in the transition from one law, or theory, to another. According to him, one can ask, for instance: 'Under what conditions the supplanted law (i.e., θ) would hold?'. The law is then explained by:

(2.5)   (a)   showing under what conditions it would hold;
        (b)   contrasting those conditions with the conditions which actually obtain.

Here (2.5b) yields an answer to the question 'Why does the law not hold?', and it may increase our understanding of the law, but what its explanatory import is is not very clear. On the other hand, if ψ is the only counterfactual special assumption, as we have assumed, (2.5a) evidently amounts to showing that a conditional of the following form is true:

(2.6)   If it were the case that ψ, then it would be the case that θ.

Glymour [1970] seems to suggest, then, that by means of a relevant deduction, of the form (2.3), that is, which is formally in conformity with the model of deductive-nomological explanation, one can establish (2.6).

Glymour's idea is closely related to an aspect of counterfactual correspondence, but it needs at least the following, essential corrections. Instead of including possible translations in the special assumptions $\psi_1,\ldots,\psi_n,\psi$, the conclusion θ is to be replaced by its translation $I(\theta)$, and, therefore, we must consider the following conditional instead of (2.6):

(2.7)   If it were the case that ψ, then it would be the case that $I(\theta)$;

more formally,

(2.8) $\psi \ \Box\!\rightarrow I(\theta)$.

Thus Glymour is close to maintaining that the argument of the form (2.3) makes the conditional of the form (2.8) true. But contrary to what Glymour seems to say, if (2.3) holds, it does not necessarily follow that (2.8) is true, as we know from theories of counterfactual conditionals [see, e.g., Lewis, 1973]. In order to find out whether this is the case, the relation of (2.3) and (2.8) must be studied more closely, and then it appears that the conditions under which (2.3) makes (2.8) true provide pragmatic constraints in the sense discussed earlier, and as noted above, the presence of such constraints is necessary for the adequacy of explanation.

It is pointed out in Rantala [1989, 1995], however, that Glymour's model and the deductive-nomological model, in particular, are too simple to describe adequately intertheory relations where counterfactual assumptions are needed; and that a more realistic model emerges if Glymour's model, the notion of counterfactual correspondence and Lewis' semantics of counterfactual conditionals are amalgamated. A *counterfactual explanation* of θ from θ' (relative to a possible world w, say) consists of the following steps:

(i)   Constructing a counterfactual correspondence $<F,I>$ of θ to θ' (with respect to a special assumption ψ);
(ii) Deriving from $<F,I>$ an argument (2.3);
(iii) Showing that (2.3) makes (2.8) true (at w).

A number of pragmatic constraints are embedded in this model, either explicitly or implicitly, for instance, to the effect that the explanation is only good relative to a properly chosen context (w), to the individual or scientific community who or which is studying the intertheory relation in question, and to adopted scientific principles. In particular, if the explainer holds the explaining law θ' rather than θ, then what scientific principles are adopted is determined by θ' and by the respective paradigmatic assumptions, rather than θ and its paradigm.

This model, and its applications in the first place, seem to support what Kuhn [1962] argues. An explanation of the above kind of a superseded law cannot be context-independent. The pragmatic constraints involved show that one has to look at the matter from the point of view of the new paradigm, and this is how we saw Kuhn arguing. But there may even be individual constraints and contexts which are effective. For instance, when the limiting case correspondence between classical particle mechanics and relativistic particle mechanics is studied, one has to consider the question whether it is in some sense *possible* that the velocity of light is infinite; and similarly one has to consider ψ in the general case. If ψ is not considered possible, the counterfactual explanation is trivialized since on this assumption (2.8) is trivially true. Some philosophers, or at least physicists, hold that this assumption is so much contrary-to-fact that it is not even possible, whereas others consider it possible. There are also other contextual conditions involved.

In short, the model is not extensional and its applications involve interpretive elements. When applying it to the relation of symbolic and subsymbolic representations one has to ask such questions as whether it in some sense is considered possible that the number of units of a neural network is infinite (see the next section). As we noticed above, Smolensky, for instance, accepts this possibility − it is even crucial for his theory − but, on the other hand, in actual research on neural networks researchers are often aiming at smaller nets, not bigger.[4] But this practical work has nothing to do with our problem concerning the conceptual relationship between the symbolic and subsymbolic.

How is the above counterfactual model related to idealization? It can be readily seen that the conditional (2.7) (i.e., 2.8) is crucial here; it says that if a certain counterfactual condition were true, then the more idealized law would (almost) be true as well. This is another way to express the fact that the law is idealized in the sense that its truth presupposes a contrary-to-fact condition [see, e.g., Niiniluoto, 1990].

III
Symbolic vs. Subsymbolic Representation

In this section, we try to point out tentatively that there is a type of symbolic representation which can be construed as being a limiting case of a type of subsymbolic representation, but, unlike Smolensky, we do not claim that the symbolic in general is a limiting case of the subsymbolic. Smolensky's conjecture is so general that it cannot have a very exact meaning and it cannot be verified in such a general form. On the other hand, we are aware of the heavy abstractions and special designs used to obtain the result, but there are two reasons why we in the face of the artificiality of our definitions still want to present the result. First, we hope that exact results even on a very small scale could help one understand and make sense of the more sweeping and less precise claim that the symbolic as a whole is a limit of some of the subsymbolic. Second, one has to start somewhere and we feel that even if the result is limited it is at least well grounded. We shall investigate a case where the number of units of a connectionist network is infinite; the case that weights are infinite can be studied in an analogous way. The relation of symbolic and subsymbolic processing will not be investigated here.

---

4 The latter fact was brought to our attention by John Haugeland.

Turing Representation

Consider a 1-way infinite Turing machine [see, e.g., Kleene, 1964] such that each square of its tape is capable of having printed upon it any one of the primitive symbols of a combinatorial language $\mathscr{L}$ and symbols 0 and 1. Let N be the set of natural numbers and let $n \in N$ be given. We let natural numbers $1,\ldots,n$ label the n first squares of the tape and, for a technical reason, we let 0 be the label of the $(n+1)$st square. (Other labels could of course be used.) By 'p' we denote a function which determines how the $n+1$ first squares are printed; hence, p is a mapping from $\{0,1,\ldots,n\}$ to $\mathrm{Prim}\mathscr{L} \cup \{0,1\}$, where $\mathrm{Prim}\mathscr{L}$ is the set of primitive symbols of $\mathscr{L}$. Thus p determines a situation (i.e., condition) of the tape up to square 0.

Since $\mathscr{L}$ is combinatorial, its expressions, considered as sequences of primitive symbols of $\mathscr{L}$, can be efficiently represented by a Turing machine in the following sense. Let $\mathrm{Exp}\mathscr{L}$ be the set of all expressions of $\mathscr{L}$ of length n belonging to some recursively defined category. For instance, if $\mathscr{L}$ is a formal language, $\mathrm{Exp}\mathscr{L}$ might be the set of its terms of length n, the set of its well-formed formulas of length n, etc. If $\mathscr{L}$ is an appropriate (recursive) fragment of natural laguage, $\mathrm{Exp}\mathscr{L}$ could be the set of all words of length n belonging to the fragment, etc. Given such an $\mathrm{Exp}\mathscr{L}$, a machine can be defined which is able to check of any printing on squares $1,\ldots,n$, that is, of any sequence of the form $s_1,\ldots,s_n$ (more precisely, $\{<1,s_1>,\ldots, <n,s_n>\}$) – where $s_i = p(i) \in \mathrm{Prim}\mathscr{L} \cup \{0,1\}$ – considered as an *input*, whether or not the sequence belongs to $\mathrm{Exp}\mathscr{L}$. In the positive case, let the machine print '1' on square 0 and then stop, that is, let the *output* be $p(0)=1$; in the negative case, let the output be $p(0)=0$. (It is assumed here for simplicity that square 0 is originally blank.)

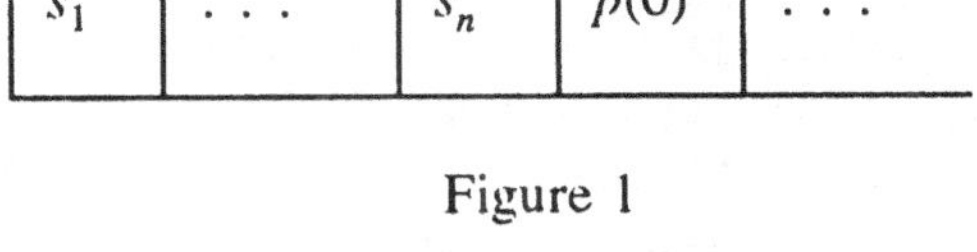

1	...	n	0	
s_1	...	s_n	$p(0)$	...

Figure 1

Thus we have for all mappings p as above:

(3.1)   if $p \upharpoonright \{1,\ldots,n\} \in \mathrm{Exp}\mathscr{L}$, then $p(0) = 1$;
   if $p \upharpoonright \{1,\ldots,n\} \notin \mathrm{Exp}\mathscr{L}$, then $p(0) = 0$,

where $p \upharpoonright \{1,\ldots,n\}$ is the restriction of p to $\{1,\ldots,n\}$.

Since the machine is able to decide of any printing on the tape whether or not it is an expression in Exp$\mathscr{L}$, we say that Exp$\mathscr{L}$ is *represented* in the machine. We may also say, if we are willing to ignore the philosophical difficulties involved, that the machine is able to represent the concepts to which the expressions in Exp$\mathscr{L}$ refer by displaying the constituent features of the concept. We call the above description *the theory of Turing representation* (of Exp$\mathscr{L}$); hereafter, TR. It only is a theory which describes the surface structure of representation, for many important and deepest details concerning general principles of Turing machines are here dispensed with. It seems, however, that this is the level at which representation in Turing machines (symbolic reresentation) and representation in connectionist networks (subsymbolic representation) approximate each other.

Network Representation

Next we consider a multi-layer connectionist network which is trained to recognize the expressions belonging to Exp$\mathscr{L}$ and to recognize whenever a sequence (of length n) of symbols from Prim$\mathscr{L}$ is not in Exp$\mathscr{L}$. We are not here concerned with questions of learning, but we may simply assume that theoretically such a training can be done. Since we shall only consider the last two layers of the network — the output layer and the one preceding it — and the other layers, whether input or hidden, may be as many as needed and have whatever connectionist architecture, it is obvious that this assumption is even realizable as soon as Exp$\mathscr{L}$ and the other relevant sets are finite and small. But if the sets are recursively infinite, the system is purely theoretical in the same way as Turing machines are purely theoretical. In this case the recognition assumption concerning the network is, of course, much more hypothetical than the same assumption (in the previous section) concerning the Turing machine, since, presumably, it is not supported by anything that would be analogous to Turing's (or Church's) thesis.

Let the (possibly hidden) layer — hereafter called the λ-layer — that is closest to the output layer be composed of two disjoint sets of units, λ_+ and λ_-, both containing m units. λ_+ and λ_- are, in turn, divided into n ensembles such that each ensemble as a whole represents a symbol of Prim$\mathscr{L}$, that is, the elements of such an ensemble represent microfeatures of the symbol. If the λ-layer is hidden, one may not know the exact structure of the latter division. Assume, for simplicity, that there is only one output unit, call it '0', which is taught to respond with a high activation value whenever λ_+ is presented with a sequence from Exp$\mathscr{L}$ and with a low activation value when λ_- is represented with a sequence of length n of symbols not in Exp$\mathscr{L}$.

Because the λ-layer is divided into λ_+ and λ_-, many problems of correlation are avoided, since we assume, furthermore, that the units in λ_+ are not feeded simultaneously with those in λ_-; that is, whenever the former units are activated, the latter ones are not, and conversely. Let the activation values of the output unit be in the open interval (0,1), whence we may adopt the logistic activation function for it.

We take the logistic activation function in its simplest form

$$a_0 = 1:(1 + e^{-\text{Inp}}),$$

where Inp is the total input feeded by λ_+ or λ_- (as the case may be) into the output unit. In the literature, it is usually taken to be of the form Inp= $\Sigma_i w_i a_i$. In the present case, w_i would be the weight of the connection of unit i to 0 and a_i the activation value (output) of i, assumed to be positive, e.g., belonging to the interval (0,1). Since in the case of λ_+ the system is trained to maximize a_0, we could assume that after the training, $w_i > 0$ for every $i \in \lambda_+$; and in the case of λ_-, to minimize a_0, whence $w_i < 0$ for every $i \in \lambda_-$.

Now, however, if the λ-layer is a hidden layer, one may not exactly know how it feeds the output unit. Therefore, we make a somewhat weaker assumption that in the case of λ_+, Inp=$f_+(m)$, where f_+ is a real-valued, monotonically increasing and positive function on the natural numbers such that $f_+(m) \to \infty$ when $m \to \infty$; and in the case of λ_-, Inp=$f_-(m)$, where f_- is a monotonically decreasing real-valued and negative function on the natural numbers such that $f_-(m) \to -\infty$ when $m \to \infty$. This much can be plausibly assumed even if the structure and behaviour of the λ-layer are not known in greater detail. It follows that when the units in λ_+ are activated after the training mode (but those in λ_- are not), $a_0 = 1:(1 + e^{-f_+(m)})$ and when the units in λ_- are activated (but those in λ_+ are not), $a_0 = 1: (1 + e^{-f_-(m)})$. Furthermore, if we let the number of units in the λ-layer increase, then in the case of λ_+, a_0 increases, and if $m \to \infty$, $a_0 \to 1$; but in the case of λ_-, a_0 decreases and $a_0 \to 0$. (See Fig. 2, where the dots depict units and connections not drawn.)

Now let p be a mapping on $\{0,1,...,n\}$ such that $p(0)=a_0$ and $p(i) \in$ Prim$\mathscr{S}$ for $i=1,...,n$. $p \restriction \{1,...,n\}$ is assumed to indicate with what symbols the system is presented. More precisely, if the sequence $p(1)...p(n)$ (i.e., $\{<1,p(1)>,...,<n,p(n)>\}$) belongs to Exp$\mathscr{S}$, the units in λ_+ are activated in the fashion described above, but if not, those in λ_- are activated. Thus we have for all such mappings p,

(3.2)   if $p \restriction \{1,...,n\} \in$ Exp$\mathscr{S}$, then $p(0)=1:(1+e^{-f_+(m)})$;
         if $p \restriction \{1,...,n\} \notin$ Exp$\mathscr{S}$, then $p(0)=1:(1+e^{-f_-(m)})$.

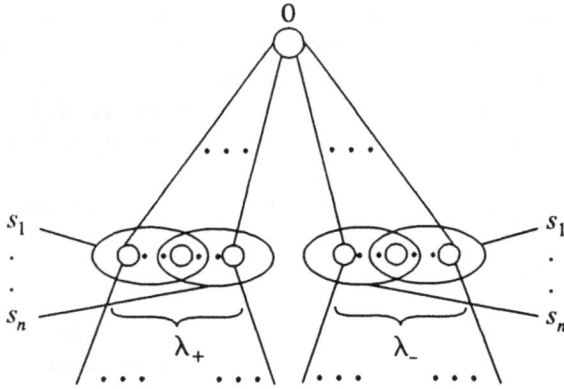

Figure 2

The above description will be called *the theory of network representation* (of Exp$\mathscr{L}$); in short, NR.

Limiting Case Correspondence

It is clear in what sense TR and NR are approximations of each other. The more units the λ-layer of the network contains, that is, the greater m is, the more closely (3.2) approximates (3.1), and if m approaches infinity, (3.2) goes to (3.1). Thus one may say that in the mathematical sense TR is a limiting case of NR. Whether there obtains a limiting case correspondence between the two theories, in the sense discussed above, is not as obvious as the existence of the mathematical correspondence. If we look at the way in which axioms (3.1) and (3.2) were reached, we may see that the meanings of the respective terms (particularly 'p') occurring in the axioms are radically different, apart from their abstract mathematical meanings. One may ask, then, whether there is here a case of incommensurability in the sense of Kuhn [1962]. This question will be discussed below by studying whether an appropriate limiting case correspondence of TR to NR can be defined and whether it results in an instance of counterfactual explanation as described in Part II.

To study the existence of a limiting case correspondence, we have to translate the above descriptions into a model-theoretic language. We shall only give an outline; for technical details the reader should consult Pearce & Rantala [1984a] and text-books of model theory and nonstandard analysis. To facilitate certain relations between formal and set-theoretic languages, we shall assume that the mathematical and set-theoretic entities needed in order to describe model-theoretically the theories of Turing and network representation are included in what is called in Pearce & Rantala [1984a] a *model of analysis*.

Consider the model $A = <A; \in, a>_{a \in A}$ where A is the *superstructure* over the set of natural numbers N, $\in$ is the membership relation (restricted to A), and each $a \in A$ is a distinguished element. Let τ_0 be the type of A, having a two-place predicate symbol $\in$ for $\in$ and an individual constant a for each $a \in A$. Then a model of analysis is a model B of $Th(A)$, i.e., a model of the set of all $L_{\omega\omega}(\tau_0)$-sentences true in A. B is *standard* if B is identical to A or isomorphic to A, otherwise *nonstandard*. It is known that A can be isomorphically and uniquely embedded in any model of $Th(A)$. In what follows, we denote by $B|A$ the unique submodel of B which is isomorphic to A and by $B|A$ its domain. Then an arbitrary model of analysis can be denoted as $B = <B; \in^B, a>_{a \in B|A}$. Here $\in^B$ is the interpretation of $\in$ in B; for instance, if B is identical to A, then $\in^B$ is $\in$, and if B is a nonstandard extension of A, then $\in^B$ is the respective extension of the relation $\in$ to B. Though $\in$ is (part of) the ordinary membership relation, $\in^B - \in$ need not be. It is assumed that if B is a nonstandard extension of A, then its domain B contains nonstandard natural and real numbers [i.e., that B satisfies an appropriate infinitary sentence to this effect; see Pearce & Rantala, 1984a].

The following notation will also be used. If R is the set of all real numbers (which is an element of A) and B is A or some of its extensions, we let $R_B = \{b \in B: b \in^B R\}$. R_B is R if B is identical to A and the respective extension of R in B if B is nonstandard. Thus the (set-theoretic) difference $R_B - R$ is the set of all nonstandard real numbers in B. If n is a standard or nonstandard natural number, then let $N_n = \{1, ..., n\}$, i.e., $N_n = \{k \in^B N: 1 <_B k <_B n\}$, where $<_B$ is the ordinary relation of order $<$ if n is standard and its respective extension to B if n is nonstandard. If f is any function and X a set, then $f \upharpoonright X$ is the restriction of f to X. If $a, b \in R_B$ are such that their difference $a - b$ (in the nonstandard sense) is infinitesimal, we say that a and b are *infinitesimally close* to each other and write '$a \approx b$'.

Even if B is a model of analysis which is different from A and its extensions, we shall, for simplicity, refer to appropriate individuals of B as (standard or nonstandard) natural and real numbers. We shall in a nonstandard case continue to use ordinary mathematical notation and the notation and terminology introduced above; no confusion will result therefrom.

Let B be a model of type τ_0. Consider a four-sorted expansion of B (of type τ, say)

196

(*) $M = <B;S,E,P,n>$

where S,E, and P are new domains and n is a distinguished individual, such
that

(3.3)   **B** is a model of analysis;

(3.4) $n \in^B N$;

(3.5) S is a nonempty subset of $B|A$;

(3.6) E is a set of mappings from N_n to S;

(3.7) P is a set of mappings from $N_n \cup \{0\}$ to $S \cup \{0,1\}$;

(3.8)   For all $p \in P$,
         if $p \restriction N_n \in E$, then $p(0) = 1$;
         if $p \restriction N_n \notin E$, then $p(0) = 0$.

Condition (3.6) says that E is a set of sequences of length n of elements
of S, and (3.4) that n is a standard natural number if **B** is standard; otherwise n is a standard or nonstandard (infinite) natural number. As we can
see, the conditions correspond − in a more abstract form − to the above
description of a Turing representation of expressions of $\mathscr{L}$. Hence we call
M a *model of Turing representation*. It is *standard* if **B** is a standard model
of analysis, otherwise *nonstandard*. If a language $\mathscr{L}$, as above, is given, we
assume that its primitive symbols are sets [cf., e.g., Barwise, 1968], even
that they are specified standard elements, i.e., elements of the superstructure A; and therefore each expression of $\mathscr{L}$ is a sequence of elements of A.
Then an *intended model* of Turing representation of Exp$\mathscr{L}$ would be a
standard model of the above form such that **B** is **A** and $S = \text{Prim}\mathscr{L}$ and $E =$
Exp$\mathscr{L}$. The above conditions axiomatize TR.

Conditions $(3.3) - (3.8)$ can be formalized in an appropriate four-sorted
(of type τ) infinitary logic.[5] The kind of quantification in (3.8) notwith-
standing (which essentially is a quantification over functions), the logic is
first order in the sense that the functions in P are in fact considered as
individuals of the respective sort (elements of the domain P). For the
translation to be involved in the forthcoming correspondence relation, one
only needs to know the types of the respective logics and the kinds of
quantification; no actual formalization is needed. One advantage of the
axiomatization of the above kind is that one can present it in ordinary
mathematical (meta)language instead of a formal language − which one

---

5 It was pointed out by Johan van Benthem that the logic cannot be $L\omega_1\omega$, as was
claimed in Pearce & Rantala [1984ab], but a much larger one (in so far as the size of
conjunctions and disjunctions is concerned). This, however, will have no effect on the
main results of these articles.

knows to be there, however, whenever needed.

Next we consider a model (of type τ', say) of the form

$$(**) \qquad \mathbf{M} = \,\langle\,\mathbf{B};S,E,P,N_{\mathbf{B}},R_{\mathbf{B}},f_+,f_-,m,n\,\rangle,$$

where $\mathbf{B}$ is as above, $S,E,P,N_{\mathbf{B}}$, and $R_{\mathbf{B}}$ are domains, f_+ and f_- are functions from $N_{\mathbf{B}}$ to $R_{\mathbf{B}}$, and m and n are distinguished individuals, such that $(3.3)-(3.6)$ and the following hold:

(3.9) f_+ (f_-) is monotonically increasing (decreasing) such that $f_+ \restriction N$ ($f_- \restriction N$) is standard and unbounded;

(3.10) $m \in^{\mathbf{B}} N$;

(3.11) P is a set of mappings from $N_n \cup \{0\}$ to $S \cup R_{\mathbf{B}}$;

(3.12)   For all $p \in P$,
   if $p \restriction N_n \in E$, then $p(0) = 1:(1 + e^{-f_+(m)})$;
   if $p \restriction N_n \notin E$, then $p(0) = 1:(1 + e^{-f_-(m)})$.

As can be seen, part of what (3.11) expresses is the condition, in an abstract form, that (for each $p \in P$) units $1, \ldots, m$ are grouped into n subsets so that each subset is a distributed representaton of some symbol in S. A model of the above kind is called a *model of network representation;* and whether it is standard or nonstandard, or intended, is defined as above. The conditions axiomatize NR.

Next we sketch a limiting case correspondence of TR to NR; since the formal proofs are analogous to (but somewhat simpler than) those in Pearce & Rantala [1984a], they will be omitted. Let K be the class of all standard models of TR and let K' be the class of all nonstandard models of NR (as in (**)), such that m is infinite (i.e., a nonstandard natural number), and n is finite (a standard natural number). Let a mapping F be defined as follows. For each $\mathbf{M} = \,\langle\,\mathbf{B};S,E,P,N_{\mathbf{B}},R_{\mathbf{B}},f_+,f_-,m,n\,\rangle$ in K', $F(\mathbf{M}) = \,\langle\,\mathbf{B}\,|\,\mathbf{A};S, E,^\circ P;n\,\rangle$ where $^\circ P$ is the set of all mappings $^\circ p$ such that for some $p \in P$, $^\circ p = (p \restriction N_n) \cup \{\,\langle 0,r\rangle\,\}$, where $r = 1$ if $p(0) = 1:(1 + e^{-f_+(m)})$, and $r = 0$ if $p(0) = 1:(1 + e^{-f_-(m)})$. It is obvious that $F(\mathbf{M}) \in K$ (i.e., that $F(\mathbf{M})$ is a standard model of Turing representation) and, conversely, that for each $\mathbf{M} \in K$, there is a model $\mathbf{M}'$ in K' such that $F(\mathbf{M}') = \mathbf{M}$. Therefore, F is a mapping from K' onto K.

In accordance with Pearce & Rantala [1984a], we may say now that $F(\mathbf{M})$ is a *standard approximation* of $\mathbf{M}$. It follows from condition (3.9) that in a model $\mathbf{M}$ as above, where m is infinite, $f_+(m)$ is an infinite (nonstandard), positive real number and $f_-(m)$ is an infinite, negative real number. Therefore, $p(0) \approx 1$ in the former case of (3.12), and since $^\circ p(0) = 1$, $p(0)$ is infinitesimally close to $^\circ p(0)$. In the latter case of (3.12), $p(0) \approx 0$, wherefore $p(0)$ is infinitesimally close to $^{\circ}p(0)$. Since F thus defines a limit procedure which is analogous to that defined in Pearce & Rantala [1984a], by means of which the models in K are obtained from those in K', we can in an analogous way define a truth preserving (that is, condition (2.1) is satisfied) translation I from the infinitary language (logic) L in which TR is characterized into the respective language L' for NR. It follows that F and I determine a limiting case correspondence of TR to NR.

If certain logical, pragmatic (and paradigmatic) conditions [which cannot be discussed here; but cf. Part II above and Rantala, 1989, 1995] are assumed to be satisfied, then, as pointed out in Part II, the existence of the limiting case correspondence of TR to NR guarantees that a counterfactual explanation of TR can be derived, that is, that this case exemplifies the general theory of counterfactual explanation, as described in part II, and, in particular, that the conditional (2.8) is true (at an appropriate world) where ψ is now the sentence 'm is infinite' (or, rather, its formal counterpart in L'), θ is (3.8), and its translation $I(\theta)$ is:

(3.13)   For all $p \in P$,
          if $p \upharpoonright N_n \in E$, then $p(0) \approx 1$;
          if $p \upharpoonright N_n \notin E$, then $p(0) \approx 0$

(which is expressible in L'). Informally, the conditional might be read, for instance, as follows:

(3.14)   If a network had infinitely many units, then a symbolic Turing representation would almost take place.

Hence, if the conditions are satisfied, TR can be considered as an idealization of NR in something like the sense suggested by Smolensky and made more exact in Part II. This case may look basically different from the physical examples mentioned in Part II since it is not perhaps appropriate to say that (3.8) and (3.12) express scientific laws (or are lawlike sentences) in the same way as the respective sentences of the physical examples. Furthermore, only if considered from an appropriate perspective, one could say something like that TR is supplanted by NR; this much might say, e.g., a connectionist who believes that the latter provides a better account of (human) cognitive representation so that the version which is provided by the former is even more idealized. Only from such a point of view does the above informal paraphrase (3.14) of the formal counterfactual conditional make sense. It is clear that it is not meaningful if looked at from a purely metamathematical or logical point of view since from that perspective the role of the notion of Turing machine, and, more generally, the role of symbolic representation and symbol manipulation, is fundamental, not that of subsymbolic.

Department of Mathematical Sciences
Philosophy
University of Tampere
P.O. Box 607, 33101 Tampere, Finland

REFERENCES

Adams, E. W. [1955]. *Axiomatic foundations of rigid body mechanics.* Unpublished diss. Stanford University.

Barwise, J. [1968]. Implicit definability and compactness in infinitary languages. In: J. Barwise (Ed.). *The syntax and semantics of infinitary languages.* Lecture Notes in Mathematics, vol. **72** (pp. 1–35). Berlin–Heidelberg–New York: Springer–Verlag.

Bohr, N. [1920]. Über die Serienspektra der Elemente. *Z. Phys.,* **2**, 423.

Fodor, J. & Pylyshyn, Z. [1988]. Connectionism and cognitive architecture: A critical analysis. *Cognition,* **28**, 3–71.

Glymour, C. [1970]. On some patterns of reduction. *Philosophy of Science* 7, 340–353.

Heisenberg, W. [1958]. *Physics and philosophy.* New York: Harper & Row.

Kleene, S. C. [1964]. *Introduction to metamathematics.* Amsterdam: North-Holland Publishing Co.

Krajewski, W. [1977]. *Correspondence principle and growth of science.* Dordrecht: D. Reidel Publishing Co.

Kuhn, T. [1962].*The structure of scientific revolutions.* Chicago: University of Chicago Press.

Kuhn, T. [1983]. Commensurability, comparability, communicability. In: P. Asquith & T. Nickles (Eds.). *PSA 1982,* vol. 2. East Lansing: Philosophy of Science Association.

Lahti, P. J. [1984]. Quantum theory as a factualization of classical theory. In: W. Balzer, D. Pearce, & H-J. Schmidt (Eds.). *Reduction in science. Structure, examples, philosophical problems* (pp. 381–396). Dordrecht: D. Reidel Publishing Co.

Lewis, D. [1973]. *Counterfactuals.* Oxford: Blackwell.

Niiniluoto, I. [1990]. Theories, approximations, and idealizations. In: J. Brzezinski, F. Coniglione, T.A.F. Kuipers, & L. Nowak (Eds.). *Idealization I: General problems.* Poznan Studies in the Philosophy of Sciences and the Humanities, vol. **16** (pp. 9–56). Amsterdam–Atlanta, GA: Rodopi.

Pearce, D. [1987]. *Roads to commensurability.* Dordrecht: D. Reidel Publishing Co.

Pearce, D. & Rantala, V. [1983]. New foundations for metascience. *Synthese* **56**, 1–26,

Pearce, D. & Rantala, V. [1984a]. A logical study of the correspondence relation. *Journal of Philosophical Logic* **13**, 47–84.

Pearce, D. & Rantala, V. [1984b]. Limiting-case correspondence between physical theories. In: W. Balzer, D. Pearce, & H-J. Schmidt (Eds.). *Reduction in science. Structure, examples, philosophical problems* (pp.153–185). Dordrecht: D. Reidel Publishing Co.

Pearce, D. & Rantala, V. [1985]. Approximate explanation is deductive-nomological. *Philosophy of Science* 52, 126–140.

Rantala, V. [1978]. The old and the new logic of metascience. *Synthese* 39, 233–247.

Rantala, V. [1979]. Correspondence and non-standard models: A case study. In: I. Niiniluoto & R. Tuomela (Eds.). *The logic and epistemology of scientific change*. Acta Philosophica Fennica, vol. 30 (pp. 366–378). Amsterdam: North-Holland Publishing Company.

Rantala, V. [1989]. Counterfactual Reduction. In: K. Gavroglu, Y. Goudaroulis, & P. Nicolacopoulos (Eds). *Imre Lakatos and theories of scientific change* (pp. 347–360). Dordrecht: Kluwer Academic Publishers.

Rantala, V. [1995]. Understanding scientific change. In: P. Bystrov *et al.* (Eds.). *Festscrift in Honour of V. Smirnov*. Dordrecht: Kluwer Academic Publishers. Forthcoming.

Rantala, V. & Vadén, T. [1995]. Minds as connoting systems: Logic and the language of thought. Forthcoming.

Smolensky, P. [1987]. The constituent structure of connectionist mental states: A reply to Fodor and Pylyshyn. *The Southern Journal of Philosophy* XXVI, Supplementary, 137–161.

Smolensky, P. [1988]. On the proper treatment of connectionism (with open peer commentary and author's response). *Behavioral and Brain Sciences* 11, 1–74.

Smolensky, P. [1991]. Connectionism, constituency and the language of thought. In: B. Loewer & G. Rey (Eds.). *Meaning in mind: Fodor and his critics* (pp. 201–227). Oxford–Cambridge, MA: Blackwell.

Suppe, F. [1977]. The search for philosophic understanding of scientific theories. In: F. Suppe (Ed.). *The structure of scientific theories* (pp. 3–232). Urbana–Chicago–London: University of Illinois Press.

Waerden, B.L. van der [1967]. Introduction. In: B.L. van der Waerden (Ed.). *Sources of quantum mechanics* (pp. 1–59). New York: Dover Publications Inc.

Poznań Studies in the Philosophy
of the Sciences and the Humanities
Vol. 42, pp. 201–216

Matti Sintonen and Mika Kiikeri

IDEALIZATION IN EVOLUTIONARY BIOLOGY

I

Idealization in Science and Philosophy of Science

Evolutionary theory has been a tough case for philosophers of science, because it resists rational reconstruction along the usual lines. One possible reason is that the theory has not yet entered a mature stage which makes such reconstruction possible. A rival explanation, which has enjoyed some support during the past decades, is that evolutionary theory is fundamentally different from physical theories. On this view it consists of a batch of heterogeneous elements which simply do not conspire closely enough to form a unitary entity.

However, one thing is clear on both views, namely, that evolutionary theory is a highly idealized one. The core theory, if there is one, is highly abstract in that its direct contacts with the observable reality are virtually non-existent. But also the less fundamental theory-elements, the evolutionary subtheories and especially the theoretical models which have been designed to live closer to the empiria, are characteristically laden with counterfactual idealizing assumptions.

Idealization challenges not just the structure of evolutionary theory but all aspects of biological inquiry. When theories are used to explain singular events or generalizations, or when they are employed to predict facts, the resultant explanations and predictions are only approximately accurate. Similarly, when experiments and measurements are performed, the resultant values are only approximations of the real ones. To deal with these situations, mathematicians and statisticians have developed a rich variety of exact approximation methods. Learning to use some of these techniques is, then, an important part of scientific training.

This state of affairs is a simple consequence of the complexity of biological phenomena. Unable to deal with, or even identify, all parameters and variables which affect or might affect a phenomenon, an evolutionist must choose the most essential factors and ignore the rest. But although this is a fact of life for any practicing evolutionary geneticist or ecologist, available metatheories have had difficulties in coming to grips with it.

With increasingly close contacts between philosophers and scientists we can ask: How do philosophers now view the discrepancy between the maddeningly complex natural systems and their idealized theoretical representations? There are a number of possibilities. One would be to ignore the issue. For the traditional hypothetico-deductivist view (H-D view, for short) of the logic of science this was the preferred way out, because the inquirer was not in the business of recording actual complexities, but rather in that of creating rational reconstructions. As it turned out, these were reconstructions of finished products in the so-called context of justification, and little or no attention was paid to the scientific activity which yields these products.

But with the downfall of the received view, and with alternatives in the horizon, this stance now seems mere stubborness. An enlightened metatheoretician must admit that idealizations are here to stay. But what else should she or he say? One could take it that the highly idealized scientific theories deliberately lie. Or one could try to make a virtue out of a seeming shortcoming. It is true, the latter course has it, that highly explanatory theories sacrifice detailed fit, truth at the level of observable consequences, for explanatory power. But this they do with the laudable aim of catching the deeper but essential connections or invariances behind empirical clutter. Rather than lie, fundamental theories are simply being economical with truth.

The Poznań School is a good representative of the latter view. It aims at a realistic description of actual scientific activity, and insists on truth as a meaningful though perhaps in practice unattainable goal. The guiding idea is to present idealization, and the reverse process called concretization or factualization, in terms of simple logical concretization schemata.[1] Although the basic ideas of the Poznań school is highly suggestive, and the motivation laudable, we shall argue for some amendations. First, it is not clear that the theory as a whole has a structure which is amenable to treatment along Poznań lines. Secondly, its notion of idealized laws and their application process does not do full justice to the model-constructing

---

[1] For a review of Poznań school, see Nowak 1992.

activities found in many sciences. Apart from difficulties already on record we shall single out one crucial feature of model building, viz. the interplay between general laws on the one hand and model-specific assumptions on the other. We shall try to show that this is more complex than Poznań philosophers have assumed, and requires us to give up the rigid idea that there is, in the concretization process of all laws or theories, a fixed order of importance.

The example of theory construction and model building we have chosen comes from the domain of evolutionary biology. We shall start with a brief account of evolutionary theory and the Poznań view and distinguish several senses in which a theory may be idealized (and need to be concretized) (part II). Part III argues that evolutionary theory at large does not fit the view in which theories have a unique core, and part IV that if we want to adopt a realistic picture of model building in evolutionary subtheories such as population genetics we have to make alterations in the Poznań account.

II
The Poznań Schema and Evolutionary Theory

To focus on idealization in evolutionary theory, let us first take a look on how it is handled in the Poznań account. On this view scientific disciplines or fields characteristically start as non-theoretical fact gathering enterprises. When they mature they go beyond fact collecting and inductive generalizations to a theoretical phase, and formulate idealizing theories with often highly abstract regularities and laws. These laws specify functional dependencies between a phenomena and their most important determinants. They are then concretisized, to accommodate factors which influence the phenomena studied to a lesser degree, in accordance with what are called concretization schemas.

In the Poznań account, a concretization schema is closely related to a certain conception of scientific laws. The basic structure of scientific laws is represented by Poznań methodologists as follows [Nowak 1992, p. 12]:

$$T^k: [G(x) \ \& \ p_1(x) = 0 \ \&...\& \ p_{k-1}(x) = 0 \ \& \ p_k(x) = 0] \rightarrow$$
$$F(x) = f_k[H(x)].$$

In this schema, F denotes some magnitude or phenomenon under study and G is a class of objects among which F occurs. H and $p_1,...,p_k$ denote all the relevant parameters influencing F. H is the *principal parameter*, thought to be the most important factor influencing F. $p_1,...,p_k$ are *secondary parameters* which are ordered according to their importance, so that if $i > j$, p_i is more important to F than p_j. All secondary factors are abstracted away from the T^k because it is assumed that each $p_i(x) = 0$, although in reality their values are not equal to zero. Consequently, T^k is called an idealizational law.

The concretization schema describes a stepwise process where each p_i is taken into consideration according to its importance (the ordering of p_i's). The most important factor p_k is considered first, the second important factor p_{k-1} secondly and so on. At every step, the functional dependence between F and H is adjusted by taking into consideration the relevant p_i. As a result, the concretization schema is presented by the following hierarchical structure:

$$T^{k-1}: [G(x) \& p_1(x) = 0 \&...\& p_{k-1}(x) = 0 \& p_k(x) \neq 0] \rightarrow$$
$$F(x) = f_{k-1}(H(x), p_k(x))$$
$$\vdots$$
$$\vdots$$
$$T^0: [G(x) \& p_1(x) \neq 0 \&...\& p_{k-1}(x) \neq 0 \& p_k(x) \neq 0] \rightarrow$$
$$F(x) = f_0(H(x), p_k(x),...,p_1(x))$$

The assumption $p(x) = 0$ is an idealizing assumption if and only if, for every member of G it holds that $p(x) \neq 0$, and 0 is the limit value of p. The complete form of an idealizing assumption is the following universal sentence:

(1) $\forall x: G(x) \rightarrow p(x) = 0.$

And now to some critical comments. It is easy to see that the assumptions of the Poznań school concretization schema are problematical in many ways. One critical point is related to universal quantification in (1) [see Kuokkanen & Tuomivaara 1992]. There are natural applications where $p(x) = 0$ for *some* $x \in G$ and $p(x) \neq 0$ for some other $x \in G$. The dichotomous on/off presumption concerning the values of x in the putative idealization assumption is too restrictive. Another critical point is related to the use of material implication in (1). As Niiniluoto [1986] suggests, it should be replaced by counterfactual implication.

But matters are more complicated still, as we can see when we attempt to apply the concretization schema to evolutionary theory as a whole.

Lastowski [1982] has applied the Poznań schema to evolutionary biology. In his account it is assumed, if perhaps tacitly, that evolutionary theory amounts to the theory of natural selection. Lastowski then gives an idealizational reconstruction of the theory of natural selection in which the idealized law describes the adaptive mechanism. The law (NS) takes the form

> If (1) there are no changes in the natural environment, and (2) the species lasts one generation, under such environmental conditions those populations of the species survive whose adaptive values are higher than the definite adaptation threshold. [Lastowski 1982, p. 362]

Here we can distinguish two idealizing assumptions, viz., that there are no changes in the natural environment, and that a species lasts one generation.

The methodologically crucial feature in this account is that, clearly, evolutionary theory as a whole is thought to fit the model designed for physical theories. An analogous view has been developed by Michael Ruse, and it may be instructive to make the parallel explicit. The idea is to think of evolutionary theory as a structure with a fundamental but abstract core theory which is then concretized for more specific applications. The parallel is here: the theory of natural selection was first presented in a qualitative form. Later developments, then, brought about quantitative subtheories and models, so that one could think of this as reinterpretation of a theory in accordance with the principle of the idealizational conception of science [Lastowski 1982, p. 354].

In Ruse's view the core of evolutionary theory is not the theory of natural selection but rather, population genetics itself. He argues that the axiomatic core theory, population genetics, contains its own theoretical entities and bridge principles. This theory, in its turn, is the explanatory core element which unifies all subsidiary disciplines.[2] In the picture which Ruse draws the architecture of the theory is as follows [Ruse 1973, pp. 48–49]:

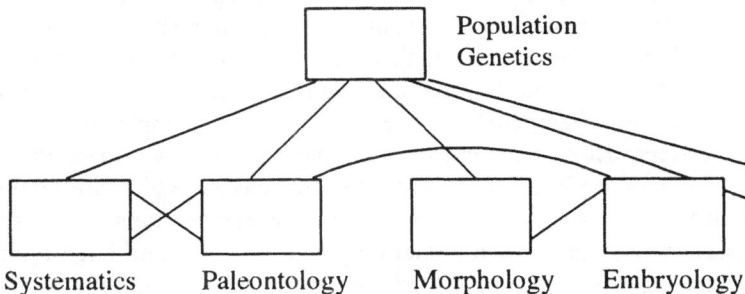

Systematics   Paleontology   Morphology   Embryology

---

2 See Ghiselin 1984, p. 65, and Ruse 1979, p. 198. For discussion, see also Recker 1987 and Sintonen 1990.

The rectangles stand for the various disciplines from systematics to embryology, the thick lines for links between the core discipline (population genetics) and the subsidiary disciplines, and the thin lines for links between the subsidiary disciplines. Now Ruse does admit that despite this parallel evolutionary theory as a *whole* does not possess "the deductive completeness possessed, say, by Newtonian mechanics."

Why is this? The answer is that the picture is idealized. Instead of explicitly formulated strong links between the various levels we find "suggestions, hypotheses, extremely weak inductive inferences, and frequently, outright guesses." Writes Ruse, because of many factors — the newness of the theory, the fact that many pertinent pieces of information are irretrievably lost, the incredible magnitude and complexity of the problem, and so on — many of the parts of evolutionary theory are just 'sketched in'". In short, the H-D model is an ideal, although, to wit, the evolutionists "are far from having it as a realized actuality".

To appreciate the merits of Ruse's proposal, consider his example, explanations in systematics. Systematics deals with the nature and the distribution of animal and plant groups, and Ruse focuses on the distribution organisms on islands some distance from the shores of large land-masses. A typical explanation starts with the observation that organisms on islands and the mainland are often similar to one another, and that the islands in turn often have a number of species of a particular genus of organism similar to each other (and to the mainland organisms). Furthermore, the organisms, especially on Oceanic islands, which bear such similarity seldom if ever include large mammals but rather comprise birds, reptiles, and insects.

The explanations that evolutionary theory gives characteristically refer to a putative common descent, i.e., that the observed similarities are best explained by assuming that there once was a group of ancestral mainland organisms from which both the contemporary mainland and the island populations have descended. They then proceed to establish that commerce between the mainland and the island, as well as commerce between islands, are and have been hampered by oceanic currents, strong winds and like natural barriers. However, when occasional passages do open, the newcomers to the islands tend to evolve into new species which are similar to but yet different from those already there. The explanations then resort to the subsidiary evidence which is that only birds, insects and like organisms typically manage to cross the barriers, or manage to drift or be blown to the islands, which is why indeed large land animals seldom exhibit these patterns.

However, this general argument pattern and its more concrete instantiations, like Darwin's argument from finches, is highly idealized and fails to mention many facts needed to yield a complete explanation. We also want to know why and how an isolated group of organisms turns into a new species. Ruse's answer is that here genetical considerations enter the scene. Quoting Dobzhansky he says that the appeal to population genetics materializes through the realization that under the different conditions that prevail on the different islands the new founding populations will be subject to new and possibly severe selective forces. Since conditions on the mainland and on the islands are different, fitness is cashed out differently, and divergent lines secure a bridgehead.

The point remains nevertheless, says Ruse, that this supposition of divergence presupposes genetic laws akin to those proposed by Mendel, for otherwise blending would quickly wipe out new forms, and selective pressure would be unlikely to lead to speciation. Thus Mendelian population genetics, and its latter day refinement, the Hardy-Weinberg law, provides the foundation of modern evolutionary theory.

Now Ruse admits that this or other evolutionary explanations do not reach deductive completeness. Yet, he says, "we can see how evolutionary phenomena are in some sense explained by assumptions, chief amongst which is the theory of population genetics." Furthermore, the genetic principles not only give the main force of evolutionary explanations but also suggest supporting details.

III

Evolutionary Explanations and Concretizations

But evolutionary theory as a whole cannot be identified with the theory of natural selection, nor with population genetics. Rather, these elements are but components in the batch of theories commonly referred to as the theory of evolution. Furthermore, evolutionary theory does not seem to form a neat hierarchical structure with any one theory as *the* structure. Logically and historically speaking it is possible to think of evolution without natural selection, as can be seen from Lastowski's account of population genetics: if by (micro)evolution we mean changes in the gene frequences of a population, selection is but one of the possible causally efficatious forces. Similarly, population genetics can hardly be a deductive explanatory core for evolution theory. There are few if any nomic generalizations which the theory could claim to be its laws. Consequently, although evolutionary theory is an idealizing theory in some sense, the Poznań type idealization schemata is not the best way to decribe it.

A useful metatheoretic model is Morton Beckner's [1959] rather early but still essentially correct *reticulationist* view. Rather than being a unitary entity with a fixed structure, evolutionary theory is more like a consortium or a treaty of relatively independent theory elements. According to Beckner evolutionary theory differs from the paradigmatic physical theories in that the various generalizations are not consequences of "one or more hypotheses of greater generality" [Beckner 1959, pp. 159−160].

On this view, evolutionary theory is not an undeveloped branch of natural science but simply different. The main problem with it is not its recent arrival among disciplines nor its undeniable complexity but rather its fundamentally different nature. Biologists are keen on building what they call theoretical models, and although population genetics often (if not always) looms behind such models, the links between the genetic core theory and the models are not deductive. In the modern case the gap is deeper still, for it does not appear that there is even interest in the construction of deductive theories.

The most remarkable distinguishing feature of evolutionary theory is that it is interested in giving what could be called quasi-historical narratives. These narratives tell the story − the emergence and spreading − of a trait by referring not just to nomic generalizations but also to historically speaking contingent facts. Furthermore, many of these facts belong rather to geology and other strictly speaking non-biological disciplines, so that the theories cannot be regarded as specifications of biological theories. Methodologically speaking the crucial difference is that the information to be obtained to tell the stories cannot be derived from any of the theories.

Evolutionary theory, then, operates on several levels simultaneously, on the level of the fundamental theories and on the level of particular quasi-historical narratives. Darwin's variant, the theory of descent with modification presented a highly abstract and idealized blueprint for a variety of subtheories from paleontology and biogeography to morphology. These subtheories were not precise enough to deliver detailed descriptions and explanations, and therefore needed concretization. And when we then trace the career of the theory from, say, Darwin's *Species* book to the first edition of *the Origin* and beyond that to later editions, it is clear that concretization of a sort took place. Without going into details, we can take it that Darwin provided a more or less cohesive package of theory-elements, with the theory of descent with modification, and the theory of natural selection as the most central ones.[3] The main claim of the former theory was the set of logically speaking singular claims that existing forms of organisms have descended from previously existing but now possibly extinct species. The latter is the more theoretical claim that the mechanism of evolution, both on the micro and the macrolevel, is natural selection. The former theory was easy to accept already at Darwin's time, with some doubts about the place of man in the scheme of things, whereas the latter was sentenced to the limbo of candidate ideas for almost a century.

Nor is this the end of the difficulties. For even if we assume, quite correctly, in our view, that Darwin initiated a revolutionary research program (one which wasn't, as we saw, the only or even the leading paradigm for evolutionary studies), there were difficulties in two dimensions. First, the claims that could be made by help of the evolutionary theory, or more precisely its selectionist theory-element, could be taken to express different degrees of commitment. Secondly, the selection theory-element involved a fuzzy quantifier in its basic claim. Let us expand on these a little.

It is useful to view the nature of the commitments along the lines suggested by Jonathan Hodge [1977, 1987, 1989]. Hodge's idea is that Darwin's strategy was this. Before one could advance this or that particular selectionist story behind an adaptation one must demonstrate both that natural selection is a phenomenon (a *vera causa* in Newtonian terms), and that it is at least in principle able to bring about effects of that type. Having done that, the selection theory-element can be used to make specific claims to the effect that such and such a generalization or fact in fact is due to natural selection.

The other difficulty is the fuzzy quantifier. The evolutionary overall claim naturally splits into a set of more specific claims about various generalizations and singular events or, rather, states of affairs. Quite a few of the generalizations, or laws as Darwin simply called them, formulated functional dependencies between features of species or populations and ecological variables, such as the law which says that organisms in colder climates tend to have larger body sizes than organisms in warmer climates. The singular facts in turn have to do with the various claims about the emergence and fixation of characteristics.

But what is the nature of the overall claim? It is often said that it is the claim that (natural) selection is the major or most important agent of

---

3 Apart from this there was also, e.g., the theory of sexual selection (treated more fully in *Descent of Man*) and other subtheories. See Mayr 1991, Sintonen 1993a, 1993b.

evolutionary change, accounting for the majority of evolutionary changes. But what is "most important" and what is "the majority"? Darwin wavered on the answer. Here we need not worry about what the right answer is, for it can be anywhere between the weak claim that some such facts and generalizations are so explained, or the strong one that all are.

Our concern is not with any particular answer. Rather, we wish to point out that there are literally myriads of possible types of theories of evolution, and that even if we focus on some particular type, such as Darwinian evolutionary theory, the basic theory can be given different interpretations. Secondly, just because evolutionary theory is in the business of constructing quasi-narrative stories about the emergence of facts and generalizations, the theory cannot be conceived along the lines suggested in the Poznań schema.

IV

Idealizations in Model Building

The grand picture that emerges is that evolutionary theory at large has a reticulate structure in which it is hard or impossible to specify a unique dominating core. However, each one of the evolutionary subtheories such as evolutionary genetics and evolutionary ecology does have an articulate structure, viz. that of an hierarchically arranged net structure. Within these subdisciplines concretization amounts to designing theoretical models with ever more accurate laws to deal with the various phenomena. The result is a reticulate picture along the lines suggested by Juha Tuomi [1981, 1992].

It is obvious, then, that concretization was needed to take down to earth the abstract core of evolutionary theory. First, special laws for the various areas of evolutionary studies had to be discovered (or created), to redeem the promise made by Darwin's theory to deal with paleontology, morphology, etc. Secondly, concretization was needed within subtheories to produce empirically adequate theories and theoretical models. Both types of concretization are important, but, as we have argued, only the latter seems to fit the intent of the Poznań school account of idealization.

But does it fit the Poznań picture in closer detail? To settle this question conclusively, we would have to examine examples from all evolutionary subtheories, such as selection theory or population genetics, the most mature and quantitative subtheory of modern evolutionary and population biology. Here we have to be satisfied with only one example. It concerns the use of equilibrium models in population genetics. It is an illuminating example for our purposes, because these models are based on the quantitative Hardy-Weinberg law, which is subject to many idealizational assumptions. It has also been a standard example in the philosophical discussions of modern evolutionary biology. In particular, Lastowski [1982] gives it an explicit Poznań-style idealizational form.

But first, what is model building? In population genetics, the goal of inquiry is usually to construct dynamical mathematical models (state space and laws of transformation) that predict and explain the changes of gene frequencies in populations as a function of time. Because empirical populations are extremely complex entities, the problem is to find a dynamically sufficient description of a population which makes it possible to find laws for its evolution. For instance, Richard Lewontin [1984, pp. 6−7] describes the problem as follows:

> [T]he problem of theory building is a constant interaction between constructing laws and finding an appropriate set of descriptive state variables such that laws can be constructed. ... We do not really know what a sufficient description of a community is because we do not know what the laws of transformation are like, nor can we construct those laws until we have chosen a set of state variables.... [T]here is a process of trial and synthesis going on in community ecology, in which both state descriptions and laws are being fitted together.

As seen from this quote, Lewontin describes the process of model constructing as a kind of trial and error activity in which the problem is to fit different but intertwined goals together. Similar points are raised by Elisabeth Lloyd [1984, 1988] in her account of population genetics in terms of so called semantic view of theory structure. In Lloyd's view there are also two *sine qua nons* to the definition of a model. On the one hand there is the necessity of specifying the state space, employing variables and parameters hoped to be adequate for the synchronic and diachronic description of the target. On the other hand there are two types of laws, viz. what Lloyd calls coexistence laws the purpose of which is to describe the system's structure, and succession (or transition) laws whose task it is to record structural changes.

On the basis of Lloyd's account, it is possible to distinguish several problems involved in the construction of theoretical models. Each particular solution to them could be conceived as requiring different sets of idealizational factors in the sense used in this paper [cf. Lloyd 1988, pp. 33−41]. The problems Lloyd discusses includes

(i)   the problem of state space description which includes (a) the choice of state variables, and (b) the choice of parameters for laws,

(ii)   the problem of state space types: the type chosen depends on the level the evolutionary processes are thought to work; the units of selection-problem.

Genetical modelling of populations presupposes that problems like these are solvable when the model is constructed. The crucial question, then, is to determine whether the Poznań schema is suitable to take into account different sets of idealizational factors which depend on the solutions given to these fundamental theoretical and practical problems. There are some reasons to think the answer is negative for the Poznań account is mainly devoted to idealizations emerging from the choice of the parameters for genetical laws. This can be seen, for instance, from Lastowski [1982] in which he gives a brief presentation of the application of Poznań schema to population genetics.

To appreciate the bearing of this observation, consider the basic law of modern population genetics, the Hardy-Weinberg law. Lastowski [1982, p. 357] discusses the case in which this idealizational law is concretized step by step. These steps include such idealizational factors as (1) the migration phenomenon (m), (2) the mutation phenomenon (n), (3) the selection phenomenon (s), and (4) genetic drift (F). Lastowski [1982, p. 358] symbolizes the idealized form of Hardy-Weinberg law as follows:

$R(x)$ & $m(x)=0$ & $n(x)=0$ & $s(x)=0$ & $F(x)=0$ →
$p^2(A,x) + 2p(A,x)\, q(a,x) + q^2(a,x) = 1$
where
$R(x)$: an idealized population (cf. below),
$m(x) = 0$: migration does not occur,
$n(x) = 0$: mutation does not occur,
$s(x) = 0$: selection does not occur,
$F(x) = 0$: genetic drift does not produce an effect,
p = frequency of allele A in some locus,
q = frequency of allele a in some locus, and
$q = 1-p$.

The initially abstract form of the H-W law is then concretized by taking the effects of these factors explicitly into consideration in the H-W law schema. The consequent part of the law becomes more complex after each of these steps.

All these idealizational factors are, however, parameters in the particular law of genetical equilibrium models. Thus it seems that the Poznań-style account leaves no room to the other types of idealizing assumptions needed in actual model-building activity. These assumptions concern factors which are involved in the solutions to problems (i)(a) and (ii): the basic structure of populations and assumptions about the levels or units of evolutionary process.

Consider, for instance, factors influencing the choice of state variables. The structure of a population is reflected in the structure of state space used to represent it. In this sense, the choice of state variables is prior to the choice of parameters. For example, Elisabeth Lloyd [1988, p. 38] writes:

> The actual state space used in each instance depends on the genetic characteristics of the system, and not usually on the parameters. For instance, the succession of a system at Hardy-Weinberg equilibrium and one that is not at equilibrium but is under selection pressure, could both be modeled in the same state space, using different laws.

According to Lloyd, the same state space could be used in many situations, depending on the genetic characteristics of the system. Once the state space with the appropriate variables has been constructed, parameters could be chosen in many ways. This gives the approach flexibility and enables modelling of different populations with the same state space, as Lloyd points out. Furthermore, the choice of parameters can be relatively independent of the construction of the state space: different parameters could be used in the same state space [see Lloyd 1988 and the quote above]. The state variables must be chosen before the parameters and laws of the model could be specified.

So, the inqury in population biology proceeds by assuming some variables to have zero values, and by supposing that some other set of variables would lead to a dynamically sufficient description of the population. All these assumptions are idealizational in nature: the effects of some factors are ignored while some are thought to be more important for the description of populations. Yet, these kinds of idealizations play no role in the Poznań account of theoretical biology (at least, not in the account of Lastowski 1982). It concentrates on the form of laws (set of parameters) although idealizations began already at the level of variable specification. Thus, it seems that the Poznań style idealized law schema does not account for the actual complexity of the idealizational practise in the population genetics.

The difficulties of Poznań style account are due to the fact that the structure of a population as presented in the variables of a model is typically ignored in the Poznań account. As Lastowski admits the population in his formalization of the Hardy-Weinberg law is an idealized one, characterized by the following condition [ibid., p. 357]:

$R(x)$ is the population that is a collection of individuals of a given species inhabiting a definite territory and living in similar conditions of environment.

But as seen from the previous discussion, this condition is clearly insufficient from the point of view of actual research activity, for it leaves out the most difficult phase of model construction in population genetics.

Another lacuna is in the assumption that there is a fixed order of importance among the idealizational factors. When we look at model building strategies in all their complexity, it is easy to see that there could be models of the same subtheory with different idealizing assumptions. In model M we may have that $p_1(x) \neq 0$ and $p_2(x) = 0$, while in model N the situation is reversed. Furthermore, individuals or subpopulations may be treated differently with respect to some idealizing assumptions of model. In this case, we could have partial idealizations and concretizations which the Poznań schema cannot handle [cf. Kuokkanen & Tuomivaara 1992].

In spite of these problems, Lastowski's treatment of Hardy-Weinberg law is very interesting as far as it goes. The problem is that it does not go far enough. It assumes that population genetics and the H-W law form the core of evolutionary biology as a whole. Because this is not the case, the idelizational factors of Lastowki's account, viz. the parameters of H-W law, form only one subset of the set of possible idealizational factors. And it must be remembered that, besides population genetics, we have not even mentioned the other subtheories of evolutionary biology in which matters could be worse still.

V

Conclusion

Our starting point in this paper was that while idealizational assumptions are important in evolutionary biology, their methodological role and logical form is seldomly discussed in available metatheories. The most notable exception, Poznań school account of idealization, turned out to be insufficient in certain respects. In particular, we argued that the structure of evolutionary theory is not amenable to treatment along Poznań lines, and that its notion of idealized laws does not do full justice to the model-constructing activities found in evolutionary subtheories such as population genetics.

Our discussion in part IV gives us a reason to suspect that the Poznań account is not even in principle suitable to deal with all the intricacies of model building in evolutionary biology. This reason is related to our discussion of the structure of evolutionary theory in parts II and III.

There we argued that evolutionary theory as a whole has a reticulate structure with no common basic laws or common theory core. But it seems that the Poznań style account of idealization requires a hierarchical theory structure in which the form of the core laws somehow determines the basic idealizational factors and their order of importance. This kind of theory structure is manifested in certain physical theories. For instance, Newton's laws are such basic laws, forming the core of classical (Newtonian) mechanics. The idealizational factors in Newton's laws are involved in every application or subtheory of Newtonian mechanics. As we argued, this is not the case in evolutionary theory. Theoretical models with different background assumptions may have to use different sets of idealizational factors.

Are there, then, any positive suggestions as to how idealizations could be handled more realistically? One such suggestion is to treat them as a part of question-theoretic model of inquiry[4]. In this approach, idealizations are thought to be relativized to some problem or question whose solution is the goal of a particular piece of inquiry. As was already seen from the discussion of model-construction above, such a problem-sensitivity of idealizations seems to be the case in population genetics. However, the more fuller and detailed treatment of this point have to wait for another occasion.

Department of Mathematical Sciences
Philosophy
University of Tampere
P.O. Box 607
33101 Tampere, Finland

REFERENCES

Beckner, Morton. [1959]. *The Biological Way of Thought*. New York: Columbia University Press.
Ghiselin, Michael. [1984]. *The Triumph of the Darwinian Method*. Chicago: University of Chicago Press.

---

4 The approach we have in mind is Hintikka's interrogative model of inquiry. See Hintikka 1992, for some suggestions of how the role of idealizations and approximations could be seen in the light of the interrogative model.

216

Hintikka, Jaakko. [1992]. The Concept of Induction in the Light of the Interrogative Approach to Inquiry. In: John Earman (Ed.). *Inference, Explanation, and Other Philosophical Frustrations.* Berkeley: University of California Press.

Hodge, Jonathan. [1977]. The Structure and Strategy of Darwin's Argument. *British Journal for the History of Science* 10, pp. 237–46.

Hodge, Jonathan. [1987]. Natural Selection as a Causal, Empirical and Probabilistic Theory. In: L. Kruger (Ed.). *The Probabilistic Revolution, vol. 2.* Cambridge, Mass.: MIT Press, pp. 233–70.

Hodge, Jonathan. [1989]. Darwin's Theory and Darwin's Arguments. In: M. Ruse (Ed.). *What the Philosophy of Biology Is: Essays Dedicated to David Hull.* Dordrecht: Kluwer, pp. 163–82.

Kuokkanen, Martti and Tuomivaara, Timo. [1992]. On the Structure of Idealizations. *Poznań Studies in the Philosophy of Sciences and the Humanities* 25, pp. 67–102.

Lastowski, Krzysztof. [1982]. The Idealizational Status of Theoretical Biology. In: Wladyslaw Krajewski (Ed.). *Polish Essays in the Philosophy of Natural Sciences.* Dordrecht: D. Reidel, pp. 353–64.

Lewontin, Richard. [1984]. The Structure of Evolutionary Genetics. In: Elliot Sober (Ed.). *Conceptual Issues in Evolutionary Biology.* Cambridge, Mass.: The MIT Press, pp. 3–13.

Lloyd, Elizabeth. [1984]. The Semantic Approach to the Structure of Population Genetics. *Philosophy of Science* 51, pp. 242–64.

Lloyd, Elizabeth. [1988]. *The Structure and Confirmation of Evolutionary Theory.* Westport, CT: Greenwood Press.

Mayr, Ernst. [1991]. *One Long Argument.* Allen Lane: The Penguin Press.

Niiniluoto, Ilkka. [1986]. Theories, Approximations, and Idealizations. In: R.B. Marcus, G.J.W. Dorn, and P. Weingartner (Eds.). *Logic, Methodology and Philosophy of Science VII.* Amsterdam: North-Holland, pp. 255–89.

Nowak, Leszek. [1992]. The Idealizational Approach to Science: A Survey. *Poznań Studies in the Philosophy of Sciences and the Humanities* 25, pp. 9–63.

Recker, Doren. [1987]. The Structure of Darwin's Argument Strategy in the *Origin of Species. Philosophy of Science* 54, pp. 147–175.

Ruse, Michael. [1973]. *The Philosophy of Biology.* London: Hutchinson.

Ruse, Michael. [1979]. *The Darwinian Revolution: Science Red in Tooth and Claw.* Chicago: University of Chicago Press.

Sintonen, Matti. [1990]. Darwin's Long and Short Arguments. *Philosophy of Science* 57, pp. 667–89.

Sintonen, Matti. [1993a]. The Emergence of Evolutionary Theory: An Interrogative Proposal. Forthcoming.

Sintonen, Matti. [1993b]. From Principles to Theories to Models. In: M. Sintonen and S. Sirén (Ed.). *Evolutionary Theory: In Need of a New Synthesis.* Philosophical Studies from the University of Tampere.

Tuomi, Juha. [1981]. The Structure and Dynamics of Darwinian Evolutionary Theory. *Systematic Zoology* 30, pp. 22–31.

Tuomi, Juha. [1992]. Evolutionary Synthesis: A Search for the Strategy. *Philosophy of Science* 59, pp. 429–38.

Poznań Studies in the Philosophy of the Sciences and the Humanities
Vol. 42, pp. 217–241

Timo Tuomivaara

ON IDEALIZATION IN ECOLOGY

Introduction

There has been much dispute concerning the use of the method of idealization in ecology. Its use has been defended especially by those ecologists who — following the exemplar of mechanistic physical science — believe that the methods of physical sciences are applicable to ecology as well. On the other hand its use has been criticized by those who have abandoned the mechanistic approach on the grounds that it is incompatible with the holistic, unique, and historically changing nature of ecological entities.

My thesis is that all sciences, from the most theoretical to the most empirical and practical, need and use idealization and abstraction. This does not mean, however, that all sciences would neccesarily be commited to the mechanistic approach. I believe, that there exists a more viable intermediate position between the extremities of mechanism and anti-mechanism. I call it the approach of interactive particularism. It is the starting point in my analysis of the use of the method of idealization in ecological theory construction and testing.

The two important additional theses upon which I base my analysis are the realistic conception of theory and the theoryladenness of all data. On the basis of the first I separate sharply the questions concerning the theoretical and the empirical content of theory, respectively. On the basis of the second I state that all observing, measuring and experimenting in ecology is founded on the theoretical ideas concerning the processes of the data generation in question. These theories of data generation — like all other theories — are abstracting and idealizing their real objects.

I present theory construction as a process in which the theoretical content of theory is gradually developed and explicated by using the methods of idealization and concretization. Theory construction proceeding in this way results in theoretical models with variable degrees of generality, realism and accuracy.

Before the constructed theory can be tested empirically, its empirical content in some designed test situation must be specified. This is done by constructing a theory of data generation which connects the theoretical content of the theory with the observable or measurable quantities of the designed test. The construction of this theory proceeds gradually, too: The relevant properties of the process of data generation are defined step by step by using the methods of isolation, idealization, and concretization.

I
Idealization Controversy in Ecology

Use of idealizations in theory construction has been an established practice in physical sciences since the days of Galileo Galilei and Isaac Newton. This method is called by different names: *the method of analysis and synthesis* (Newton), *the method of resolution and composition of causes* (Mill), *the method of successive approximations* (Jevons), *the method of idealization and concretization* (Nowak). Its basic steps may be presented as follows:

First, the objects of research are analysed or reduced to their elementary parts with the help of experimental isolation and/or conceptual abstraction. Secondly, the basic laws and theories are formulated by using idealizations and abstractions: only the most essential parts of objects are taken into account, all others being idealized away as secondary. Thirdly, the behaviour of the complex whole is explained by combining relevant parts and their laws, i.e. by the method of synthesis, in which the basic theory of essential parts is supplemented or concretized by one secondary part after another. Fourthly the concretization of the basic theory is stopped when its predictions approximate sufficiently the observed behaviour of the object studied. (See the different formulations for example Newton [1704], Mill [1836 and 1844], Jevons [1879], Poincaré [1902], Such [1978], Nowak [1980], McMullin [1984].)

Mechanistic Approach in Ecology

When ecology was establishing itself as an autonomous field of biology in the second half of the nineteenth century, physiology was for many the only true form of biology. This physiological concept of biology was usually connected with a strongly mechanistic view of nature and science: Just as in the case of inorganic nature complexities of living nature are reducible to simpler parts or factors, following some universal, deterministic, and experimentally verifiable laws (Coleman [1982]). This point of view was adopted by many early ecologists, too. They practiced ecology following the exemplar of the mechanistic physical sciences. Later on in this century the mechanistic approach has been applied especially in ecophysiology, population ecology and system ecology (McIntosh [1985], Kingsland [1985]).

The most important ontological and methodological theses of the mechanistic approach in ecology may be formulated as follows:

M1) *Reductivism* or *atomism*: Causally complex wholes in nature are reducible to more simple or atomistic parts. The properties of these wholes are deducible from the properties of their parts.

M2) *Universalism*: Parts follow universal laws, i.e. laws which are not restricted in space or time. Wholes, being the aggregates of parts, follow resultant laws, which are also universal.

M3) *Methodological unity of sciences*: Ecology and biology use the same experimental, quantitative, mathematical, and analytical methods as the physical sciences do.

Anti-Mechanistic Approach in Ecology

If one foot of early ecology was in the physiological biology of the nineteenth century and in its mechanistic approach, its other foot was in the *naturalistic biology* of the eighteenth and nineteenth centuries and its anti-mechanistic approach. Naturalists, especially in their biogeographical studies, applied the descriptive and comparative methods of field biology instead of the quantitative and experimental methods admired by physiological biologists. Naturalists were impressed by the individuality, uniqueness, variability and wholeness in living nature. On this bases many of them questioned the validity of the mechanistic approach in biology. They believed that biology needs more holistic, qualitative or descriptive, and historical approach than that used in physical sciences (McIntosh [1985], Hagen [1989]).

The basic ontological and methodological theses of the anti-mechanistic approach may be formulated as follows:

AM1)   *Holism*: Organic nature consists of holistic wholes, which cannot be reduced to more simple atomistic parts. These wholes have emergent properties, which must be investigated on their own hierarchical level.

AM2)   *Singularism*: The variability of living nature is not merely noise or appearance hiding the immutable order of universal laws. It is an intrinsic property of living. Consequently the laws of living nature are not universal but singular, i.e. they are time and space bound.

AM3)   *Methodological autonomy of ecology*: In ecology the methods of physical sciences must be replaced or supplemented by more holistic, descriptive, comparative and qualitative methods. They are needed in ecology because of the holistic, unique, and historically changing nature of its entities.

According to holism, living nature is characterised by complex webs of interactions. While a physicist can analyze or reduce the complexities of inorganic nature into some simpler and more basic parts and study them in isolation, the ecologist is unable to do so. In the systems he is studing "everything is affecting everything else". This means, that the traditional analytical or reductive ways of scientific thinking practiced in physical sciences must be replaced in ecology by *the holistic way of thinking*, which does not deny the irreducible complexity and interrelatedness of nature.

According to singularism, defended recently by Mayr [1988] and Simberloff [1980] for instance, the variability of living nature is not merely noise or appearance that may and must be eliminated away by idealizing abstractions and/or experimental arrangements. Instead it is its intrinsic and irreducible property. It is cause and effect of its evolutionary origin. Consequently the generalizations made about living entities are not universal laws, "but simply facts about particular spatio-temporally localized objects" says Mayr [1988, 348]. This variability concerns also the basic mechanisms of inheritance, evolutionary chance etc. Because of this "there is hardly a theory in biology for which some exceptions are not known" [Mayr 1988, 19].

According to the anti-mechanistic view, ecology is not a generalizing science like physics, looking the individual cases in its domain only as instances of universal laws. Instead it is more like a historical science, or *scientific natural history*, describing and catologing singularities, individualities, and irregularities appearing in the historically and geographically changing nature.

Variations of Mechanistic and Anti-Mechanistic Approaches in Ecology

It is important to note that the mechanistic and anti-mechanistic approaches are applied with many variations in ecology. For example the theoretical, quantitative, and analytical ways of thinking, characterizing the approach of classical mechanism in physical sciences, have been supplemented or replaced in ecology often by descriptive, qualitative, and numerical simulation ways of thinking (Quinn and Dunham [1983], Levins [1966], Haila and Levins [1993], Jorgensen [1988]).

In addition, some variations combine elements from both approaches. One recent example of this is the system theoretical ecology. In some its formulations it combines holism from anti-mechanism with the ideals of universalism, mathematical analysis and quantitative accuracy, which are traditional elements of mechanism (L. von Bertalanffy [1968], E. Odum [1977], and B.C. Patten [1971], Onstad [1988]).

In this paper I cannot, however, go into the details of these variations and combinations. I limit myself to only a single comment: As can be seen, my above definitions of mechanism and anti-mechanism, doesn't take into account all these variations and combinations. But from my point of view in this paper this doesn't matter. It is sufficient that my definitions include the ontological and methodological theses commonly associated with mechanistic and anti-mechanistic approaches in ecology, and that they throw light on the deeper conceptual disagreements behind the idealization controversy in ecology.

Interactive Particularism: an Alternative

I defend the use of the method of idealization in ecology. I don't want, however, to deny the validity of criticism the holists and naturalists are directing against reductivism and universalism connected with this method in the physical sciences. Instead I argue that the validity of this method in ecology is not dependent on the correctness of the mechanistic approach.

I believe that there exists a more viable intermediate position between the extremities of mechanism and anti-mechanism. I call this intermediate position *interactive particularism*. Its basic ontological and methodological tenets may be formulated as follows:

IP1) *Interactivism*: There exist parts and there exist wholes. Wholes have emergent properties lacking their parts. The parts effect their wholes

and the wholes effect their parts. This means that there exist interactions between entities at different hierarchical levels.

IP2) *Particularism*: All living entities are time and space bound. But their
laws hold universally in all time and space regions where the relevant
conditions of their lawlike behaviours are fulfilled.

IP3) *Methodological plurality of ecology*: Ecology needs the analytical,
mathematical, quantitative and experimental methods of the mechanistic approach. But because its objects are historically changing
entities, it needs the descriptive, comparative, qualitative, and historical methods of the anti-mechanistic approach as well.

Interactivism in my view consists of the following ideas: I admit to
holism, that nature is not a heap of disconnected parts. It includes systems
of interconnected biotic and abiotic entities. These systems have a hierarchical organization, where the entities at one level are compounded into new
entities at the next higher level. Entities at a higher level have emergent
properties lacking in the entities at a lower level. In addition I assume that
the higher level entities can causally affect their lower level parts. This may
be called the assumption of *downward causation*. (Popper and Eccles
[1977], Bunge [1979], Mayr [1988].)

On the other hand I admit to reductivism, that nature is not a seamless
block or totality. It can be dissected into parts and subsystems because there
exist joints in its organization across which forces binding parts together are
weaker than elsewere. In addition, I admit that there exists also *upward
causation* from lower levels to higher levels. This means that an entity at a
higher system level can be partially explained by identifying its lower level
components and their interactions with one another and with entities in the
environment of the system. The partiality of this reductive explanation
follows from emergent properties at the higher levels (Bunge [1979]).

Particularism in my view consists of the following ideas: I admit to
singularism and naturalism, that living entities are time and space bound.
They are subject to variation and change produced by evolutionary forces
and random factors. Because of this, in the case of ecological entities the
existence of universal laws − laws not subject to any exceptions or limitations in their applicability − seems very unlikely. On the other hand I
admit to universalism, that searching for laws and patterns is essential in
ecology from its most basic forms to its more applied ones, and that, in
addition, true laws are universal.

My solution of this apparent dilemma is as follows: The lawlike behaviour of ecological entities at different hierarchical levels depends on their
structure and their environment. Of course if their structure and environ-

ment are changing, laws lose their validity or applicability in the sense that they are unable to describe correctly the behaviour of an entity in its changed conditions. But this doesn't deny nor exclude the possibility that the laws might have retained their validity or applicability if the relevant structural and environmental conditions had remained unchanged. Law statements, as such, do not include the assumption that the relevant conditions for the realization of laws exist or persist. This is an additional question, which concerns the applicability or testability of the law statement but not its validity as such. (Ereshefsky [1991], Wigner [1987].)

It is this more limited form of universalism, called here particularism, which can direct ecology in its search for laws in a living nature full of irregularities and historical contingencies. Ecological entities are particular systems in the sense that their existence is time and space bound. We must, of course, avoid the error of accepting as real such a contingency or singularity which in the end is only apparent. Particularism means, however, that we must take seriously the possibility that in the historical contingencies and singularities appearing in living nature there exists an irreducible remainder. In such cases the best and the most that ecologists can do is to describe the states and systems they are witnessing, that is: to do natural history. This, however, does not exclude the possibility of theoretical or generalizing ecology: The aim of generalizing ecology is to find, what is universal in the behaviour of such entities, existing in some particular conditions of time and space.

II
On the Methodology of Theory Construction and Testing

Standard Conception

I start my methodological comments from the standard conception, as it is called in the philosophy of science (Suppe [1977]). Its basic theses can be formulated in this connection as follows:

S1) *Empirical character of scientific theories*: Scientific theories have and must have clear empirical meaning or content. Theories lacking it are not included in science but in mathematics or metaphysics.

S2) *Neutrality of empirical data*: Empirical data, produced by observation, measurement and experiment, form a neutral and reliable foundation for science. Hypotheses and theories — on the contrary — are uncertain constructions of the human mind.

S3) *Theory construction by the method of induction and/or hypothesis*:
    Scientific theory construction is directed by data: Theories are induc-
    tive generalizations or summaries of data or hypotheses testable and
    tested by data.

The fundamental idea of the standard conception is that scientific theories are systems with empirical meaning or content, that is: they are able to
describe, explain or predict the observable behaviour of their objects. This
means that the law statements of a theory are interpreted as describing or
corresponding to universal regularities in the object's observable behaviour.

Disputes Concerning Theory Construction and Testing in Ecology

Much of the discussion concerning the adequacy of theory construction and
testing in ecology has revolved round the standard conception. Robert P.
McIntosh [1985, 245] writes that in the 1960's and 1970's many — especially in the fields of population and system ecology — viewed ecology as
finally achieving scientific status. The fulfilment of the requirements of the
standard conception was seen as a central point in this new, scientific
ecology.

Many ecologists, however, have critiziced the theoretical results of
population and system ecology as contradicting the standard conception
(Slobodkin [1965], Peters [1976], Simberloff [1980], Hall [1988], McIntosh
[1980]). Some of the critics have required that the rules of standard conception — especially its requirement of testability — should be carried out
more rigorously and consistently (Simberloff [1980, 1983]; Strong [1980,
1983], Hall [1988]). A more radical reaction has been to question the
adequacy of the hypothetical-deductive approach in ecology and to demand
a more inductive or empirical approach (Quinn and Durham [1983], Peters
[1976, 1980]).

A third reaction had been to defend the adequacy of theoretical ecology
on the grounds that the strict testability requirement is inappropriate in the
case of ecology (Levins [1966], Levin [1981], Caswell [1988]). Further,
some have reacted by retreating from the realistic interpretation of scientific
theories to the instrumentalistic one: theories in ecology should not be
viewed as truthful representations of their objects but merely as conceptual
devises or instruments useful in the systematizing of data (Jorgenssen
[1988]).

The latest reactions in this controversy have been based on the recent
anti- or postempirical approaches in the philosophy of science. As is well

known, the origin of these anti-empirical approaches is in the criticism which Thomas Kuhn [1962] launched against the standard conception at the beginning of the sixties. These critics of the standard conception state that scientific theories in general do not have any clear empirical content or meaning; that empirical data are not neutral but theoryladen and because of this uncertain and fallible just like theories are; and that theory construction is not directed by data but by research paradigms, consisting of ontological, methodological and theoretical assumptions and exemplars.

The most radical anti-empiricists − such as Feyerabend [1975], Barnes [1977], Bloor [1976], Collins [1985], Woolgar [1988] − have argued that the development of science cannot be analyzed at all on the basis of the classical realistic idea that there exists an independent outer reality confronting theories in empirical tests. Instead, the images of reality and the results of research must be seen as *social constructions* determined in the end by such unepistemic factors as ideological, aesthetical, and social interest. In ecology this radical anti-empiricism is supported by Fagerström [1987], for example.

Realistic Conception

I accept the critical points of anti-empiricists, without, however, abandoning the realistic conception of the scientific theories and tests. By utilizing and synthetizing the ideas of Lakatos [1971], Bunge [1967, 1970, 1985], Bhaskar [1975], Haré [1970], Haré and Madden [1975], Nowak [1980], Suppe [1989], Giere [1988] and Popper [1983] I formulate the following realistic alternative to the standard conception as well as to the radical anti-empiristic conception:

R1)   *Realistic interpretation of theory*: Scientific theories do not describe the world as it is observed by us, but as it is in itself and independently of us and of our observations.

R2)   *Importance of theories of data generation*: Because all data and tests are dependent on theoretical ideas, it is essential to develop these ideas into the form of a explicitly and systematically presented theory of data generation.

R3)   *Theory construction by the method of idealization and concretization in the context of the research paradigm*: The theories of research object and data generation are constructed by the method of idealization and concretization in the framework of the research paradigm.

In my realistic conception I make a sharp distinction between theoretical and empirical content of theory: The auxiliary statements needed in the specification of a theory's empirical content are not included in the essential structure of the theory. In it are included only those concepts and statements which are needed in the specification and development of its theoretical content. According to my realism, theories speak in primarily on the research object itself: on its supposedly real and essential properties. As such they don't necessarily say anything about how these supposedly real properties appear in the observations, measurements, and experiments of scientists.

Although I am critizing empiricism and the standard conception, I don't want to deny their requirement of empirical testability as such. I argue, however, that this requirement is formulated by them in a very misleading way for three reasons: First, it is not qualified with the necessary distinction between questions concerning the theoretical and the empirical content of theory. Consequently, autonomous theoretical thinking is often criticized as unscientific metaphysics, and the process of theory construction is seen predominantly as a process aiming at the specification of the empirical content of a theory.

Secondly it gives an oversimplified and misleading description of the process by which the empirical content of a theory is specified. It is not done − as the standard conception suggests − simply by adding to the basic statements of a theory some auxiliary statements (called operational definitions) which connect the variables of the theory with some observable quantities. It is a much more complicated process. For instance, in addition assumptions are needed which describe the nature and functioning of instruments used, give the other relevant factors affecting the behaviour of the quantities observed, fix the conditions of observation and experiment etc. I call the totality of all assumptions needed in the specification of the empirical content for a theory in a given test the theory of data generation. According to my view, a theory which in its basic form does not have any empirical content is given the form in which it can be empirically tested with the help of an auxiliary theory defining some relevant process of data generation.

Thirdly, it is connected with a questionable view, that data have an epistemological priority in relation to theory. I agree with the critics of the standard conception that the data are theoryladen etc. I disagree, however, when the radical anti-empiricists imply that empirical tests and data cannot − even in principle − have any qenuine empistemic role as evidence in questions of the truth and falsity of theory. From the theoryladenness and fallibility of data I conclude only that the attainment of data deserving this role or status as evidence may be and often is very difficult, and that because of this, the process of empirical testing must be based on a clearly formulated and well grounded theory of data generation.

Structuralistic Conception

There is much to recommend a structuralistic approach (Suppes [1959], Stegmuller [1976], Giere [1985], Suppe [1989], Balzer, Moulines and Sneed [1987], Kuokkanen and Tuomivaara [1992 and 1993]) with regard to the exact analysis of the content of a theory and its relation to data. But in some formulations (for instance Stegmuller [1976]), it is connected with theses which are incompatible with my realistic view of theories. There exist, however, some other formulations of structuralistic conception (Suppe [1989] and Giere [1985], for instance), which are more in accordance with it.

According to Giere [1985] a scientific theory consists of 1) a set of concepts and statements (expressed with words and sentences of some language), 2) of a set of models defined by these concepts and statements, and 3) of a set of statements, called theoretical hypotheses, which fix the scope of the intended applications of the models defined in this theory. When applied in my realistic framework, Giere's analysis says that during the process of theory construction models are defined which represent better and better the supposedly real objects under study. This view is explicated in more detail in the following chapter.

III
On the Method of Idealization
in the Construction of Ecological Theory

The construction of an ecological theory proceeds within the framework of a research paradigm consisting of more or less clearly formulated ontological, theoretical and methodological assumptions and exemplars. Its aim is to explicate what is the theoretical content of these framework assumptions and exemplars in the question under consideration. This is done step by step by using the methods of isolation, idealization, concretization, and instantiation.

Formulation of Conceptual Model

At the very beginning of the theory construction lies the process of conceptualization of the entity under consideration. On the bases of research problem and framework assumptions, ecologists draw the boundaries of their systems, i.e. isolate or identify them. The result of this process is a conceptual model for the object of study.

Conceptual models used often in ecological modelling represent objects under study as systems consisting of a set of compartments or subsystems with processes or flows between them, or in more general terms: of a set of elements and relationsphips between them. From the environment of the identified system the models usually mention some factors influencing the system (called forcing functions), and some others, influenced by it (called output functions) (Jorgenssen [1988], Botkin [1993], Beeby [1993]).

A conceptual model includes all elements and relations which, according to the ontological and theoretical framework assumptions, are supposed to be relevant. All others are omitted as irrelevant. This means that already the conceptualization or identification of the research object is based on idealizations and abstractions.

Formulation of Core Model

A core model is constructed by applying the method of idealization to the conceptual model. This results in a model representing only those elements and relations which — according to the ontological and theoretical assumptions — are the most essential in the set of all relevant elements mentioned in the conceptual model. In this connection it is usual that a more accurate verbal or mathematical formulation is given to the elemets choosen as essential.

This is the old strategy of analytical thinking of trying first the most simple models. There are several reasons, why ecologists follow this strategy, even if they — in many cases — have abandoned the mechanistic view. First, in some cases a simple model in terms of some basic elements may be sufficient. Secondly, there are cases where the limitations of our knowledge restrain us from going far beyond simple models.

Thirdly, the correct degree of the complexity may be unknown at the beginning of theory construction, or it may vary from one application of the theory to another. Forthly, there always exist an upper limit to the useful degree of complexity beyond which the added complecity does not improve the realism or accuracy of a model (Jorgenssen [1988]). All these are good reasons to start from as simple and general model as possible and to proceed to more complex ones only after it is shown as insufficient.

Formulation of Theoretical Model

A core model usually strongly simplifies and idealizes the research object. It takes into account only the most essential elements omitting all others. This, of course, makes the derivation of the object behaviour easier. But when the core model is applied in a situation where the omitted elements have significant effect, it must be specified or developed into the form of a theoretical model, which is able to represent these omitted elements as well.

The enrichment or development of a core model is done by the method of concretization or specification, that is: by adding, in the order of their importance, one omitted element after another to the core model. Under favorable conditions this procedure would result in a succession of theoretical models in which the latter models approximate the real properties and behaviours of the object studied better than the earlier ones. (Nowak [1980], Krajewski [1974], Lakatos [1970].)

Formulation of Special Case Model or Solution

A special case model or solution represents the effects which are generated or caused by the elements and relationships (i.e. by mechanisms) described in the theoretical model when these mechanisms are supposed to operate in some specified conditions. This model is developed from the theoretical model by supplementing it with parameter values and initial and boundary conditions characterizing these conditions and by making the calculations needed in order to solve the equations of the model.

Structure of Ecological Theory

Theory construction proceeds through different structural levels consisting of models with variable degrees of realism, generality, and accuracy. In the conceptual model, at the beginning of theory construction, the researcher formulates as realistic and general a picture of the research object as possible. It is, however, in most cases, too complicated for it to be formulated accurately enough. In the search for models with adequate degrees of accuracy, realism, and generality the content of the conceptual model is explicated further:

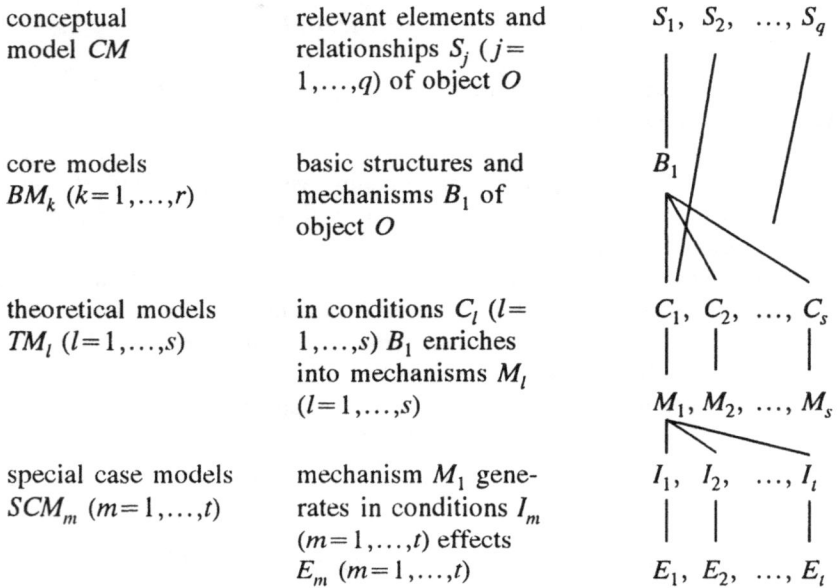

conceptual model *CM*	relevant elements and relationships S_j ($j=1,\ldots,q$) of object O	$S_1,\ S_2,\ \ldots,\ S_q$
core models BM_k ($k=1,\ldots,r$)	basic structures and mechanisms B_1 of object O	B_1
theoretical models TM_l ($l=1,\ldots,s$)	in conditions C_l ($l=1,\ldots,s$) B_1 enriches into mechanisms M_l ($l=1,\ldots,s$)	$C_1,\ C_2,\ \ldots,\ C_s$ $M_1,\ M_2,\ \ldots,\ M_s$
special case models SCM_m ($m=1,\ldots,t$)	mechanism M_1 generates in conditions I_m ($m=1,\ldots,t$) effects E_m ($m=1,\ldots,t$)	$I_1,\ I_2,\ \ldots,\ I_t$ $E_1,\ E_2,\ \ldots,\ E_t$

Fig. 1. Process of explicating the theoretical content of a theory.

The conceptual model is first simplified into the form of a core model. This model gives an accurate and general description of the most essential mechanisms of the research object. But because it omits all other relevant factors, its degree of realism is low. It does form, however, the most general level in the explicitly and accurately formulated structure of the theory.

Secondly, in order to attain models with better degrees of realism the basic model is concretized into the form of a theoretical model. It represents how the basic mechanisms of the research object are enriched and complicated in the different assumed conditions. Theoretical models form the second level in the explicitly formulated structure of a theory. The third level in this structure is defined by the special case models, instantiating a theoretical model with definitive parameter values and initial and boundary conditions.

An ideal of theory construction, especially in physical sciencies, has been that theories have a compact and hierarchical structure. The attainability of this ideal depends on 1) the existence of a theoretical framework, which is strong enough to unify different approaches of research adopted in the area; and 2) how coherently an adopted theoretical approach can be developed in its different applications. It is clear that if every theoretical model in the area inherits the most essential parts of its content from a core model common to all of them, then the resulting set of theoretical models has a very unified structure.

In ecology neither of these condition is generally fulfilled. For example, it is not uncommon that two ecologists facing a more or less similar problem begin from different theoretical backgrounds and formulate very different and incompatible conceptual, core or theoretical models of the entities under study: what is relevant or essential for one, is irrelevant or inessential for the other, and vice versa. In addition it is usual that in the development of the adopted theoretical approach its core model must be supplemented, modified and corrected so significantly in its different applications that the resulting set of theoretical models is broken into several subsets, more or less independent of one another.

Some Important Types of Idealization in Ecology

One important type concerns interactions between ecological entities at different hierarchical levels. When a system is described only at one hierarchical level its dependency on the entities at lower and higher levels are idealized away. A model generated by such idealizations may be viewed as a black box in two directions: in down- and upward directions. Many times, conceptual, core or theoretical models in ecology are heterogenous assemblies of subsystems, some treated as grey or white boxes, that is: by describing some of their dependencies down- and/or upwards, and some other treated as black boxes, that is: by describing merely their overall performance at one level.

Another important class of idealizations used in ecology concerns the variability and the time and space boundness, that is: the degree and nature of particularity of ecological entities. Concerning the question of particularity the researcher must answer two interrelated questions.

First, what is the extent of variation in the currently existing ecological entities? Usual assumption is that the variability of the properties studied is insignificant, limited, constant or homogenous in the entities belonging to the intended domain of application of a theory. In some cases this may be true or a good approximation. But in many cases it is shown to be an idealization.

The other question requiring an answer is: What is the stability of the environmental conditions and genetical structures on which the properties and the variability of currently existing ecological entities depend? Usual assumption is that these conditions are stable or remain unchanged or that they are changing so slowly that their effects can be omitted as insignificant. For some ecological entities and for some limited time spans and areas this is often true. But for other entities and for longer or wider stretches of time and space it is only idealization. Holdgate and Beament [1990, 412], for example, write on this question: "There is obviously something of a paradox here. The ecologist studies the interactions that are the very cause of evolution: in doing so he describes his system in terms of units which an understanding of evolution demonstrates are at best transitory".

Some of the idealizations or abstractions used in the formulation of ecological theories are concretized later in the process of theory construction. For example a black box in the original model may be concretized by replacing it with a grey or white box illuminating some entities at a lower or higher level affecting it. On the other hand a static box assumed in the original model as time and space independent is concretized by replacing it with a dynamic box specifying some of the the time and space dependencies in the behaviour of the box.

Verification of Theoretical Model

An important phase in ecological theory construction is model verification. According to one definition "a model is said to be verified, if it behaves in the way the model builder wanted it to behave" (Jorgenssen [1988, 40]). Model verification proceeds by testing or trying out a model in different situations and comparing the behaviour generated by the model to the expected behaviour. The methods used in the verification differ depending on whether the aim of the theory or model is 1) the qualitative modelling of an ecological entity or 2) its numerical simulation.

In qualitative modelling an important verification method used is called by Levins [1966] *robustness analysis*. Its aim is to check that the obtained result depends "on the essentials of a model" and not "on the details of the simplifying assumptions". This is done by applying the essential assumptions of the model with variable parameter values and simplifying assumptions. If irrespective of these variations the result is always qualitatively the same, "we have what we can call a robust theorem that is relatively free of the details of the model". Also, the robustness of the model's predictions shows that the qualitative correctness of the model is not jeopardized because of its ingnorance of the quantitative details of its object. (Wimsatt [1981], Kingsland [1985].)

In the approach of numerical simulation the starting models are usually much more complex and "realistic" than the basic models in the analytical approach. A central method used in their verification is called *sensivity analysis*. It is carried out by simulating the behaviour of the model over a range of conditions, defined by different values of relevant external variables and parameters intrinsic to the model. It tells how great a change in the value of a relevant state or output variable results when a change in the value of an external variable or an intrinsic parameter is made. As a result of this analysis the components in the relation to which the performance of the model is most sensitive or insensitive are identified. The former are included in the final model and the latter are elimated from it as superfluous. (Jorgensen [1988], Botkin [1993], Beeby [1993].)

Methods of Verification and the Method of Idealization and Concretization

I view the robustness and sensivity analyses as variants of the method of idealization and concretization. Irrespective of their differences they have this in common: In both cases a core model is tried out in different theoretical and special case models and the sensitivity of essential factors of the core model to the secondary factors and initial conditions mentioned in the theoretical models and their solutions are noted. This is like the process of concretization. But its bases and aims differ from the bases and aims of concretization described by Nowak [1980], for instance.

By using the terminology of Nowak [1980], we can say that the methods of verification are used in situations where the researchers doesn't have any clear or definite picture of "the essential structure" of the object or phenomena under study. In the case of qualitative modelling they have − in the face of the complexity of living nature − abandoned the aim of giving an accurate quantitative representation of this structure or of the behaviour of the object they are studing. Their aim is more modest: to show that their theories are able to identify some relevant "primary factors" or mechanisms in the essential structure of the object and to give qualitatively correct predictions of the behaviour of the object, even if the knowledge of the other primary or secondary factors in the essential structure of the object is uncertain, inaccurate, incomplete, or impossible.

In the approach of numerical simulation the aim of quantitative accuracy has been retained. But the researchers are ingnorant of which are the most important primary and secondary factors in the essential structure of the object. By applying sensitivity analysis to the hypothetical and alternative sets of primary and secondary factors, they try to test that all relevant factors are included in the model and all irrelevant ones are exluded from it. In this case the various concretizations applied to the original model may result in further idealizations (irrelevant or secondary factors are omitted) and/or concretizations (new relevant factors are included).

IV
On the Method of Idealization in the Testing of Ecological Theory

Data Generation in the Context of Testing

In the context of testing the aim of data generation is to test if the theoretical model corresponds to its object with the desired accuracy or not. This demands that the data be 1) *relevant*: it describes such features of the object which are connected with the processes and factors of the theoretical model; 2) *valid*: it gives correct and unbiased representations of these features; 3) *reliable*: it can be checked and repeated; and 4) *evidentially powerful*: it provides evidence for or against the theory. In other words, in the context of testing the researcher's aim is to design and implement a system of observation, measurement and/or experiment which is able to produce relevant, valid, reliable, and evidentially powerful data.

I believe, that the aim described above is attainable − especially in more complex test situations often encountered in ecology − only with the help of a clearly formulated and well grounded theory of data generation. Such a theory is needed in order to specify correctly the empirical content of the theory in the designed test. It connects the theoretical content of the theory with the quantities observed in the test. It tells which other systematic and random factors − in addition to the factors given by the theory − are in operation in the process of data generation. Further it tells how the conditions of observation and experiment have to be designed so that the systematic effects predicted by the theory would not be lost in the disturbing variation produced by other systematic and random factors in operation. (Views, more or less similar to my views on the nature and importance of the theoretical modelling of the system of data generation, can be found in Bunge [1970], Blalock [1984], Lakatos [1971], Spanos [1986], Suppes [1962 and 1977] and Suppe [1974 and 1989]).

The System of Data Generation

The system of data generation may be analysed into the three components: the object of experiment or data generation, consisting of the object of the theory connected with some observable or measurable conditions; the instrumentation, consisting of the instruments used in the observation and measurement; the experimental arrangements, consisting of the operations of control, blocking, homogenizing, matching, variation of treatment levels, randomization etc., and of the conditions of data generation produced by these operations.

A main problem in the design of the system of data generation is that every real system of data generation is open to disturbances. The source of these disturbances may be located: 1) in the object of data generation: the observable quantities in this object may react to some other factors in addition to the factors described in the theoretical model; 2) in the instrumentation: measured quantities may be affected by unintended factors besides the intended object of measurement; and 3) in the experimental arrangements: although a main task of the experimental arrangements is to minimize the effects of disturbing factors they themselves may generate errors as well.

There are three prinicipal strategies to deal with the problem of disturbing factors in data generation. 1) Controlling: a factor is controlled or eliminated by an experimental arrangement. 2) Measuring and introducing into analysis: a disturbing factor is taken into account, its effects are measured, and included in the analysis and in the models of data used as the basis of analysis. 3) Randomizing: The idea of randomization is to convert the uncontrolled and unmeasured disturbances into random effects which would leave the systematic effects predicted by a theoretical model discernable in the behaviour of the mean values of relevant quantities.

Conceptual Model of Data Generation

The theory of data generation describes or defines the system by which the data is generated. Its construction proceeds step by step towards a more and more accurate and detailed definition of the designed system of data generation.

In the first step the system of data generation is conceptualized or identified on the basis of the theory under test and all other relevant assumptions concerning the object, instruments and arrangements of data generation. This step results in a conceptual model of data generation, including all things assumed to be relevant in the designed system. This model, athough it is already simplifying a real system, is usually too complicated for exact formulation. So, in the search for a model which is able to describe the designed system with adequate degrees of accuracy, realism and generality, the conceptual model is processed further.

Core Model of Data Generation

In the second step the conceptual model is simplified or idealized into the form of a core model of data generation, which defines the object of data generation by using only the most essential measurable quantites of data generation. This model is more commonly known as the operational model.

The core model gives an idealized and simplified picture of the designed system: It represents only those of its properties, which are most essential from the point of view of the theoretical model under test; that is, properties which are connected with the factors or mechanisms of this model. It omits all other properties of the object, instrumentation, and arrangements of data generation. It describes an ideal experimental situation in which the behaviour of the object is seen without any disturbances of instrumentation or experimentation. The significance of this idealizing model of data generation is that it describe as clearly as possible, what is the empirical content of the theoretical model in terms of the observable quantities choosen.

Theoretical Model of Data Generation

The third step in the defining of the system of data generation is to fix the arrangements and instruments of data generation and to describe all the relevant new quantities they introduce into the core model of the designed system. This results in a specification or concretization of the core model of data generation and is called here the theoretical model of data generation.

A theoretical model of the designed system describes what if the observed behaviour of the object of data generation in the situations characterized by the arrangements of data generation, which generate such and such systematic and random disturbances in its behaviour, and by the instrumentation of data generation, which disturb the observation of its behaviour by such and such systematic and random errors of measurements.

Special Case Model of Data Generation

The fourth step in the design of data generation is to fix a particular situation in which the system defined in the theoretical model of data generation is to be applied and to describe the expected results to be produced by this system in this particular situation. This is done by adding to the theoretical model of data the parameter values and initial and boundary conditions characterizing some particular situation. This results in a special case model of data generation.

In a special case model of data generation the desired degrees of accuracy of results, the expected degrees of errors in measurements, the size of the experiment or sample, the levels or values of treatment variables and controlled boundary conditions etc. are fixed. When a theoretical model of data defines a whole class of data generations, a corresponding special case model of data identifies one member from this class.

Importance of the Theory of Data Generation

I admit that in the case of ecological research the multiple causality of the object under study cannot be decomposed as easily as in physical research into a handful of essential factors of the theory, the effects of which can clearly be seen in the data generated. The systematic factors given by an ecological theory are usually able to explain only a very limited part of the actual variation in the data. Thus, other systematic and random factors acting in the data generation are prone to produce vast amounts of unexplained variation. This situation has discouraged the theoretical approach in ecology in general and in the generation and analysis of ecological data in particular.

I think, however, that a partial solution to this problem is that the generation of ecological data and its statistical modelling and analysis be based on more clearly formulated and better grounded theories of data generation. With such theories it may be possible to reduce considerably the amount of unexplained variation and to bring the systematics of living nature more clearly into view. (See Hari, Häkkinen, Rita and Tuomivaara [1995], for a more detailed analysis of the problems of testing, statistical analysis and inference in ecology.)

V
Conclusion

I believe that the methodological approach outlined in this paper — beginning with interactivistic and particularistic ontology, and proceeding through the realistic conception of theory to the critical analysis of theoryladenness in data — is able to clarify many of those problems which have confronted theory construction and testing in ecology, and which have divided ecologists into deeply disagreeing factions with regard to the possibility and utility of idealizing theoretical thinking in ecology.

Department of Philosophy
University of Helsinki
P. O. Box 24
00014 University of Helsinki, Finland

REFERENCES

Balzer, W., Moulines C.-U. & Sneed, J.D. [1987]. *An Architectonic for Science. The Structuralist Program.* Dordrecht/Boston/London: Reidel.

Barnes, B. [1977]. *Interests and the Growth of Knowledge.* London: Routledge & Kegan Paul.

Bloor, D. [1976]. *Knowledge and Social Imagery.* London: Routledge & Kegan Paul.

Beeby, A. [1993]. *Applying Ecology.* London/Glasgow/New York: Chapman & Hall.

Bertalanffy, L. von [1968]. *General System Theory. Foundations, Development, Applications.* New York: George Braziller.

Bhaskar, R. [1975]. *A Realist Theory of Science.* Leeds/Bristol: Leeds Books Ltd.

Blalock, Jr., H. [1984]. *Basic Dilemmas in the Social Sciences.* Baverly Hills/London/New Delhi: Sage Publications.

Botkin, D.B. [1993]. *Forest Dynamics. An Ecological Model.* Oxford/New York: Oxford University Press.

Bunge, M. [1967]. *Scientific Research I—II.* New York: Springer-Verlag.

Bunge, M. [1970]. Theory Meets Experiment. In: H. Kiefer and M. Munitz (Eds.). *Mind, Science, and History.* New York, pp. 138—165.

Bunge, M. [1979]. *Treatise on Basic Philosophy. Vol. 4. Ontology II: A World of Systems.* Dordrecht/Boston/London: Reidel.

Bunge, M. [1985]. *Treatise on Basic Philosophy. Vol. 5. Epistemology & Methodology I: Exploring the World.* Dordrecht/Boston/London: Reidel.

Caswell, H. [1988]. Theory and Models in Ecology: A Different Perspective. *Ecological Modelling*, **43**, 33—44.

Coleman, W. [1982]. *Biology in the Nineteenth Century.* Cambridge/London: Cambridge University Press.

Collins, H. [1985]. *Changing Order: Replication and Induction In Scientific Practice.* Beverly Hills: Sage.

Ereshefsky, M. [1991]. The Semantic Approach to Evolutionary Theory. *Biology and Philosophy,* 6, 59–80.

Fagerström, T. [1987]. On Theory, Data and Mathematics in Ecology. *Oikos,* 50, 258–261.

Feyerabend, P. [1975]. *Against Method.* London: New Left Books.

Giere, R.N. [1988]. *Explaining Science. A Cognitive Approach.* Chicago/London: The University of Chicago Press.

Hagen, J. [1989]. Research Perspectives and the Anomalous Status of Modern Ecology. *Biology and Philosophy,* 4, 433–455.

Haila, Y. and Levins, R. [1992]. *Humanity and Nature. Ecology, Science and Society.* London: Pluto Press.

Hall, C.A.S. [1988]. An Assessment of Several of the Historically Most Influential Theoretical Models Used in Ecology and of the Data Provided in their Support. *Ecological Modelling,* 43, 5–32.

Haré, R. [1970], *The Principles of Scientific Thinking.* London: Macmillan.

Haré, R. and Madden, E.H. [1975]. *Causal Powers.* Oxford: Blackwell.

Hari, P., Häkkinen, R., Rita, H. and Tuomivaara, T. [1995]. Guide-Dog Approach: A Methodology for Quantitative Ecology. Forthcoming.

Jevons, W.S. [1879]. *The Principles of Science: A Treatise on Logic and Scientific Method.* London: Macmillan and Co.

Jorgenssen, S.E. [1988]. *Fundamentals of Ecological Modelling.* Amsterdam/Oxford/New York/Tokyo: Elsevier.

Kingsland, S.E. [1985]. *Modeling Nature. Episodes in the History of Population Ecology.* Chicago/London: The University of Chicago Press.

Kitching, R.L. [1983]. *Systems Ecology.* St Lucia: University of Queensland Press.

Krajewski, W. [1974]. *Correspondence Principle and the Growth of Knowledge.* Dordrecht/Boston/London: Reidel.

Kuhn, T. [1962]. *The Structure of Scientific Revolutions.* Chicago: University of Chicago Press.

Kuokkanen, M. and Tuomivaara, T. [1992]. On the Structure of Idealizations. In: J. Brzeziński and L. Nowak (Eds.). *Idealization III: Approximation and Truth.* Poznań Studies in the Philosophy of the Sciences and the Humanities, 25, Amsterdam/Atlanta, GA: Rodopi, pp. 67–102.

Kuokkanen, M. and Tuomivaara, T. [1993]. The Threshold Model of Scientific Change and the Continuity of Scientific Knowledge. *Journal for General Philosophy of Science,* 115.

Lakatos, I. [1970]. Falsification and the Methodology of Scientific Research Programs. In: I. Lakatos and A. Musgrave (Eds.). *Criticism and the Growth of Knowledge,* Cambridge: Cambridge University Press, pp. 91–196.

Levin, S. [1981]. The Role of Theoretical Ecology in the Description and Understanding of Populations in Heterogeneous Environments. *American Zoologist,* 21, 865–875.

Levins, R. [1966]. The Strategy of Model Building in Population Biology. *American Scientist,* 54, 421–431.

240

Mayr, E. [1988]. *Toward a New Philosophy of Biology. Observations of an Evolutionist.* Cambridge/Massachusetts/ London: Harvard University Press.

McIntosh, R. [1980]. The Background and Some Current Problems of Theoretical Ecology. In: E. Saarinen (Ed.). *Conceptual Issues in Ecology*, Dordrecht/Boston/London: Reidel, pp. 1–62.

McIntosh, R. [1985]. *The Background of Ecology: Concept and Theory.* New York: Cambridge University Press.

McMullin, E. [1985]. Galilean Idealization. *History and Philosophy of Science*, **16**, 247–273.

Mill, J.S. [1836]. On the Definition of Political Economy; and on the Method of Investigation Proper to it. In: *Collected Works of John Stuart Mill, Vol. IV.* London/Toronto: Routledge & Kegan Poul and University of Toronto Press, 1967, pp. 309–339.

Mill, J.S. [1844]. *A System of Logic.* London: Longmans, Green and Co.

Newton, I. [1704]. *Optics; Or a Treatise of the Reflections, Refractions, Inflections, and Colour of Light.* London: Printed for S. Smith and B. Walford.

Nowak, L. [1980]. *The Structure of Idealization. Towards a Systematic Interpretation of the Marxian Idea of Science*, Dordrecht/Boston/London:Reidel.

Odum, E. [1977]. The Emergence of Ecology as a New Integrative Discipline. *Science*, **195**, 1289–1293.

Onstad, D.W. [1988]. Population-Dynamics Theory: the Roles of Analytical, Simulation and Supercomputer Models. *Ecological Modelling*, **43**, 111–123.

Patten, B.C. [1971]. *System Analysis and Simulation in Ecology. Vol. I.* New York: Academic Press Inc.

Peters, R. [1976]. Tautology in Evolution and Ecology. *American Naturalist*, **110**, 1–7.

Peters, R. [1980]. Useful Concepts for Predictive Ecology. In: E. Saarinen (Ed.). *Conceptual Issues in Ecology*, Dordrecht/Boston/London: Reidel, pp. 215–227.

Poincaré, H. [1902]. Science and Hypothesis. In: H. Poincaré. *The Foundations of Science.* Lancaster, PA.: The Science Press, 1913 and 1946.

Popper, K. [1983]. *Realism and the Aim of Science.* Totowa, New Jersey: Rowman and Littlefield.

Popper, K. and Eccles, J. [1977]. *The Self and Its Brain.* New York: Springer-Verlag.

Quinn, J. and Dunham, A. [1983]. On Hypothesis Testing in Ecology and Evolution. *American Naturalist*, **122**, 602–617.

Simberloff, D. [1980]. A Succession of Paradigms in Ecology: Essentialism to Materialism and Probabilism. In: E. Saarinen (Ed.). *Conceptual Issues in Ecology*, Dordrecht/Boston/London: Reidel, pp. 63–99.

Simberloff, D. [1983]. Competition Theory, Hypothesis-Testing, and Other Community Ecological Buzzwords. *American Naturalist*, **122**, 626–635.

Slobodkin, L.B. [1965]. On the Present Incompleteness of Mathematical Ecology. *American Scientist*, **53**, 347–357.

Spanos, A. [1986]. *Statistical Foundations of Econometric Modelling.* Cambridge/London/New York: Cambridge University Press.

Stegmüller, W. [1976]. *The Structure and Dynamics of Theories.* New York: Springer.

Strong, D. [1980]. Null Hypotheses in Ecology. In: E. Saarinen (Ed.). *Conceptual Issues in Ecology*, Dordrecht/Boston/London: Reidel, pp. 245–259.

Strong, D. [1983]. Natural Variability and the Manifold Mechanisms of Ecological Communities. *American Naturalist*, **122**, 636—660.

Such, J. [1978]. Idealization and Concretization in Natural Sciences. *Poznań Studies in the Philosophy of the Sciences and the Humanities*, **4**. Amsterdam: Gruner, 49—73.

Suppe, F. [1974]. Theories and Phenomena. In: W. Leinfellner and E. Köhler (Eds.). *Developments in the Methodology of the Social Sciences*. Dordrecht: Reidel, pp. 45—92. (Also in: Suppe [1989].)

Suppe, F. [1977]. *The Structure of Scientific Theories*. 2nd ed. Urbana: University of Illinois Press.

Suppe, F. [1989]. *The Semantic Conception of Theories and Scientific Realism*. Urbana and Chicago: University of Illinois Press.

Suppes, P. [1957]. *Introduction to Locig*. New York: Van Nostrand.

Suppes, P. [1962]. Models of Data. In: E. Nagel, P. Suppes and A. Tarski (Eds.). *Logic, Methodology and Philosophy of Science: Proceedings of the 1960 International Congress*. Stanford, Calif.: Stanford University Press, pp. 252—261.

Suppes, P. [1977]. The Structure of Theories and the Analysis of Data. In: F. Suppe [1977], pp. 266—283.

Wigner, E. [1987]. The Unreasonable Effectiveness of Mathematics in the Natural Sciences. In: J.H. Weaver (Ed.). *The World of Physics*. New York: Simon and Schuster, pp. 82—96.

Wimsatt, W. [1981]. Robustness, Reliability and Multiple Determination in Nature. In: M. Brewer and B. Collins (Eds.). *Knowing and Validating in the Social Sciences: A Tribute to Donald T. Campbell*. San Francisco: Jossey-Bass, pp. 124—163.

Woolgar, S. [1988]. *Science: The Very Idea*. London and New York: Tavistock Publications and Ellis Horwood Limited.

Poznań Studies in the Philosophy
of the Sciences and the Humanities
Vol. 42, pp. 243–268

Martti Kuokkanen and Matti Häyry

EARLY UTILITARIANISM AND ITS IDEALIZATIONS
FROM A SYSTEMATIC POINT OF VIEW

Introduction

We shall reconstruct some basic ideas of Early Utilitarian Theorizing using
the framework of the structuralist theory of science. We first divide the core
assumptions of Early Utilitarianism into three groups:

the Greatest Happiness Principle,
the Impartiality Principle and
the assumptions on the quantitative and qualitative aspects of pleasures.

[For more details, see Häyry 1994.] Historically several substantially
different versions of the Greatest Happiness Principle have been put forth.
In the paper we consider only two of them explicitly, namely Strict Universal Altruism [Cumberland 1672, Berkeley 1712] and Classical Utilitarianism
[Hutcheson 1725, Bentham 1789]. A third one, called here Total Utilitarianism [Hutcheson 1725, Hartley 1749, Bentham 1789, Godwin 1793, J.S.
Mill 1861] is only mentioned here due to the fact that it seems to presuppose an *aggregated* happiness-concept which in turn is the focus of *modern
welfare theorizing*.

There are also several formulations of the Impartiality Principle. In this
paper we formulate two versions of impartiality, the weak and the strong
one. They reflect at least partially the views of Godwin [1793] and J.S. Mill
[1861]. Concerning the assumptions of the qualitative and quantitative
aspects of pleasures we study four different positions, among them those of
Bentham and J.S. Mill. To be sure, there are other possible positions.
However, we shall not extend our study with respect to them.

Our analysis shows that at a *very general level* the three core assumptions of Early Utilitarianism constitute three *mutually independent (and*

compatible) sets of assumptions. They make it possible to *specialize* the core assumptions in the *structuralist sense*. Using the structuralist terminology it can be said that there are two basic theory-elements, a *quantitative* and a *qualitative* one, and both constitute a theory-net. In particular, we shall show in detail that there exists a quantitative theory-net of utilitarian theorizing whose elements constitute a *lattice*. A completely analogous result can be shown for qualitative utilitarian theorizing.

We analyse the idealizations of utilitarian theorizing at two levels. Using the Poznań School Theory of Idealization it is shown that three of the four positions with regard to the relation of the qualitative and quantitative aspects of pleasures contain idealizing assumptions. Moreover, it is shown that *applying* utilitarian theorizing via a *theory of hierarchies of human needs* for example, presupposes idealizations. Using the Poznań School Theory of Idealization the *general form* of such rather concrete idealizations can be crystallized. It turns out that the idealizations of the qualitative and quantitative aspects of pleasures are special, stronger cases of the general form of idealizations, presupposed in applications of utilitarian theorizing. To be sure, there are other idealizations contained in utilitarian theorizing, for example assumptions connected with the "measurability" of pleasures. They are, however, outside the scope of this paper.

The Core Assumptions of Early Utilitarianism

The core assumptions or axioms of Early Utilitarianism can be divided into three groups: the Greatest Happiness Principle, the Impartiality Principle and the assumptions on the relation of quantitative and qualitative aspects of happiness. In fact there are several substantially different versions of each principle. In particular, the Greatest Happiness Principle has several formulations.

We start from the following nominal definition: Happiness equals net pleasure (= "pleasure−pain"). Using this definition, at least four different formulations of the Greatest Happiness Principle in Early Utilitarianism emerge:

(P1) Morally right act, omission etc. maximize the net pleasure for each sentient human being.

(P2) Morally right act, omission etc. maximize the net pleasure for the greatest possible set of sentient human beings.

(P3) Morally right act, omission etc. maximize the number of sentient human beings who may utilize the pleasure.

(P4)   Morally right act, omission etc. produce the greatest (sum of) pleasure for the greatest set of sentient human beings.

Principle (P1) constitutes the core of Strict Universal Altruism [Cumberland 1672, Berkeley 1712], principles (P2) or (P3) constitute Classical Utilitarianism [Hutcheson 1725, Bentham 1789] and principle (P4) constitutes a utilitarian doctrine, called here Total Utilitarianism [Hartley 1749, Hutcheson 1725, Bentham 1789, Godwin 1793 and Mill 1861]. Principle (P4) seems to presuppose an *aggregated* concept of pleasure contrary to principles (P1) − (P3) and for this reason we will not consider any version of utilitarianism which contains (P4).

Consider next the Impartiality Principles. At least the following two can be differentiated, but they seem to be substantially the same. However, the principles have at least two formally different explicates as will be seen later:

(I1)   Everybody should be taken equally into account in pleasure calculations.

(I2)   Everybody to count for one, and nobody to count for more than one.

The latter comes from J.S. Mill [1861] and, for example, Godwin [1793] strongly emphasizes the former.

Concerning the assumptions on the qualities and quantities of pleasure, at least four different positions can be differentiated:

(Q1)   Pleasures contain only quantitative differences.

(Q2)   Pleasures contain both quantitative and qualitative differences, but only quantitative differences are relevant.

(Q3)   Pleasures contain both quantitative and qualitative differences, and both are relevant.

(Q4)   Pleasures contain both quantitative and qualitative differences, but only qualitative differences are relevant.

For example, Bentham is a representative of (Q2), and J.S. Mill is a representative of (Q3). *Prima facie* principles (Q1) − (Q4) contradict one another. However, we shall show that there are cases where principles (Q1) − (Q4) plus all versions of principles (P) and (I) yield the *same* act, omission etc. as a morally right act or omission, i.e., any combination of principles selected from P, I and Q is compatible in this sense.

The Basic Concepts

In the following we first define the needed concepts formally. Then we axiomatize several versions of Early Utilitarianism and study their intra- and intertheoretic properties.

(1) S is a non-empty finite set of sentient human beings.

(2) A is a finite set of acts, omissions, rules, laws, policies, reforms etc. which are available in a given situation, such that for *every* $a_i \in A$ *there is at least one* $s_j \in S$, and (some of) the consequences of a_i have effects of the happiness of s_j.

(3) $Ar \subseteq A$, $Ar \neq \emptyset$ is the set of realized acts, omissions etc.

Let $\mathbb{R}$ be a suitable set of real numbers. Then

(4) P is a function of form $A \times S \rightarrow \mathbb{R}$.
P is called a *net pleasure function*.
The case $P(<a_i,s_j>)=0$ denotes the case where act, omission etc. a_i does not have any effect on the happiness of s_j.

Let $\mathbb{R}^+$ be a suitable set of positive real numbers. Then

(5) W is a function of form $A \times S \rightarrow \mathbb{R}^+$.
W is called a *weight function*.

Let $\mathbb{N}=\{1,2,...\}$ be a suitable finite set of natural numbers. Then

(6) Q is a function of form $A \times S \rightarrow \mathbb{N}$.
Q is called a *quality function*.

An Axiomatization of Strict Universal Altruism

We axiomatize three versions of Strict Universal Altruism. Consider principle (P1). Using the formal concepts defined above it would read as

(PA1) An act, ... $a \in A$ is morally right if and only if for all $a' \in A$, for all $s \in S$: $P(<a,s>) \geq P(<a',s>)$.

Principles (I1) and (I2) can be formalized in two ways:

(IA1)   For all $a,a' \in A$, for all $s,s' \in S$:
$$W(<a,s>) = W(<a',s'>),$$

and

(IA2)   For all $a \in A$, for all $s,s' \in S$:
$$W(<a,s>) = W(<a,s'>).$$

The former implies the latter. The latter says "the consequences of actions for any individuals together are taken equally into account although the weights of the consequences of actions for individuals in themselves may vary". The stronger version states individuals in the same position relative to one another, as does the weaker one. However, it excludes the possibility that the weights of the consequences of actions in themselves may vary.[1]

We next define some auxiliary concepts which make it possible to combine axioms (PA1) and (IA1) or (IA2).

D1 $A_S = \{a \mid a \in A$ and for all $a' \in A$, for all $s \in S$:
$$W(<a,s>) \cdot P(<a,s>) \geq W(<a',s>) \cdot P(<a',s>)\}$$
is the *set of morally right actions in Strict Universal Altruism*.

D2 $W_W = \{z \mid z \in W$ and for all $v,v' \in D_I(W)$: if
$$v = <a,s>, \ v' = <a,s'> \text{ then } W(v) = W(v')\}$$
is the *set of weaker impartiality weight functions*.

D3 $W_S = \{z \mid z \in W$, and for all $v,v' \in D_I(W)$: $W(v) = W(v')\}$
is the *set of stronger impartiality weight functions*.

Next we define three versions of Strict Universal Altruism (SUA). The first is SUA without impartiality, the second and the third are SUA with weak and strong impartiality.

D4   Let S, A, P, W, Q, Ar satisfy the definitions $(1)-(6)$ of the basic concepts. Then
(1) $M(SUA) = \{<S,A,P,W,Q,Ar> \mid Ar \subseteq A_S\}$
is the *set of models of Strict Universal Altruism without impartiality*,

---

1 In fact there is logically a third alternative to formalize the impartiality principle, namely
For all $a,a' \in A$, for all $s \in S$: $W(<a,s>) = W(<a',s>)$.
It says that "the consequences of any actions are taken equally into account for any individuals *taken separately*", i.e., the formalization allows the case where the consequences of a given action for separate individuals are not taken equally into account. For this reason it seems that the third formalization does not contribute much in the proper sense of impartiality.

(2) $M(\text{SUA})_W = \{ <S,A,P,W,Q,Ar> \mid Ar \subseteq A_S \text{ and } W \subseteq W_W \}$
is the *set of models of Strict Universal Altruism with weak impartiality*,

(3) $M(\text{SUA})_S = \{ <S,A,P,W,Q,Ar> \mid Ar \subseteq A_S \text{ and } W \subseteq W_S \}$
is the *set of models of Strict Universal Altruism with strong impartiality*.

The motivation and adequacy of D4 should be obvious.[2]

Theorem 1: $M(\text{SUA})_S \subset M(\text{SUA})_W \subset M(\text{SUA})$.
Proof: Theorem follows trivially from definitions D1 – D3.
Theorem 2: Strict Universal Altruism with strong impartiality is consistent.
Proof: We have to show that $M(\text{SUA})_S$ is not empty. We construct a simpliest possible model which proves the theorem.

---

2 Related to the "orthodox" structuralist programme [cf. Balzer, Moulines and Sneed 1987] our approach is rather heavily simplified. We consider neither constraints nor the sets of partial potential models (and consequently nor the theoreticity-problem) although they may be relevant, too. The *set of potential models*, presupposed by the sets of models of different utilitarianisms are taken to consist of structures of the form

$z = <S,A,P,W,Q,Ar>$

where the components satisfy the defining clauses of the basic concepts.

Strictly speaking any *purely quantitative* utilitarian theory contains no reference to the qualities of pleasures, nor does any *purely qualitative* utilitarian theory refer to quantities of pleasures. Consequently the *set of potential models* of *purely quantitative* utilitarian theorizing consists of structures of the form

$z' = <S,A,P,W,Ar>$

and the *set of potential models* of *purely qualitative* utilitarian theorizing consists of structures of the form

$z'' = <S,A,W,Q,Ar>$.

Thus, the related structures are of set-theoretically different form. However, both structures of form z' and structures of form z'' are *sub-structures (reducts)* of structures of form z [cf. Balzer, Moulines and Sneed 1987]. Due to the fact that any purely quantitative utilitarian theory *in no way restricts* the quality-aspect of pleasures (Q-component) relative to the quantity-aspect of pleasures (P-component) in a given structure z, any *model* $z' = <S,A,P,W,Ar>$ of any purely quantitative utilitarian theory can be *expanded* to a structure $z = <S,A,P,W,Q,Ar>$, which again is a structure where the relevant assumptions are satisfied, i.e., z is a model of any purely quantitative utilitarian theory if z' is. The same holds for the relation of a *model* z'' of any purely qualitative utilitarian theory and its *expansion relative to the P-component*. Thus, the *mutual independence* of purely quantitative and qualitative utilitarian theorizing, as shown later in the text, makes it completely possible to fix the set of potential models in the way we have fixed it.

Consider a structure $y=\langle S,A,P,W,Q,Ar\rangle$ with $S=\{s\}$, $A=\{a\}$, $P=\{\langle\!\langle a,s\rangle,+1\rangle\}$, $W=\{\langle\!\langle a,s\rangle,+1\rangle\}$, $Q=\{\langle\!\langle a,s\rangle,+1\rangle\}$ and $Ar=\{a\}$. Condition $Ar\subseteq A_S$ is trivially satisfied in y (see D1). Also condition $W\subseteq W_S$ is trivially satisfied in y (see D3). Thus, y is a model of SUA.

Theorem 3: Condition $Ar\subseteq A_S$ is independent of conditions $W\subseteq W_S$ and $W\subseteq W_W$ and vice versa.

Proof: Note first that relation $W_S\subset W_W$ holds due for definitions D2 and D3. So, to show that conditions $Ar\subseteq A_S$ and $W\subseteq W_W$ are independent is sufficient to demonstrate the claim.

Consider a structure $y=\langle A,S,P,W,Ar\rangle$ with $A=\{a,a'\}$, $S=\{s,s'\}$, $P=\{\langle\!\langle a,s\rangle,+1\rangle$, $\langle\!\langle a',s\rangle,0\rangle$, $\langle\!\langle a,s'\rangle,+1\rangle$, $\langle\!\langle a',s'\rangle,0\rangle\}$, $W=\{\langle\!\langle a,s\rangle,+1\rangle$, $\langle\!\langle a',s\rangle,\tfrac12\rangle$, $\langle\!\langle a,s'\rangle,+1\rangle$, $\langle\!\langle a',s'\rangle,\tfrac12\rangle\}$ and let $Ar=\{a\}$. Then $\{a\}\subseteq A_S$ and $W\subseteq W_W$ hold in y.

Replace W in y by $W'=\{\langle\!\langle a,s\rangle,+1\rangle$, $\langle\!\langle a',s\rangle,\tfrac12\rangle$, $\langle\!\langle a,s'\rangle,\tfrac12\rangle$, $\langle\!\langle a',s'\rangle,\tfrac12\rangle\}$. A structure y' results such that $\{a\}\subseteq A_S$ holds in y' but $W'\subseteq W_W$ does not.

Replace P in y by $P'=\{\langle\!\langle a,s\rangle,+1\rangle$, $\langle\!\langle a',s\rangle,0\rangle$, $\langle\!\langle a,s'\rangle,0\rangle$, $\langle\!\langle a',s'\rangle,+1\rangle\}$. A structure y'' results such that $W\subseteq W_W$ holds in y'' but $\{a\}\subseteq A_S$ does not.

So far we have shown that there are three versions of Strict Universal Altruism if the impartiality principle is used in reconstructing SUA. In fact SUA is really very strong. We illustrate its content by the following example after Godwin [1793]. The example is typically used to illustrate a radical version of impartiality principle. However, in the following we show that the example is incompatible with principle (P1), which obviously is the core of SUA.

The example goes as follows: Suppose that the palace of Archbishop Fénelon (a known benefactor) is on fire and only one person, either the archbishop or his valet, can be saved. Then justice would demand that Fénelon ought to be saved. The reason Godwin gave for the resolution is simply that the learned benefactor and reformer would be more useful to other people than the humble valet.

Consider now the example from the point of view of the present formalism. Let S be a finite set of individuals, $F=$ Fénelon and $V=$ the valet among them. The set of available actions would become $A=\{a,a'\}$, a meaning that Fénelon is saved, the valet not; a' meaning that the valet is saved, Fénelon not. The Q-function is irrelevant and the related W-function should satisfy at least the weaker impartiality principle $W\subseteq W_W$ because Godwin strongly emphasized impartiality. Thus, W satisfies conditions for all $s,s' \in S$: $W(<a,s>) = W(<a,s'>)$ and $W(<a',s>) = W(<a',s'>)$. For simplicity denote $W(<a,s>) = W(<a,s'>)$ by w^* and $W(<a',s>) = W(<a',s'>)$ by w^{**}.

Consider next the related P-function. The argument obviously presupposes

(1) $w^*(P(<a,s>)) \geq w^{**}(P(<a',s>))$ for all $s \in S$, $s \neq F$, $s \neq V$.

On the other hand, looking the situation from the point of view of the valet, the following obviously holds:

(2) $w^{**}(P(<a',V>)) > w^*(P(<a,V>))$.

But result (2) shows that $\{a\} \nsubseteq A_S$ because $V \in S$. An analogous argument relative to Fénelon shows that $\{a'\} \nsubseteq A_S$. Thus, neither a nor a' is a morally right action in SUA.

An Axiomatization of Classical Utilitarianism

We next characterize Classical Utilitarianism (CU) along the same lines as SUA above. We first show that principles (P2) and (P3) are independent of one another. Taking their conjunction characterizes *at least partially* the core of Classical Utilitarianism.[3] Adding the weaker and stronger versions of the impartiality principle to the conjunction yields three versions of Classical Utilitarianism analogously to SUA. We are then able to prove some results for CU, analogous to the results of SUA. Finally, we show that SUA implies CU.

Consider principles (P2) and (P3). Their formalizations would read as follows:

D5   Let $a \in A$. Then
    (1) $g(a)$ denotes the number of individuals in S for which action a maximizes individuals' W-weighted net pleasure;

---

3 The independence of principles (P2) and (P3) combined with impartiality assumptions opens *six different ways* to characterize Classical Utilitarianism from a systematic point of view. For simplicity's sake we consider here only the conjunction of principles (P2) and (P3).

(2) $g'(a)$ denotes the number of individuals in S for which action a gives a positive, non-zero value of individuals' W-weighted net pleasure function.

In clause (2) the predicate "may utilize the pleasure" occurring in (P3) is interpreted as "a positive, non-zero value of the W-weighted P-function". There are certainly other alternatives, however, the selected one seems to be the most simple and "natural". In any case, nothing essential depends on this stipulation.

Using D5 we formalize (P2) and (P3) as follows:

(PA2)   An act $a \in A$ is morally right if and only if for all $a' \in A$: $g(a) \geq g(a')$;

(PA3)   An act $a \in A$ is morally right if and only if for all $a' \in A$: $g'(a) \geq g'(a')$.

Theorem 4: (PA2) and (PA3) are independent.

Proof: Consider a structure $y = <S,A,P,W,Ar>$ with $S = \{s_1,s_2,s_3\}$, $A = \{a,a'\}$, $P = \{\ll a,s_1>,+1>, \ll a,s_2>,+1>, \ll a,s_3>,0>, \ll a', s_1>,\frac{1}{2}>, \ll a',s_2>,\frac{1}{2}>, \ll a',s_3>,\frac{1}{2}>\}$, $W(<a,s>) = +1$ for all $<a,s> \in A \times S$ and $Ar = \{a\}$. Principle (PA2) is satisfied in y because $g(a) = 2 > g(a') = 1$. Principle (PA3) is not satisfied in y because $g'(a) = 2 < g'(a') = 3$.

Replace $Ar = \{a\}$ in y by $Ar = \{a'\}$. A structure $y' = \{S,A,P,W\{a'\}>$ results. Principle (PA3) is satisfied in y' because $g'(a) = 2 < g'(a') = 3$, but principle (PA2) doesn't because $g(a) = 2 > g(a') = 1$.

The next definition makes it possible to combine principles (PA2) and (PA3).

D6 $A_C = \{a \mid a \in A$ and for all $a' \in A$: $g(a) \geq g(a')$ and $g'(a) \geq g'(a')\}$.

Using D6 we next define three versions of Classical Utilitarianism analogous to Strict Universal Altruism.

D7   Let S, A, P, W, Q, Ar satisfy the definitions $(1) - (6)$ of the basic concepts. Then

(1) $M(\text{CU}) = \{<S,A,P,W,Q,Ar> \mid Ar \subseteq A_C\}$

is the *set of models of Classical Utilitarianism without impartiality*,

(2) $M(\text{CU})_W = \{<S,A,P,W,Q,Ar> \mid Ar \subseteq A_C$ and $W \subseteq W_W\}$

is the *set of models of Classical Utilitarianism with weak impartiality*,

(3) $M(CU)_S = \{ <S,A,P,W,Q,Ar> \mid Ar \subseteq A_C \text{ and } W \subseteq W_S \}$
is the *set of models of Classical Utilitarianism with strong impartiality*.

We next prove some results analogous to SUA.

Theorem 5: $M(CU)_S \subset M(CU)_W \subset M(CU)$.
Proof: Theorem 5 follows trivially from definitions D2, D3 and D6.
Theorem 6: Classical Utilitarianism with strong impartiality is consistent.
Theorem 7: $M(SUA)_S \subset M(CU)_S$.
Proof: We prove Theorem 7, which is sufficient to demonstrate Theorem 6.
Consider a structure $y = <S,A,P,W,Q,Ar>$. Let $y \in M(SUA)_S$. Then $Ar \subseteq A_S$ and $W \subseteq W_S$. We have to show that $A_S \subset A_C$.

Let a be an arbitrary element in A_S, i.e., the condition "for all $a' \in A$, for all $s \in S$: $W(<a,s>) \cdot P(<a,s>) \geq W(<a',s>) \cdot P(<a',s>)$"
is satisfied in y.

Contrary to the claim assume that $a \notin A_C$. Then there is an a' in A such that $g(a) < g(a')$ or $g'(a) < g'(a')$. Suppose $g(a) < g(a')$ holds. Then the number of individuals S for which a' maximizes individuals' W-weighted net pleasure is greater than the related number associated with a. But then a cannot maximize the W-weighted net pleasure function for all s in S. Thus, $a \notin A_S$, which contradicts assumption $a \in A_S$.

Suppose $g'(a) < g'(a')$. Then the number of individuals in S for which action a' gives a positive, non-zero value of individuals' W-weighted net pleasure function is greater than the related number associated with a. Hence, there exist $g'(a')-g'(a)$ individuals in S such that the weighted net pleasures of a' for these individuals are more than the weighted net pleasures of a. But then a cannot maximize the weighted net pleasure for all individuals in S. Thus, $a \notin A_S$, which once again contradicts assumption $a \in A_S$.

We next show that $A_S \neq A_C$. Let $S = \{s,s'\}$, $A = \{a,a'\}$ and let $W(<a,s>) \cdot P(<a,s>) = +1, W(<a,s'>) \cdot P(<a,s'>) = 0, W(<a',s>) \cdot P(<a',s>) = 0$ and $W(<a',s'>) \cdot P(<a',s'>) = +1$. Counting g and g' in A yields $g(a) = g(a') = 1$ and $g'(a) = g'(a') = 1$. Thus, $\{a,a'\} \subseteq A_C$. However, $\{a,a'\} \cap A_S = \emptyset$ because there exists no action in A which maximizes the W-weighted net pleasure for both s and s' in S.

Theorem 8: Condition $Ar \subseteq A_C$ is independent of conditions $W \subseteq W_S$ and $W \subseteq W_W$ and vice versa.
Proof: Analogous to the proof of theorem 3.

Let us now illustrate the content of Classical Utilitarianism via the Fénelon example. Consider Classical Utilitarianism with weak impartiality. Count g and g' in $A=\{a,a'\}$, ($a\sim$ Fénelon is saved and $a'\sim$ the valet is saved) for individuals in S. It is obvious that the following conditions hold in the example:

$$(3) \quad w^*(P(<a,s>)) \geq w^{**}(P(<a',s>)) \text{ for all } s \in S, \ s \neq V$$

and

$$(4) \quad w^{**}(P(<a',V>)) > w^*(P(<a,V>))$$

because the argument clearly presupposes that "to save Fénelon" is better than "to save the valet" for the *other* in S except Fénelon and the valet, and on the other hand it is obvious that "to save Fénelon" is better than "to save the valet" for *Fénelon*, but "to save Fénelon" is worse than "to save the valet" for *the valet*. Counting g in A yields

$$(5) \quad \begin{cases} g(a) = \text{the number of individuals in } S\text{-1 (the valet)} \\ g(a') = 1 \text{ (the valet).} \end{cases}$$

The example clearly presupposes that S contains more individuals than only Fénelon and the valet. Thus, condition

$$(6) \quad g(a) > g(a')$$

holds.

Make the following assumptions, which are completely reasonable in the light of Godwin's example:

$$(7) \quad w^*(P(<a,s>)) > 0 \text{ for all } s \text{ in } S, \ s \neq V$$

and

$$(8) \quad w^{**}(P(<a',V>)) > 0.$$

Counting g' in A yields

$$(9) \quad \begin{cases} g'(a)=g(a) \\ g'(a')=g(a'). \end{cases}$$

Combining conditions (6) and (9) yields that condition

(10) $g'(a) > g'(a')$

holds. But then results (6) and (10) imply that in the example

(11) $A_C = \{a\}$.

Thus the structure $<S,A,P,W,Q,\{a\}>$ is a model of Classical Utilitarianism with weak impartiality.

The Relations of the Three Versions of Strict Universal Altruism and Classical Utilitarianism

Above three versions of Strict Universal Altruism and three versions of Classical Utilitarianism are formulated; versions with strong impartiality, with weak impartiality and without impartiality respectively. The following figure summarizes the relations of these theories.

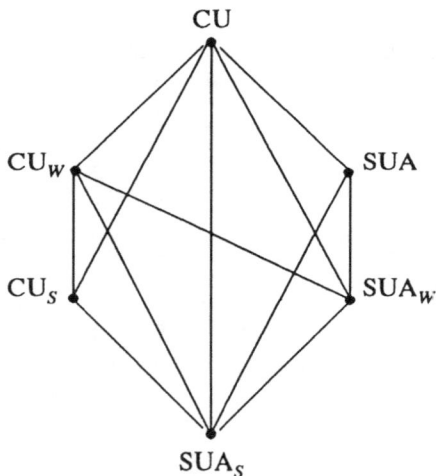

Figure 1.

Here the lines between the theories indicate the proper sub-set relation between the sets of models of the theories, i.e., between the sets (connected by one or more lines) of models of any theories, the proper sub-set relation holds from a theory located lower in Figure 1 to another one located higher. Thus the relations between the six theories constitute a lattice.

Different Versions of Early Utilitarianism with
Qualitative Differences in Pleasures

Start with the consideration of the *quality function Q*, which was defined above as a function of form

(12) $A \times S \rightarrow \mathbb{N}$

where $\mathbb{N} = \{1,2,\dots\}$ is a suitable finite set of natural numbers. We posit first the following convention which fixes the content of Q. Condition

(13) $Q(<a,s>) \geq Q(<a',s>)$

says that the *quality* of pleasure of act, omission etc. a for s is at least as good as is the *quality* of pleasure of a' for s. Respectively, condition

(14) $Q(<a,s>) \geq Q(<a,s'>)$

says that the *quality* of pleasure of act, omission etc. a for s is at least as good as is the *quality* of pleasure of a for s'.

Thus, theoretically, the "goodness" of the quality of pleasure of *one* and the *same* action may vary in the set of individuals as condition (14) indicates. However, for simplicity's sake we make the following idealizing restriction

(15) for all $a \in A$, for all $s,s' \in S$: $Q(<a,s>) = Q(<a,s'>)$.

Condition (15) says that individuals in S *share a preference* among the qualities associated with the pleasures yielded by available actions. However, the preference itself may vary, as may the qualities in themselves, i.e., the quality associated with the pleasure of action a for individual s need not be identical with the quality associated with the pleasure of action a for individual s'. Condition (15) demands only that if the quality of pleasure associated with a is *better* than (or equal to) the related quality with a' for *some* s in S, then the same holds for *all* s in S.

The next simplification or idealization concerns the *weights* of *qualities* of pleasures. Theoretically it would be possible to introduce *an independent weight function for pleasures*, analogously to the weight function of quantities of pleasures, W above. However, we take it for granted that any qualities of pleasures are taken into account to the same extent. This simplification or idealization makes any weight function for qualities of pleasures unnecessary.

Now look at utilitarian doctrines from a purely qualitative perspective. Then the core of any purely qualitative utilitarian theory would contain the following principle:

(16) Comparing any two $a,a' \in A$: a is morally right if and only if
$$Q(<a,s>) \geq Q(<a',s>)$$

where the relevant condition can be simplified to the form

(17) $Q(a) \geq Q(a')$

due to restriction (15).

Principle (16) is generalizable relative to the whole A, and a set of morally right acts can be defined analogous to A_S as follows.

D8 $A_Q = \{a \mid a \in A$ and for all $a' \in A$: $Q(a) \geq Q(a')\}$
is the *set of morally right actions of Strong Qualitative Utilitarian Theory*.

Define next the set of models of Strong Qualitative Utilitarian Theory analogous to SUA without impartiality (see D4(1) above).

D9 Let S, A, P, W, Q, Ar satisfy the defining conditions $(1)-(6)$ of the basic concepts. Then
$M(SQU) = \{ <S,A,P,W,Q,Ar> \mid Ar \subseteq A_Q\}$
is the *set of models of Strong Qualitative Utilitarian Theory*.

Removing the idealizing assumptions made above and formulating the needed weak and strong impartiality principles relative to the qualities of pleasures makes it possible to explicate six different, purely qualitative utilitarian theories: Strong Qualitative Utilitarian Theory with strong impartiality, with weak impartiality and without impartiality and Qualitative Utilitarian Theory with strong impartiality, with weak impartiality and without impartiality (SQU_S, SQU_W, SQU, QU_S, QU_W and QU), completely analogously to the quantitative theorizing above. Their interrelations are similar to the interrelations of the quantitative theorizing, i.e., the proper sub-set relation between the sets of models of the qualitative theories induces a structurally similar lattice.

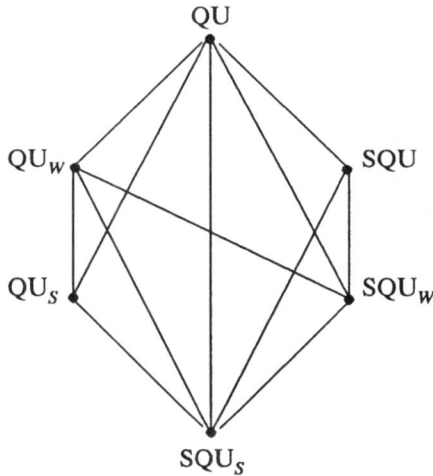

Figure 2.

The Core Assumptions Reconsidered

Consider now principles (Q1)−(Q4) given above. We next show that in utilitarian theorizing principles (Q1)−(Q4), principles (P1)−(P3) and principles (I1)−(I2) are compatible in the sense that there are actions which satisfy all the principles. This result, however, does not imply that any action which satisfies one of the principles would satisfy another, i.e., the principles are not equivalent. Our result only shows that quantitative utilitarian theorizing with or without impartiality can be combined with a qualitative theorizing in a consistent way.

We first show that A_Q is independent of A_C and vice versa. Because condition $A_S \subset A_C$ holds (Theorem 7), the result with Theorem 8 shows that Q-conditions defining A_Q are independent of P- and W-conditions defining A_C and A_S, and vice versa.

Theorem 9: Strong Qualitative Utilitarian Theory is independent of Classical Utilitarianism without impartiality and vice versa.

Proof: We have to construct two structures y and y' such that $y \in M(\mathrm{SQU})$ $-M(\mathrm{CU})$ and $y' \in M(\mathrm{CU})-M(\mathrm{SQU})$.

Consider structure $y = \langle S,A,P,W,Q,Ar \rangle$ with $S=\{s,s'\}$, $A=\{a,a'\}$; $W(\langle a,s \rangle)\cdot P(\langle a,s \rangle)=2$, $W(\langle a',s \rangle)\cdot P(\langle a',s \rangle)=1$, $W(\langle a,s' \rangle)\cdot$ $P(\langle a,s' \rangle)=0$ and $W(\langle a',s' \rangle)\cdot P(\langle a',s' \rangle)=1$; $Q(a)>Q(a')$ and

$Ar=\{a\}$. $y\in M(\text{SQU})$ because $Q(a)>Q(a')$. $y\notin M(\text{CU})$ because $g(a)=g(a')=1$ but $g'(a)=1<g'(a')=2$.

Replace $Ar=\{a\}$ in y by $Ar=\{a'\}$. A structure y' results such that $y'\in M(\text{CU})$ because $g(a')=g(a)=1$ and $g'(a')=2>g'(a)=1$, but $y'\notin M(\text{SQU})$ because $Q(a')<Q(a)$.

Thus, $A_Q\nsubseteq A_C$ and $A_C\nsubseteq A_Q$ which means that A_Q is independent of A_C (and A_S) and vice versa.

Theorem 10: Strict Universal Altruism (with strong impartiality), Classical Utilitarianism (with strong impartiality) and Strong Qualitative Utilitarian Theory are compatible.

Proof: We construct a structure which is a model of each of the three theories.

Consider structure $y=<S,A,P,W,Q,Ar>$ with $S=\{s,s'\}$; $A=\{a,a'\}$; $W(<a,s>)=W(<a',s'>)$ for all elements in $A\times S$; $P(<a,s>)=P(<a,s'>)>P(<a',s>)=P(<a',s'>)$; $Q(a)=Q(a')$ and finally $Ar=\{a\}$.

Condition $\{a\}\subseteq A_S$ holds in y because a maximizes the weighted net pleasure for all $s\in S$ in A. Thus, $y\in M(\text{SUA})_S$. Then $y\in M(\text{CU})_S$ due to Theorem 7, $y\in M(\text{SUA})_W$ and $y\in M(\text{SUA})$ due to Theorem 1, $y\in M(\text{CU})_W$ and $y\in M(\text{CU})$ due to Theorem 5.

Condition $\{a\}\subseteq A_Q$ holds in y because condition $Q(a)=Q(a')$ holds. Thus, $y\in M(\text{SQU})$.

Hence, all the three versions of SUA and CU, and SQU are compatible.

Theorem 10 shows that principles (PA1), (PA2), (PA3), (IA1), (IA2) and "quality principle": a is morally right iff for all $a'\in A: Q(a)\geq Q(a')$, are compatible. We take it for granted that (PA1)−(PA3), (IA1) and (IA2) formalize in an adequate way the formulations (P1)−(P3) of the Greatest Happiness Principle and the formulations (I1) and (I2) of the Impartiality Principle. Consider now four positions to the relation of qualities and quantities of pleasures in utilitarian theorizing, expressed by (Q1)−(Q4) above. Define $M=\{y\,|\,y=<S,A,P,W,Q\{a\}>$, and the following conditions are satisfied in y:

1) for all elements in $A\times S$: $W(<a,s>)=W(<a',s'>)$,
2) for all $a'\in A$, for all $s\in S$: $P(<a,s>\geq P(<a',s>)$,
3) for all $a'\in A$: $Q(a)\geq Q(a')\}$

Thus, M is *a part of* the *joint* set of models for SUA, CU and SQU (with or without impartiality).[4] Theorem 9 shows that A_Q is independent of A_C (and A_S) and vice versa. This result implies that conditions (1), (2) and (3) of M can be removed or changed independently of one another.

Position (Q3) says that pleasures contain both quantitative and qualitative differences and both are relevant. (Q3) is satisfied in any $y \in M$ because $y \notin M$ if the quantitative aspect fails (condition 2) *or* the qualitative aspect fails (condition 3).

Remove now condition (3) from M. A set M' results such that (Q2) is satisfied in any $y \in M'$ because conditions (1) and (2) are satisfied, i.e., the quantitative aspect, and only it, is now relevant. But note that $M \subset M'$ because any $y \in M$ satisfies conditions (1), (2) and (3), and hence y satisfies conditions (1) and (2) if $y \in M$.

Remove now conditions (1) and (2) from M. A set M'' results such that (Q4) is satisfied in any $y \in M''$ because condition (3) is satisfied, i.e., the qualitative aspect, and only it, is now relevant. As above, the result $M \subset M''$ holds because for any y: if $y \in M$ then $y \in M''$.

Combining the above results yields: for any y: if $y \in M$ then $y \in M'$ and $y \in M''$. So, $M = M' \cap M''$, i.e., (Q2), (Q3) and (Q4) are satisfied in any $y \in M$.

Consider now (Q1). It says that pleasures contain only quantitative *differences*. Define M^* replacing condition (3) of M by (3'): for all a, $a' \in A$: $Q(a) = Q(a')$, which says that pleasures cannot be differentiated with respect to their qualities. Then (Q1) is satisfied in any $y \in M^*$. Condition (3') implies condition (3) of M, but the converse does not hold. Thus, $M^* \subset M$. But then (Q1)−(Q4) are all satisfied in any $y \in M^*$ and so are (PA1), (PA2), (PA3), (IA1) and (IA2). Thus, principles (PA1)−(PA3), (IA1), (IA2) and (Q1)−(Q4) are compatible.

The above results can be expressed by the following figure.

4 Note that in general Ar in the models of SUA, CU and SQU need not be a singleton, i.e., it is possible that there are *several morally right realizable* actions.

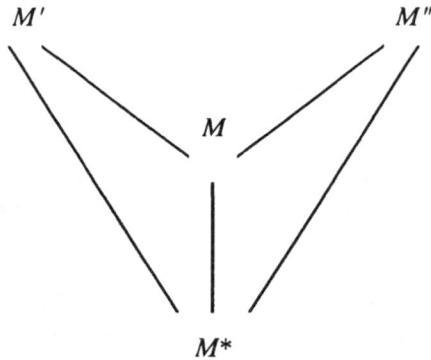

Figure 3.

Here M^* is a set of models in which principles (PA1) − (PA3), (IA1) − (IA2) and (Q1) − (4) are satisfied. In the elements of M principles (PA1) − (PA3), (IA1) − (IA2) and (Q2) − (Q4) are satisfied. In the elements of M' principles (PA1) − (PA3), (IA1) − (IA2) and (Q2) are satisfied and finally in the elements of M'' principles (PA1) − (PA3), (IA1) − (IA2) and (Q4) are satisfied. Thus, $M^* \subset M$, $M \subset M'$ and $M \subset M''$.

Some Idealizations in Utilitarian Theorizing

It is obvious that there are idealizing assumptions at different levels of utilitarian theorizing. In the following we consider first some of the *most basic* ones connected with the four different positions to the relation of qualities and quantities in pleasures (expressed by (Q1) − (Q4) above). Then we shall argue that *applications* of utilitarian theorizing presuppose somewhat weaker idealizations. The basic idealizations can be seen as special, stronger cases of the weaker ones.

In general it is very difficult to say what all a pleasure contains. However, so far as we are interested in the *differences* in pleasures, two aspects can be differentiated due to (Q1) − (Q4): a difference between pleasures can be yielded by varying the *quantity* of pleasure or varying the *quality* of pleasure (or by varying both aspects). This view implies that a pleasure, denoted here by $\mathcal{P}$, is somehow compounded with respect to its quantity and quality plus possibly with respect to some other things. We denote this feature of pleasure $\mathcal{P}$ by the following formula

(18) $\mathcal{P} = F(P,Q)$

where P and Q refer to the P- and Q-funtions defined by (4) and (6) of the definition of basic concepts above. We won't try to characterize F in (18) at all. We only assume that it is a *functional* (many−one) relation which contains the quantitative and qualitative aspects of pleasure among other things. We take it for granted that formula (18) characterizes at least partially the content of pleasures.

We next study the idealizing assumptions contained in positions (Q1)− (Q4), in particular (Q1) is in the focus. Here we utilize the Poznań School Theory of Idealization [cf. Krajewski 1977, Nowak 1980 and Kuokkanen and Tuomivaara 1992, 1993a,b].[5]

---

5 The core of the Poznań School Theory of Idealization can be presented as follows. Assume that a researcher wants to explain the phenomena F in the class of objects R. On the basis of background knowledge some factors, say $H, p_k, p_{k-1}, \ldots, p_1$, are distinguished as *essential* for F. All the factors which are not considered to be essential for F are *reduced away*. Here H denotes *the principal factor* for F and the remainder are *the secondary factors* for F. Moreover, H is the *most important* factor relative to F, and it holds for the secondary factors that if $i > j$ then p_i is *more essential* for F than p_j, $i,j = 1,\ldots,k$.

First, all the secondary factors $p_k, p_{k-1}, \ldots, p_1$ are *abstracted away* by assuming $p_i(x) = 0$, $i = 1,\ldots,k$ and postulating counterfactually that essential, secondary factors p_i (with the given value $p_i(x) = 0$) do not influence F. Then a simple dependence between F and what is believed to be its principal factor is hypothetically proposed. In this manner an idealizational scientific law

(T_k) if $R(x)$ and $p_1(x) = 0 \ldots p_{k-1}(x) = 0$ and $p_k(x) = 0$ then
$F(x) = f_k(H(x))$

is put forward.

The idealized law (T_k) or, using Nowak's [1980, p. 29] terminology, *an idealizational statement* (T_k) is *concretized with respect to the influence of the secondary factors*. Because p_k is the most essential factor of the secondary factors, condition $p_k(x) = 0$ is removed and an appropriate correction to the consequent of (T_k) is introduced. Then the first concretization of the idealized law (T_k) is of the form:

(T_{k-1}) if $R(x)$ and $p_1(x) = 0 \ldots$ and $p_{k-1}(x) = 0$ and $p_k(x) \neq 0$ then
$F(x) = f_{k-1}(H(x), p_k(x))$.

Then condition $p_{k-1}(x) = 0$, concerning the most essential secondary factor after p_k, is removed and so on. The *final concretization* relative to the least essential secondary factor p_1 yields a *factual* statement (a fully concretized law):

(T_0) if $R(x)$ and $p_1(x) \neq 0 \ldots$ and $p_k(x) \neq 0$ then
$F(x) = f_0(H(x), p_k(x) \ldots p_1(x))$.

We first define an auxiliary function f which makes it easy to express several idealizations in the context of (Q1)–(Q4).

D10 Let S, A, P and Q satisfy conditions (1), (2), (4) and (6) of the definition of basic concepts above. Then

$f(Q)=0$ if the following condition is satisfied in Q: for all $s,s' \in S$, for all $a,a' \in A$: $Q(<a,s>)=Q(<a',s'>)$,

$\neq 0$ otherwise;

$f(P)=0$ if the following condition is satisfied in P: for all $s,s' \in S$, for all $a,a' \in A$: $P(<a,s>)=P(<a',s'>)$,

$\neq 0$ otherwise.

Consider now (Q1) "Pleasures contain only quantitative differences". (Q1) obviously implies

(19) There are no qualitative differences in pleasures.

Using f-function, defined by D10 above, sentence (19) can be formalized as

(20) for all Q: $f(Q)=0$,

which is, according to the Poznań Theory of Idealization, an *idealizing* assumption (a factually false sentence), *given that pleasures in fact contain qualitative differences.*

Consider now the parallel idealization with respect to the quantitative differences in pleasures,

(21) for all P: $f(P)=0$,

which says that there are no quantitative differences in pleasures. The conjunction of (20) and (21) says:

(22) there are neither qualitative nor quantitative differences in pleasures,

which is clearly an excessively strong idealization because position (Q1) clearly presupposes that *there are* quantitative differences in pleasures. Thus, (Q1) presupposes that sentence

(23) there is at least one P such that $P(<a,s>) \neq P(<a',s>)$ or $P(<a,s>) \neq P(<a,s'>)$

holds. Thus, the conjunction of (20) and (23) formalizes position (Q1). Applying the Poznań School Theory of Idealization to the partial characterization of pleasures (formula 18) yields the following idealization:

(24) if for all Q: $f(Q)=0$, for all P: $f(P)\neq0$ then $\mathcal{P}=F(P)$.

Because condition for all P: $f(P)\neq0$ implies condition (23), sentence (24) related with respect to formula (18) says that

(25) if there are no qualitative differences in pleasures then the differences in pleasures depend only on quantitative aspects of pleasures.

Sentence (24) is an idealizing sentence in the sense of the Poznań School Theory of Idealization. Sentence (24) can be concretized or factualized precisely in the way the Poznań School Theory of Idealization presupposes. Replace condition "for all Q: $f(Q)=0$" in the antecedent of (24) by the condition

(26) for all Q: $f(Q)\neq0$

which implies the sentence "there are qualitative differences in pleasures". Then the result is sentence

(27) if for all Q: $f(Q)\neq0$ and for all P: $f(P)\neq0$ then $\mathcal{P}=F(P,Q)$,

which is the concretization (or factualization) of sentence (24) in the Poznań sense. Sentence (27) says

(28) if there are both qualitative and quantitative differences in pleasures then the differences in pleasures depend on both.

But then sentence (27) concretizes or factualizes position (Q1) to position (Q3) "Pleasures contain both quantitative and qualitative differences and both are relevant" because (Q3) obviously presupposes that the conditions "for all Q: $f(Q)\neq0$" and "for all P: $f(P)\neq0$" hold, but (Q1) in turn presupposes that the conditions "for all Q: $f(Q)=0$" and "for all P: $f(P)\neq0$" hold.

Consider now (Q2) and (Q4). Both positions share the presupposition of (Q3), i.e., there are both qualitative and quantitative differences in pleasures. However, the former *neglects* the influence of qualitative differences in pleasures, the latter *neglects* the quantitative ones. The difference between (Q2) and (Q4) and between either of these and position (Q3) can be expressed simply as follows

(29)   for all Q,Q',P: $F(P,Q) = F(P,Q')$
and
(30)   for all Q,P,P': $F(P,Q) = F(P',Q)$.

Sentence (29) says that any quality in pleasure is irrelevant and sentence (30) says that any quantity in pleasure is irrelevant. The conjunction of (27) and (29) characterizes position (Q2) and the conjunction of (27) and (30) characterizes position (Q4). In both cases it can be said that pleasure depends on both qualitative and quantitative aspects, although in a trivial way.[6]

Some Idealizations in the Applications of Utilitarian Theorizing

Consider now the following very concrete example.[7] Suppose that S contains the population in a particular area and A contains a particular set of measures available for the local social workers. Functions P and Q express then the quantity and quality of the impacts of measures $a \in A$ for $s \in S$. In typical cases it is realistic to assume that a given measure a may have *different* qualitative and quantitative impacts on *different* individuals and two given measures a and a' may have *different* qualitative and quantitative impacts on *one and the same* individual. This means that both of the following conditions hold:

(31) $f(Q) \neq 0$
(32) $f(P) \neq 0$.

Consider now condition (31). This condition is *completely realistic* but it does not help the social workers much in selecting the most efficient

---

6 Analogously to the relation of positions (Q1) and (Q3), the relations of (Q2) and (Q4) to (Q3) can be presented using the Poznań School Theory of Idealization. The essential difference, however, is that position (Q1) *denies the existence of qualitative differences* in pleasures whereas (Q2) declares that *qualitative differences are irrelevant*, and (Q4) in turn declares that *quantitative differences are irrelevant*.

7 The example is partially intended to show that the core ideas of Early Utilitarianism surely have applications still nowadays, especially is the theory is modified using the Poznań-type idealizations.

measures relative to the individuals. Consider a very simple case with $A = \{a,a'\}$ and $S = \{s,s'\}$. Condition (31) allows the following cases, all of them different from one another:

$$Q(<a,s>) \neq Q(<a,s'>) \neq Q(<a',s>) \neq Q(<a',s'>)$$
$$Q(<a,s>) = Q(<a,s'>) \neq Q(<a',s>) \neq Q(<a',s'>)$$
$$Q(<a,s>) \neq Q(<a,s'>) = Q(<a',s>) \neq Q(<a',s'>)$$
$$Q(<a,s>) \neq Q(<a,s'>) \neq Q(<a',s>) = Q(<a',s'>)$$
$$Q(<a,s>) = Q(<a,s'>) = Q(<a',s>) \neq Q(<a',s'>)$$
$$Q(<a,s>) = Q(<a,s'>) \neq Q(<a',s>) = Q(<a',s'>)$$
$$Q(<a,s>) \neq Q(<a,s'>) = Q(<a',s>) = Q(<a',s'>).$$

Moreover, note that any "$\neq$" above may be replaced by "$<$" or "$>$". Thus, the simple case generates a total of $12 \times 3 = 36$ different cases. The same holds for P. So it seems that the realistic conditions (31) and (32) are too weak and liberal because for selecting the most adequate measures from $\{a,a'\}$ for s and s', the social workers have to study a total of 72 cases (although not all of them are independent of one another). One way out of the problem is to make *idealizations* which are informative and realistic to a sufficient extent.

Consider the set of available measures A. A can be partitioned due to the effects the measures have on the individuals. Any theory of the hierarchies of human needs serves as an example. Second, S can be partitioned, based for example on a theory of social structure. Then the related idealizing assumptions are made relative to the partitions of measures and individuals, i.e., it is assumed that *inside the parts* the effects of measures *approximately* equal for the individuals. The relevance and adequacy of such idealizations depends essentially on the partition's relevance and adequacy. The related idealizations for the qualitative aspects of the effects are in general of the following form.

D11 Let S, A, P and Q satisfy conditions (1), (2), (4) and (6) of the definition of basic concepts and let S_i, $i = 1,\ldots,n$, A_j, $j = 1,\ldots,m$ constitute partitions of S and A. Then
for all $s,s' \in S_i$, for all $a,a' \in A_j$, $i = 1,\ldots,n, j = 1,\ldots,m$: $Q(<a,s>) = Q(<a',s'>)$ is the *general form of idealization relative to partitions* S_i and A_j.

We next show that for example idealizations connected with *a hierarchy of human needs* entail the above general form. Consider the following: if $i < j$ then any $a \in A_j$ contributes more to the quality of life than does any $a' \in A_i$.

The first idealization says that the possible differences between actions *in parts* A_i, $i=1,\ldots,n$ are irrelevant, i.e., if $a,a'\in A_i$ then their effects on the quality of life are approximately equal. The second idealization says that the effects of any a on the quality *of life are approximately equal for all individuals*. The three conditions read as

(33)(a)   for all $s\in S$, for all $a,a'\in A$: if $a\in A_i$, $a'\in A_j$ and $i<j$ then $Q(<a,s>)<Q(<a',s>)$.

(33)(b)   for all $s\in S$, for all $a,a'\in A$, for all i: if $a,a'\in A_i$ then $Q(<a,s>)=Q(<a',s>)$.

(33)(c)   for all $s,s'\in S$, for all $a\in A$: $Q(<a,s>)=Q(<a,s'>)$.

Consider the conjunction of (33,b) and (33,c). It is equivalent to

(34)   for all $s,s'\in S$, for all $a,a'\in A_i$, for all i: if $a,a'\in A_i$ then $Q(<a,s>)=Q(<a',s'>)$,

which is an instance of the general form of idealization *with the special partition* where S contains precisely one part, itself. Thus, the extra contribution relative to the general form of idealization is due to the *crucial* condition (33,a) entailed in the theory of hierarchies of human needs. Similar results can be shown for the quantitative aspects (*P*-functions).

We end the paper with an important result which shows that the idealization $f(Q)=0$, used to constitute the presupposition of position (Q1) (see D10), implies the general form of idealization relative to any partitions S_i and A_j, but the converse does not hold. Thus, the former idealization is a special, stronger case of the latter. The parallel result holds for the quantitative aspect P of idealizations.

Theorem 11: Let Q be any quality function. Condition $f(Q)=0$ implies the general form of idealization relative to any partitions S_i and A_j, but the converse does not hold.

Proof: Let Q be an arbitrary quality-function. Condition $f(Q)=0$ says that condition "for all $s,s'\in S$, for all $a,a'\in A$: $Q(<a,s>)=Q(<a',s'>)$" holds in Q. Consider now arbitrary s, s' and a, a'. The above condition implies that condition $Q(<a,s>)=Q(<a',s'>)$ holds independently of the fact to which parts S_i and A_j of S and A s, s', a and a' happen to belong. Thus, the implication holds.

Consider now $S_1=\{s\}$, $S_2=\{s'\}$, $A_1=\{a\}$ and $A_2=\{a'\}$. Let $Q=\{\ll a,s>,1>,\ll a',s'>,2>\}$. Then the sentence for all $s,s'\in S_i$, $i=1,2$, for all $a,a'\in A_j$, $j=1,2$: $Q(<a,s>)=Q(<a',s'>)$ is true in structure $y = <\{s\} \cup \{s'\},\{a\} \cup \{a'\},Q>$. However, $f(Q) \neq 0$ because the sentence for all $s,s' \in S_1 \cup S_2$, for all $a,a' \in A_1 \cup A_2$: $Q(<a,s>) = Q(<a',s'>)$ fails in y. Thus, the converse of the implication does not hold.[8]

Conclusions

The above analysis supports the following general conclusions:

o the core assumptions of Early Utilitarianism, the Greatest Happiness Principle, the Impartiality Principle and the assumptions on the quantitative and qualitative aspects of pleasures, constitute three sets of mutually independent (and compatible) assumptions

o from a systematic point of view there exist two structuralist theory-nets, a qualitative one and a quantitative one; and

o applying the Poznań School Theory of Idealization to utilitarian theorizing shows that the assumptions on the qualitative and quantitative aspects of pleasures contain definite *special* idealizations which are of the same *general form* of idealizations used in applying utilitarian theorizing to concrete social cases. The former imply the latter.

The Academy of Finland,
and *Department of Philosophy*
University of Helsinki
P.O. Box 24
00014 University of Helsinki, Finland

---

8 Looked at from a general logical point of view the most basic idealizations, formalized by D10, are of the form

(1) $\forall x: x \in A \rightarrow x \in B$

whereas the general form of idealization relative to a partition is of form

(2) $\forall x: x \in A_i \rightarrow x \in B$

where $A_i \subseteq A$.

Any sentence of form (1) trivially implies a sentence of the form (2) but the converse does not hold. This observation leads directly to *partial idealizations*, initiated and studied in detail in Kuokkanen and Tuomivaara [1992, 1993a].

ACKNOWLEDGEMENTS

Prof. Henry Fullenwider has kindly revised the language of the paper.

REFERENCES

A   Classical Works

Bentham, J. [1789]. *An Introduction to the Principles of Morals and Legislation*. 1982 edition by J.H. Burns and H.L.A. Hart. Reprinted with new Introduction, London and New York: Methuen.

Berkeley, G. [1712]. *Passive Obedience*. Reprinted in: A.C. Fraser (Ed.). *The Works of George Berkeley*. Vol. 3. Oxford, second edition 1901.

Cumberland, R. [1672]. *De Legibus Naturae*. Partly reprinted in: Raphael [1991], trans. by D.D. Raphael.

Godwin, W. [1793]. *Enquiry Concerning Political Justice and its Influence on Modern Morals and Happiness* (third edition first published 1798). 1985-edition by I. Kramnick. Harmondsworth, Middlesex: Penguin Books.

Hartley, D. [1749]. *Observations on Man, His Frame, His Duty, and His Expectations*. 1934 edition by J.B. Priestly, third edition, London.

Hutcheson, F. [1725]. *An Inquiry into the Original of our Ideas of Beauty and Virtue*, Treatise II. Partly reprinted in: Raphael [1991].

Mill, J.S. [1861]. *Utilitarianism*. In: J.S. Mill and J. Bentham. *Utilitarianism and Other Essays*. 1987 edition by A. Ryan. Harmondsworth, Middlesex: Penguin Books.

Raphael, D.D. [1991] (Ed.). *British Moralists 1650−1800 I−II*. Indianapolis and Cambridge: Hackett Publishing Company (first published 1969).

B   Other References

Balzer, W., Moulines, C.-U. and Sneed, J.D. [1987]. *An Architectonic for Science. The Structuralist Program*. Dordrecht: Reidel.

Häyry, M. [1994]. *Liberal Utilitarianism and Applied Ethics*. London and New York: Routledge.

Krajewski, W. [1977]. *Correspondence Principle and the Growth of Science*. Dordrecht: Reidel.

Kuokkanen, M. and Tuomivaara, T. [1992]. On the Structure of Idealizations. Explorations in the Poznań School Methodology of Science. In: J. Brzeziński and L. Nowak (Eds.). *Idealization III: Approximation and Truth* (*Poznań Studies in the Philosophy of the Sciences and the Humanities*. Vol. 25). Amsterdam: Rodopi, 67−102.

Kuokkanen, M. and Tuomivaara, T. [1993a]. On the Adequacy of the Idealization Theory of the Poznań School. *Science Studies*. Vol. 6 [1993], No. 1, 28−33.

Kuokkanen, M. and Tuomivaara, T. [1993b]. The Threshold Model of Scientific Change and the Continuity of Scientific Knowledge. To appear in: *Zeitschrift für allgemeine Wissenschaftstheorie − Journal of General Philosophy of Science*.

Nowak, L. [1980]. *The Structure of Idealization*. Dordrecht: Reidel.

IDEALIZATION, APPROXIMATION
AND MEASUREMENT

Poznań Studies in the Philosophy
of the Sciences and the Humanities
Vol. 42, pp. 271–284

Rainer Westermann

MEASUREMENT-THEORETICAL IDEALIZATIONS
AND EMPIRICAL RESEARCH PRACTICE

Introduction

In the course of a structuralist (re-)construction of an empirical theory T, there are at least three occasions for dealing with measurement problems:

(1)   The substantial axioms used to define the set of models of any theory-element T_i give rise to certain *uniqueness constraints* representing, for example, the fact that a linear relationship between two terms can be meaningfully assumed only if both variables are measured at least at interval scale level [Westermann, 1989].

(2)   Any term that has been identified as theoretical with respect to the theory T according to the (semi-formal) structuralistic criterion of theoreticity can be measured (only) by applying this theory (*theoretical measurement*, Balzer, 1983).

(3)   By means of *intertheoretical links*, the theory can be connected with other theory-elements and nets. The resulting *theory-holon* [Balzer, Moulines and Sneed, 1987] may especially cover auxiliary theories representing, for example, concrete observation and coding procedures for non-theoretical terms [Westmeyer, 1989].

In this paper, the third approach is taken up and we will discuss how substantial theories can be connected with measurement theories in order to adequately specify the theoretical basis of scores and numbers used in empirical research.

The most prominent and widely accepted measurement-theoretical approach was put forward by Suppes and Zinnes [1963], Pfanzagl [1971] and Krantz, Luce, Suppes, and Tversky [1971], Luce, Krantz, Suppes and Tversky [1990] and Suppes, Krantz, Luce and Tversky [1989]. Since both structuralism and measurement theory are based on semi-formal set-theoretical axiomatizations, the technical problems of linking substantial and measurement theories are not difficult to solve and need not be considered in this paper. The main problem dealt with here is the fact that, although there is no empirical research without some form of measurement, in analyzing and reconstructing actual scientific theories and research processes it becomes apparent that there is little or no manifest connection to a theory of measurement.

There are several reasons for this unsatisfactory state of affairs. On the one hand, most empirical researchers do not even attempt to cope with formalized measurement-theoretical approaches because they seem difficult to understand. On the other hand, experts in measurement theory have failed to make their concepts and methods as routinely usable as, say, analysis of variance. The most basic problem, however, lies in the fact that measurement theories usually involve two kinds of idealizations that make it difficult to apply them in actual empirical research practice.

Measurement theory primarily deals with an ideal form of measurement, the "representational" measurement, whereas the vast majority of variables are, at least in the social and behavioral sciences, amenable only to non-ideal forms of measurement which are called "derived" and "quasi-representational". In the body of this paper these three types of measurement are explained with the use of specific examples and some suggestions are given as to how they can be incorporated into theory-holons to form an adequate basis for deriving scale values for non-theoretical terms and assessing their scale type.

In addition, most measurement-theoretical structures are strictly deterministic and pertain to ideal error-free situations. This idealization can be reduced by probabilistic formulations and specific error theories [Falmagne, 1980; Hamerle and Tutz, 1980; Heyer and Mausfeld, 1987] or, more simply, by testing derived statistical hypotheses. A short explanation of this latter approach is given below in the section on quasi-representational measurement.

I
Representational and Fundamental Measurement

Measurement is representational if a particular empirical structure (consisting of at least one set of empirical objects A and one nonnumerical relation R) can be mapped homomorphically into an adequate numerical structure, i.e., if each object a can be assigned a number $f(a)$ so that each relevant empirical relation between the objects is represented by a corresponding numerical relation [Krantz et al., 1971, pp. 8−9; Dawes, 1972, p. 11]. If the nonnumerical relations are not contingent upon a prior numerical assignment, the resulting representational measurement is said to be also "fundamental" [Suppes, 1968; Dawes, 1972, p. 84].

Empirical relations and objects can be of very different kinds ranging from, say, a relatively clear-cut ordering of sticks according to their length to subjective judgments of loudness ratios or personal evaluations of social values. In each case, measurement is not a representation of inner attributes or subjective sensations, but of nonnumerical relations based on judgements which are more or less subjective [cf. Shepard, 1981, p. 50].

The primary aim of measurement theory is to define, by semi-formal set-theoretical axiomatizations, *measurement structures*, i.e., idealized types of empirical structures for which *representation theorems* can be proved. The axioms of a measurement structure specify, in terms of empirical relations, a set of conditions which taken together are sufficient for a homomorphic numerical representation. The most important axioms are those which are necessary for the representation and which can be tested empirically.

Take for example an individual's pairwise preference judgments (e.g. "*S.F.* is nicer than *N.Y.*", "*N.Y.* is nicer than *L.A.*", etc.). They can be consistently represented by assigning an ordered series of numbers (e.g. 5, 4, 3) only if they are transitive, i.e., if there are no *circular triads*. Moreover, it can be shown analytically that a homomorphic numerical representation always exists if the empirical preference relation is not only transitive but also asymmetric and complete, i.e., if it is a *strict simple order* relation [Roberts, 1979, pp. 32, 109]. To be sure, this "guarantees only formal representability, but neither theoretical fruitfulness nor practical utility" [Schwager, 1991, p. 625].

Having proved such a representational theorem, the uniqueness properties of the representation can be specified. In the example at hand, all numerical assignments leading to a homomorphic numerical representation of the empirical structure are related by (strictly) monotone increasing functions. Thus any assignment is unique up to monotone increasing transformations and is an ordinal scale.

How representational measurement can be achieved in actual psychological research practice is illustrated by a well-known experiment from social psychology which at first glance has no connection with measurement theory. Aronson and Carlsmith [1963] tested a hypothesis derived from Festinger's [1957, p. 586] theory of cognitive dissonance, namely, "that a mild threat of punishment for playing with a desired toy would lead to a devaluation of that toy while a severe threat would not". In their experiment, five different toys were used, one of which was declared forbidden.

To obtain each child's order of preference, ten posssible pairs of toys were presented, and each time the child had to indicate the toy he or she would rather play with. Aronson and Carlsmith ruled out minor deviations from transitivity by repeating the judgments in the case of inconsistencies, and excluded those children from their analysis for whom the preference relation was clearly intransitive. Thus, for the remaining subjects, the empirical preference relation was transitive. Since each possible pair of objects was judged and no equality judgments were allowed, it is also necessarily complete and asymmetric. As a consequence, Aronson and Carlsmith's procedure leads to a strict simple order and a representational and fundamental ordinal scale measurement.

Representational measurement at interval scale level (i.e. unique up to positive-linear transformations) can be based on nonnumerical relations that pertain to the order of differences or dissimilarities between pairs of stimuli. A person may have to judge, for example, whether the difference in attractiveness between objects *a* and *b* is greater than the respective difference between two other objects *c* and *d*. The resulting empirical dissimilarity relation is numerically representable if it satisfies the axioms of a *difference structure* [Krantz et al., 1971, Ch. 4]. In addition, special measurement structures leading to interval scales were developed for stimuli made up of different components and for bisection operations (*conjoint* and *bisymmetric structures*).

Difference structures were used by Schneider, Parker and Stein [1974] for measuring the loudness of ten tones differing in intensity. Each subject had to compare 990 combinations of two different tone pairs and indicate which pair had the greater difference in loudness. From the resulting rank order of pairs, scale values for the single stimuli were determinded by nonmetric scaling techniques [Shepard, 1966]. In addition, the large number of judgments provided many opportunities to test whether the judgments were transitive.

To reduce the amount of labor involved, the order relation between object pairs may be obtained by asking the subjects to give direct numerical estimations of the differences within each pair [Roberts, 1979, p. 135; Schneider, 1982, p. 325]. Since the final assignment of numbers to the object of interest is based on prior numerical ratings, this procedure leads to representational, but not to fundamental measurements.

Although, as our examples show, representational measurement is possible in different areas of psychology, it is only seldom applied in common research practice. This is partly due, of course, to the fact that it is relatively laborious. But there is a more fundamental reason: Most concepts of interest simply do not correspond to qualitative, relatively directly observable relations which can be taken as a basis for fundamental or representational measurement. As a consequence, researchers often have no choice but to quantify variables by direct rating procedures or by summing up scores in a test or questionnaire. These widely used kinds of assessment methods are analyzed below.

II
Quasi-Representational Measurement

Quasi-Representational Measurement is best illustrated by contrasting the representational measurement of Aronson and Carlsmith [1963] with procedures used in two follow-up studies.

(1) To test alternative explanations for the rise in attractiveness of the forbidden object, Turner and Wright [1965] used eight toys and obtained each girl's preference order by means of a process of elimination. After acquainting the child with all the toys, the experimenter asked which one she liked best. The chosen toy was put away and the procedure was repeated until the child had chosen seven out of eight toys.

Of course, it is quite possible that an elimination procedure leads to the same result as a complete pair comparison. But when the elimination procedure alone is applied, we get a rank order for each person even if the judgments were given in a completely random way. The important difference between the two procedures lies in the fact that the elimination procedure does not permit us to test whether transitivity, the necessary condition for ordinal scale representations, is satisfied. As a consequence, a rank order by elimination can not be considered a representational measurement.

(2) Freedman [1965] was concerned with the effects of mild and severe threats over a relatively long period, and he considered not only judgements of attractivity, but also actual playing behavior. The important point here is that Freedman did not content himself with preference orders alone. Instead, each child "was asked to indicate his liking of each of five toys on a scale ranging from 0 (very, very bad toy) to 100 (very, very good toy) by pointing to a place on the scale" [Freedman, 1965, p. 149]. Numerical differences between corresponding ratings in the first and the second session were used to test the predictions concerning changes in the attractiveness of forbidden and non-forbidden

toys. The ratings were thus interpreted as if they were at least interval scale measurements.

Both the use of rating scales and their (implicit) interpretation as interval scale measurements is ubiquitous in large areas of the social and behavioral sciences. There are many different types of rating scales, ranging from small sets of verbally labeled categories (e.g., agree, undecided, disagree) over numerical or graphical responses as in our example to direct numerical "magnitude estimations" [Dawes, 1972; Lodge and Tursky, 1982; Stevens, 1968]. None of these are representational measurements, because they do not include empirical tests of transitivity or of other conditions necessary for interval scale representations.

Dawes [1972] uses the term "index measurement" for such scaling procedures without consistency checks. He argues that index measurements can only be evaluated in terms of their usefulness, e.g. in terms of predictive validity, and that there is no sense in asking what scale type is attained by a particular index measurement technique. This position could, in my opinion, have detrimental consequences in that the usual (implicit) interpretation of rating scale values as interval scales can be neither justified nor criticized from a methodological point of view and there are no criteria for differentiating between a sound and a questionable use of ratings or magnitude estimations.

Fortunately, it is not necessary to exclude direct rankings, ratings, and magnitude estimations from measurement theoretical analyses. In fact, it is possible to test the hypothesis that one of these procedures has led to scale values at ordinal or interval level. Since such tests are based on an indirect, *a posteriori* application of representational measurement structures, we can speak of a *quasi-representational* measurement.

For the interval scale level, the basic idea of this approach can be summarized as follows [Westermann, 1985]: In a preliminary step, individual or group scale values $f(a)$, $f(b)$, ... for the objects and the attribute of interest are determined empirically by one of the rating or estimation methods. The hypothesis that these values are at interval scale level is tested in a subsequent step by recoursing to the necessary axioms of a finite, equally-spaced difference structure [Krantz et al., 1971, pp. 167–168]. In this part of the study, the chosen objects are presented in groups of four and the subjects are asked to judge whether the difference (with respect to the attribute of interest) is larger within the first or within the second pair of objects.

Firstly, different quadrupels *(a,b,c,d)* of objects are selected so that the absolute difference between the scale values are (at least approximately) equal: $|f(a)-f(b)| = |f(c)-f(d)|$. From the equal-spacing axiom of finite difference structures it follows that the subjective differences $|a-b|$ and $|c-d|$ must be equal if the scale values have indeed interval level. As a consequence, the probability should be 0.5 that the $|a-b|$-difference is judged the larger in a forced choice pair comparison task. When replications (over or within subjects) can be considered as independent, this statistical hypothesis is tested by a binomial or sign test. Since the interval scale hypothesis is corroborated only if the null hypothesis can be accepted, it is important that power tables [Cohen, 1977] are used to deliberately choose the number of replications so that the probability β for a Type II error is low at least for medium effect sizes.

Secondly, different sets of six objects (i,j,k,l,m,n) are selected under the constraint that $|f(i)-f(j)| > |f(l)-f(m)|$, $|f(j)-f(k)| > |f(m)-f(n)|$ and $|f(i)-f(k)| > |f(l)-f(n)|$. If the scale values have interval level, then the subjective differences $|i-j|$, $|j-k|$, and $|i-k|$ should actually be judged larger with probability $p > 0.5$. A significant and substantial deviation from the binomial null hypothesis is thus expected for these quadrupels. In addition, according to the monotonicity axiom of difference structures, any subject that judges $|i-j| > |l-m|$ and $|j-k| > |m-n|$ must also judge $|i-k| > |l-n|$. Again, a binomial test can be used to test whether this necessary condition for an interval scale representation is satisfied.

III
Derived Measurement

According to the theory of derived measurement [Suppes and Zinnes, 1963, pp. 17−22; Roberts, 1979, pp. 76−91], a derived scale can be defined as an ordered n-tupel $<A, f_1,\dots, f_k, RR, g>$ consisting of a set of objects A, k basic numerical assignments, a derived numerical assignment g, and a so-called representation relation RR. That many typical dependent variables in psychological research can be regarded as derived measurements is illustrated by another well-known dissonance-theoretical experiment. Janis and Gilmore [1965, p. 17] investigated "the influence of incentive conditions on the success of role playing in modifying attitudes". In their experiment, students' attitudes towards a policy of increased college courses in physics and mathematics were considered. Five statements concerning the proposed policy had to be answered on a five-point scale of approval−disapproval. The response categories were labeled numerically and these numbers were summed up to a combined "attitude score". Thus there are five basic numerical assignments and the representing relation consisting of a simple summation rule, the result of which is a derived numerical assignment to each subject.

Both this summation rule and the representing relations for the three most prominent general methods of attitude scaling [Likert, 1932; Osgood, Suci and Tannenbaum, 1957; Thurstone and Chave, 1929] are special cases of the following expression [Fishbein and Ajzen, 1975; Westermann, 1983]:

$$(1) \quad A_{po} = \sum_{i=1}^{n} B_{ipo} E_{ipo} \, .$$

In this equation, E_{ipo} is the positive or negative evaluation of object o expressed by an endorsement of statement i by person p, B_{ipo} is the degree of agreement or disagreement of person p with this statement i on object o, and A_{po} is the resulting score for the attitude of person p towards object o. Both in Janis and Gilmore's experiment and in most other attitude scalings the basic numerical assignments are not representational measurements, but rather only quasi-representational ratings. This further stresses the great importance of the theoretical and empirical foundation of quasi-representational measurement described above. Without the ability to assess the scale type of a quasi-representational basic numerical assignment like B_{ipo}, one cannot say anything about the scale type of derived assignments such as the attitude score A_{po}.

The scale type of a derived measurement is defined by the consequences that admissible transformations of the basic numerical assignments have for the derived assignment [Suppes and Zinnes, 1963]. If all derived assignments resulting from admissible transformations of the basic assignment are related, say, by positive linear transformations, the derived assignment is of interval scale level. In the case of product-summation models such as our general representing relation (1), the derived scores have interval scale level if both B_{ipo} and E_{ipo} are at least interval scales and if, in addition, several sets of specific conditions are satisfied [Dohmen, Doll and Orth, 1986; Mellenbergh, Molendijk, de Haan and ter Horst, 1990; Westermann, 1983]. Janis and Gilmore's attitude scores, for example, can be considered as being unique up to positive linear transformations if the numbers expressing the degree of endorsement have interval scale properties and if the sums of these numerical responses are equal for all subjects. Since these conditions are not usually satisfied for standard attitude scalings, it is worth considering a simple special case of derived measurement.

Take for example an attitude questionnaire, an intelligence test or the like with n positively evaluated items, with two response categories for each item (endorsed or rejected, passed or failed, etc.), and with a resulting test score S defined as the number of positively answered items. According to Suppes and Zinnes [1963], such sum scores, being the result of a mere counting operation, can be regarded as a fundamental measurement at the absolute-scale level ("pseudopointer measurement"). In actual research practice, however, scores of this kind are not interpreted as absolute-scales that can not be transformed in any way. Intelligence test results, for example, are usually transformed linearly in IQ-values. When we consider test scores as absolute scales, we must either condemn these operations as inadmissible, or we must exclude test results from our measurement theoretical analyses. For these reasons, I prefer to regard counting variables not as representational but as derived measurements.

For sum scores, our general representing relation (1) can be simplified considerably since E_{ipo} is equal 1 for all i and all p, and B_{ipo} equals 1 or 0 in the case of a positive or a negative response of person p to item i, respectively. As a result, we get

$$(2) \quad S_p = \sum_{i=1}^{n} B_{ip} .$$

As the same empirical information is depicted by any two ordered numbers such as 0 and 1, 0 and 2, -16 and $+27$, etc., any monotone increasing transformation of the two commonly used numerical response labels 0 and 1 is an admissible transformation according to Suppes and Zinnes [1963]. Since in the case of a variable X with only two possible realizations any monotone increasing transformation can be described by a positive linear function $X^* = aX + b$, all admissible transformations B_{ip}^* of the basic numerical assignment B_{ip} result in derived numerical assignments S_p^* that are related by linear functions:

$$(3) \quad S_p^* = \sum_{i=1}^{n} B_{ip}^* = \sum_{i=1}^{n} (aB_{ip} + b) = aS_p + nb .$$

Thus, a variable defined as the number of positive responses to some n items is a derived measurement at the interval scale level.

This result corresponds exactly to the way results of questionnaires or intelligence tests are commonly used. They are analyzed by parametric tests; product-moment correlations are used as measures of statistical association; numbers of positive answers are transformed linearly into z-

values or IQs, and so on. The approach adopted here clearly leads to the conclusion that statistical operations of this kind are meaningful in the sense of Suppes and Zinnes [1963; cf. Michell, 1986; Stevens, 1946; Stine, 1989].

Conclusion

When an empirical theory is constructed or reconstructed from the structuralist point of view, the theoretical reference points of theory-oriented experiments can be explicated as specific elements of the complex theory-net. They are a result of successive enlargements, restrictions and specializations of more general elements [Westermann, 1989]. To reconstruct adequately the measurement-theoretical bases of such experimental theory-elements, intertheoretical links to adequate measurement theories must be established.

When an experimental variable is measured representationally, we have a "measurement link" between the substantial theory and a representational measurement structure. More concretely, the corresponding term in the potential models of the experimental theory-element is linked directly, for example, to a *strict simple order* or a *difference structure* in the case of ordinal or interval scale measurement, respectively. Typically, measurement theories of this kind are not yet the proper *bedrock elements* [Balzer et al., 1987, pp. 413] of a theory-holon, for the empirical objects and relations in these representational measurement structures can not be specified without further presuppositions. All our observations are theory-laden in the sense that they are at least partially determined by our attitudes, expectations, experience, and so on [Hanson, 1969; Popper, 1975; Suppe, 1977]. Even in physics, fundamental measurements of extensive properties such as length or temperature often depend on theoretical assumptions [Fraser, 1980; Schwager, 1991; Shepard, 1981]. In particular, human observations and judgment in psychological experiments and social science surveys are usually related and dependent on the wording of instructions, the presentation order of stimuli, the response format, the situational setting, etc. In a comprehensive theory-holon, these determinating factors can be represented by linking the theories of measurement to more elementary theories of observation and data collection.

When rating scales, magnitude estimations or other quasi-representational measurements are used, the corresponding experimental theory-element can be linked with an auxiliary element representing the assumption that the scaling procedure at hand has lead to scores at interval scale level.

When this hypothesis is actually tested by deriving statistical hypotheses from a *finite difference structure* (see above), this measurement structure as well as the indirect testing procedure and its foundation in the theories of statistical tests can be reconstructed by specific theory-elements and -nets. Otherwise, the empirical supposition that ratings lead to interval scales can only be accepted by convention or by recoursing to positive empirical results with this approach for ruler-like continua [Westermann, 1985].

In the case of test scores or other derived measurements, we have to incorporate derived measurement structures in the sense of Suppes and Zinnes [1963] in the theory-holon. Usually, their representing relations must be considered as syntactical definitions that have no empirical content and can be accepted only because of their convenience, simplicity, usefulness, etc. Alternatively, it is sometimes possible to test the validity of representing relations empirically or to link them with substantial theories. The general representation relation (1) for attitude scaling, for example, is substantiated by the fact that it is in accordance with Fishbein's theory of attitude [Fishbein and Ajzen, 1975].

For a proper interpretation of derived measurements, it must be taken into account that numerical relations between derived scores do not directly represent observable relations between nonnumerical objects. The result of a derived measurement depends not only on the representation relation that has been chosen, but also on the concrete items that constitute the test or questionnaire. Of course, it is neither necessary nor advisable to choose items in a purely intuitive manner, for there are several concepts and procedures to select sets of items that are homogenous or unidimensional [Hattie, 1984].

In summary, different kinds of measurement-theoretic bases must be chosen for representational, quasi-representational and derived measurements. When adequate measurement theories become part of a theory-holon, we have a comprehensive representation of both the empirical theory of interest and auxiliary theories that are necessary or useful for its interpretation and application. It is hoped that the possibility of such a joint representation will promote the effective use of measurement-theoretical idealizations in empirical research practice.

Institut für Psychologie
Georg-August-Universität
Goßlerstr. 14
D-37073 Göttingen, Germany

REFERENCES

Aronson, E. and Carlsmith, J.M. [1963]. Effect of the severity of threat on the devaluation of forbidden behavior. *Journal of Abnormal and Social Psychology*, **66**, 584 – 588.

Balzer, W. [1983]. Theory and measurement. *Erkenntnis*, **19**, 3 – 25.

Balzer, W., Moulines, C.U. and Sneed, J.D. [1987]. *An architectonic for science. The structuralist program*. Dordrecht: Reidel.

Cohen, J. [1977]. *Statistical power analysis for the behavioral sciences*. (Rev. ed.) New York: Academic Press.

Dawes, R.M. [1972]. *Fundamentals of attitude measurement*. New York: Wiley.

Dohmen, P., Doll, J. and Orth, B. [1986]. Modifizierte Produktsummenmodelle und ihre empirische Prüfung in der Einstellungsforschung. *Zeitschrift für Sozialpsychologie*, **17**, 109 – 118.

Falmagne, J.C. [1980]. A probabilistic theory of extensive measurement. *Philosophy of Science*, **47**, 277 – 296.

Festinger, L. [1957]. *A theory of cognitive dissonance*. Evanston, Ill.: Row, Peterson.

Fishbein, M. and Ajzen, I. [1975]. *Belief, attitude, intention, and behavior*. Reading, MA: Addison-Wesley.

Fraser, C.O. [1980]. Measurement in psychology. *British Journal of Psychology*, **71**, 23 – 34.

Freedman, J.L. [1965]. Long-term behavioral effects of cognitive dissonance. *Journal of Experimental Social Psychology*, **1**, 145 – 155.

Hamerle, A. and Tutz, G. [1980]. Goodness of fit tests for probabilistic measurement models. *Journal of Mathematical Psychology*, 153 – 167.

Hanson, N.R. [1969]. Logical positivism and the interpretation of scientific theories. In: P. Achinstein and S.F. Barker (Eds.). *The legacy of logical positivism* (pp. 57 – 84). Baltimore: Hopkins.

Hattie, J. [1984]. An empirical study of various indices for determining unidimensionality. *Multivariate Behavioral Research*, **19**, 49 – 78.

Heyer, D. and Mausfeld, R. [1987]. On errors, probabilistic measurement and Boolean valued logic. *Methodika*, **1**, 113 – 138.

Janis, I.L. and Gilmore, J.B. [1965]. The influence of incentive conditions on the success of role playing in modifying attitudes. *Journal of Personality and Social Psychology*, **1**, 17 – 27.

Krantz, D.H., Luce, R.D., Suppes, P. and Tversky, A. [1971]. *Foundations of measurement: Vol. 1*. New York: Academic Press.

Likert, R. [1932]. A technique for the measurement of attitudes. *Archives of Psychology*, **140**.

Lodge, M. and Tursky, B. [1982]. The social psychophysical scaling of political opinion. In: B. Wegener (Ed.). *Social attitudes and psychophysical measurement* (pp. 177 – 198). Hillsdale: Erlbaum.

Luce, R.D., Krantz, D.H., Suppes, P. and Tversky, A. [1990]. *Foundations of measurement: Vol. 3*. San Diego: Academic Press.

Mellenbergh, G.J., Molendijk, L., Haan, W. de and Horst, G. ter [1990]. The sum-of-products variable reconsidered. *Methodika*, **4**, 37−46.

Michell, J. [1986]. Measurement scale and statistics: A clash of paradigms. *Psychological Bulletin*, **100**, 398−407.

Osgood, C.E., Suci, G.J. and Tannenbaum, P.E. [1957]. *The measurement of meaning.* Urbana: University of Illinois Press.

Pfanzagl, J. [1971]. *Theory of measurement.* (2nd ed.) Würzburg: Physica-Verlag.

Popper, K.R. [1975]. *The logic of scientific discovery.* (8th ed.) London: Hutchinson.

Roberts, F.S. [1979]. *Measurement theory with applications to decision-making, utility, and the social sciences.* Reading, MA: Addison-Wesley.

Schneider, B. [1982]. The nonmetric analysis of difference judgments in social psychophysics: Scale validity and dimensionality. In: B. Wegener (Ed.). *Social attitudes and psychophysical measurement* (pp. 317−337). Hillsdale: Erlbaum.

Schneider, B., Parker, S. and Stein, D. [1974]. The measurement of loudness using direct comparisons of sensory intervals. *Journal of Mathematical Psychology*, **11**, 259−273.

Schwager, K.W. [1991]. The representational theory of measurement: An assessment. *Psychological Bulletin*, **110**, 618−626.

Shepard, R.N. [1966]. Metric structures in ordinal data. *Journal of Mathematical Psychology*, **3**, 287−315.

Shepard, R.N. [1981]. Psychological relations and psychophysical scales: On the status of "direct" psychophysical measurement. *Journal of Mathematical Psychology*, **24**, 21−57.

Stevens, S.S. [1946]. On the theory of scales of measurement. *Science*, **103**, 677−680.

Stevens, S.S. [1968]. Ratio scales of opinion. In: D.K. Whitla (Ed.). *Handbook of measurement and assessment in behavioral sciences* (pp. 171−199). Reading, MA: Addison-Wesley.

Stine, W.W. [1989]. Meaningful inference: The role of measurement in statistics. *Psychological Bulletin*, **105**, 147−155.

Suppe, F. [1977]. *The structure of scientific theories.* (2nd ed.) Urbana: University of Illinois Press.

Suppes, P. [1968]. The desirability of formalization in science. *Journal of Philosophy*, **65**, 651−664.

Suppes, P., Krantz, D.H., Luce, R.D. and Tversky, A. [1989]. *Foundations of measurement: Vol. 2.* San Diego: Academic Press.

Suppes, P. and Zinnes, J.L. [1963]. Basic measurement theory. In: R.D. Luce, R.R. Bush and E. Galanter (Eds.). *Handbook of mathematical psychology: Vol. 1* (pp. 1−76). New York: Wiley.

Thurstone, L.L. and Chave, E.J. [1929]. *The measurement of attitudes.* Chicago: Chicago University Press.

Turner, E.A. and Wright, J.C. [1965]. Effects of severity of threat and perceived availability on the attractiveness of objects. *Journal of Personality and Social Psychology*, **2**, 128−132.

Westermann, R. [1983]. Interval-scale measurement of attitudes: Some theoretical conditions and empirical testing methods. *British Journal of Mathematical and Statistical Psychology*, **36**, 228–239.

Westermann, R. [1985]. Empirical tests of scale type for individual ratings. *Applied Psychological Measurement*, **9**, 265–274.

Westermann, R. [1989]. Festinger's theory of cognitive dissonance: A revised structural reconstruction. In: H. Westmeyer (Ed.). *Psychological theories from a structuralist point of view* (pp. 33–62). Berlin: Springer.

Westmeyer, H. [1989]. The theory of behavior interaction. A structuralist construction of a theory and a reconstruction of its theoretical environment. In: H. Westmeyer (Ed.). *Psychological theories from a structuralist point of view* (pp. 145–185). Berlin: Springer.

Poznań Studies in the Philosophy of the Sciences and the Humanities
Vol. 42, pp. 285–297

Uwe Konerding

PROBABILITY AS AN IDEALIZATION OF RELATIVE FREQUENCY:
A CASE STUDY BY MEANS OF THE BTL-MODEL

Introduction

It is common scientific experience that deterministic empirical laws are seldom exactly valid. In nearly every non-trivial empirical application there are some data which are not completely in correspondence with the law. Mainly two approaches are applied for handling this problem. The first of these approaches has a long and successful tradition in the natural sciences. It consists of interpreting the empirical law only as an idealization. This means that the law is not expected to hold exactly; instead it is only expected to approximate data well enough to serve the practical purposes for which the respective empirical application is performed. This point of view presupposed, small deviations from predicted data can easily be tolerated. The second approach is mainly applied in the social sciences. It consists of combining the law with assumptions concerning random deviations from ideal data. As a result formulations arise which refer to probabilities.

In philosophy of science and consequently also within the structuralist considerations [Sneed 1971, Balzer, Moulines and Sneed 1987] both strategies have received quite different attention. The whole conceptual framework of structuralism has been developed with regard to deterministically formulated theories. Consequently, it is primarily appropriate for analyzing and reconstructing scientific activities which are examples of the first approach. The second approach however, i.e., the formulation and empirical application of probabilistic models, has hardly ever been treated within the structuralist framework. In order to counter this deficit, the attempt is made within this article to apply the structuralist conception for reconstructing a probabilistic model together with common strategies of its empirical application. For this purpose the special version of the structuralist framework provided by Balzer, Moulines and Sneed [1987] is selected. The probabilistic model which is to be reconstructed is the so-called BTL-model, which is of great importance in psychology [Luce 1959].

Two alternative reconstructions are presented. The first reconstruction is nearer to usual expositions of the model in textbooks of mathematical psychology [Colonius 1984]. It will, however, be argued that this reconstruction provides no adequate conceptual basis for reconstructing the empirical application of the model. In contrast, the second reconstruction is less similar to usual expositions; but on the other hand it is adequate for reconstructing how the model is empirically applied. The core idea of the latter reconstruction consists of interpreting probability as an idealized relative frequency.

I

Informal Description of the BTL-Model

The BTL-model is meant to describe and explain choice-behaviour. It refers to situations in which subjects are confronted with n-tuples of objects from which they have to select the most dominant object with respect to an a-priori given attribute. The objects may be marbles, politicians, types of beer or different types of crimes; the attribute may be physical weight, competence, good taste or severeness of the offense. There is practically no limitation concerning the kind of objects or the kind of the attribute which can be considered by means of the BTL-model.

In its most generalized form the model can be applied to n-tuples of different size. In this article only the restricted version, which refers to pairs of objects, is considered. In the most common intended application of the model, each pair of objects which can be constituted out of a given sample of objects is presented at least once to the subject. The model can be applied for predicting either individual or group behaviour. In the first case, the model refers to repeated independent choices performed by the same person with respect to the same pair of objects; in the second case, it refers to choices performed by different persons with respect to the same pair of objects. In both cases the choice probabilities are reduced to a latent variable which refers to the presented objects. When individual choice behaviour is considered, this variable can be interpreted as the subjective sensation of the attribute in question; in the case of group behaviour it can be interpreted as the modal sensation of the group.

From this description it follows that there are some essential differences between the BTL-model and its applications on the one hand and a prototypical physical theory-element, as for example the basic element of classical particle mechanics (CPM), on the other. One difference concerns the set of intended applications. In contrast to CPM there is, in the case of the BTL-model, no clearly defined set of intended applications for which the model is generally expected to be empirically valid. Consequently, the whole set of intended applications of the BTL-model is by far more amorphous than the set of intended applications of CPM. A second difference concerns the nature of the terms involved. In the case of CPM the term 'mass' can easily be considered as the same regardless of whether CPM is applied to systems of planets or to atoms. In the case of the BTL-model it may, however, be questionable whether the indivual sensation of weight is the same concept as the modal group sensation of political competence.

In view of these differences it may even be questionable whether the BTL-model and its intended applications can be reconstructed as one single theory-element at all. For the purpose of this article, i.e., a special consideration of those problems which arise when a probabilistic model is reconstructed, the answer to this question is rather irrelevant. For this purpose it may be assumed that a reconstruction of the BTL-model and its intended applications in one single theory-element is feasible.

First Reconstruction

To give an overview of all involved terms, the reconstructions presented here start with the definition of the potential models. In the first possible reconstruction of the BTL-model this is

D1 x is a *potential BTL-model* $(x \in M_p(BTL))$ iff there exist A, p, τ, so that

 (1) $x = \; <A, [0, 1]$ subset of $\mathbb{R}, \mathbb{R}^+, p, \tau>$;

 (2) A is a finite set with at least two elements;

 (3) $p: A \times A \rightarrow [0, 1]$ subset of $\mathbb{R}$;

 (4) $\tau: A \rightarrow \mathbb{R}^+$.

In this reconstruction A is the sample of objects out of which the presented pairs of objects are constituted; this set is the only non-auxiliary base-set in this definition. The function p is the probability that the first object of the respective pair is selected when it is presented together with the second object of this pair; this probability will also be refered to as probability of preference. The function τ is the subjective sensation of the attribute in question.

In order to discriminate non-theoretical and theoretical terms, a very pragmatic criterion of theoreticity is applied. According to this criterion a term is theoretical with respect to a model iff it is usually estimated from data under presupposition of this model. The use of such a vague term as 'usually' becomes necessary because of the vast set of very different intended applications of the BTL-model. But again, with respect to the purpose of this article the application of such a vague criterion will do no harm. Anyway, according to this criterion the discrimination between non-theoretical and theoretical terms can easily be performed. The values for the sensation are usually estimated from data under presupposition of the model; therefore, this term is theoretical with respect to the model. In contrast, the probability of preference is usually not determined under presupposition of the model; therefore, this term is non-theoretical with respect to the model.

Consequently, the definition for the partial potential models is

D2　　x is a *partial potential BTL-model* ($x \in M_{pp}(BTL)$) iff there exist A, p, so that
　　　(1)　　$x = \,<A, [0, 1]$ subset of $\mathbb{R}, p>$;
　　　(2)　　A is a finite set with at least two elements;
　　　(3)　　$p: A \times A \to [0, 1]$ subset of $\mathbb{R}$;

The definition of the actual models is

D3　　x is an *actual BTL-model* ($x \in M(BTL)$) iff
　　　(1)　　$x \in M_p(BTL)$;

　　　(2)　　$p(a,b) = \dfrac{\tau(a)}{\tau(a) + \tau(b)}$　　for all $a,b \in A$.

Here the second axiom describes the already mentioned relationship between the probability of preference and the subjective sensations of the two objects which are presented together.

Up to this point the reconstruction may seem easy. Difficulties arise, however, when the conceptual apparatus developed thus far is applied for reconstructing how the model is empirically applied. The reason for these difficulties is the concept of probability. Social scientists usually interpret probability as the limiting value of relative frequency when sample size tends to infinity. Data sets, however, are always finite. This means that the term 'probability of preference' is not suitable for describing data and, furthermore, that the BTL-model in the reconstruction just presented contains no function at all, which is suitable for describing data. As a consequence the empirical claim cannot be reconstructed. Taken at a whole this means that the reconstruction just presented turns out to be a conceptual dead end.

Second Reconstruction

In the second possible reconstruction the definition of the potential models is

D4 x is a *potential BTL-model* ($x \in M_p(BTL)$) iff there exist A, n, k, τ, so that
  (1) $x = <A, \mathbb{N}_0, \mathbb{R}^+, n, k, \tau>$;
  (2) A is a finite set with at least two elements;
  (3) $n: A \times A \to \mathbb{N}_0$;
  (4) $k: A \times A \to \mathbb{N}_0$;
  (5) $\tau: A \to \mathbb{R}^+$.

As far as the non-auxiliary base-sets are concerned, the second reconstruction is the same as the first. As in the first reconstruction, there is only one set of this kind, namely the sample of objects which is again denoted by A. Both reconstructions differ, however, with respect to their functional terms. The term 'probability of preference' is omitted in the second reconstruction. Instead two new terms are introduced. The first new term, the function n, is the number of times the respective pair of objects has been presented; the second new term, the function k, is the number of times the first object has been selected. The last term, the function τ, is, exactly as in the first reconstruction, the sensation of the attribute in question.

According to the theoreticity-criterion which has been proposed in the preceding part, the terms 'number of presentations' and 'number of times the first object has been selected' can easily be identified as non-theoretical with respect to the model. The term sensation again is theoretical. In contrast to the first reconstruction, all non-theoretical terms in the second reconstruction can be directly applied for describing data.

The definition for the partial potential models is

D5 x is a *partial potential BTL-model* ($x \in M_{pp}(BTL)$) iff there exist A, n, k, so that
  (1) $x = <A, \mathbb{N}_0, n, k>$;
  (2) A is a finite set with at least two elements;
  (3) $n: A \times A \to \mathbb{N}_0$;
  (4) $k: A \times A \to \mathbb{N}_0$;

The definition of the actual models is

D6   x is an *actual BTL-model* $(x \in M(BTL))$ iff
   (1) $x \in M_p(BTL)$;
   (2)   for all $a,b \in A$: if $n(a,b) > 0$ then
         $$\frac{k(a,b)}{n(a,b)} =_{\text{IDEALIZED}} \frac{\tau(a)}{[\tau(a) + \tau(b)]} \, .$$

This last definition is quite analogous to the corresponding definition in the first reconstruction. Again the second axiom describes the 'fundamental law' of the model. In contrast to the first reconstruction, however, there is no relationship stated between sensation and probability but instead a relationship between sensation and relative frequency. This relationship can, of course, only be understood as an idealization. Especially for small data sets it will usually be impossible to find theoretical terms so that this relationship holds exactly. The general idea of probability as the limiting case of relative frequency can now be taken as a kind of metaprinciple which determines how this idealization has to be understood. In other words, if the right hand side of the equation is interpreted as probability, then the derivation of strategies for empirically applying the model can be based upon all that is known from mathematical statistics about the relationship between probability and relative frequencies.

II
Reconstruction of the Empirical Application Procedure

The next step is to elaborate how the actually performed procedure of empirically applying the BTL-model can be reconstructed within the structuralist framework. This reconstruction will be performed in two stages: firstly, the conception which will serve as a reconstructional frame will be discussed; secondly, the procedures which are actually applied when the BTL-model is tested empirically will be subsumed under this frame.

Reconstructional Frame

The reconstructional frame will be produced partly by selecting relevant concepts from the structuralist conception and partly by adequately modifying these selected concepts. The part of the structuralist considerations which is most promising in terms of providing relevant concepts is the part concerned with empirical claims. Here, the attempt is made to find a general characterization of the claim which can be made by means of the mathematical structure of a theory element. Most of these considerations refer to exact empirical claims made by means of an idealized theory element. Balzer, Moulines and Sneed, however, also discuss so-called approximative claims, i.e. claims in which a certain degree of inaccuracy is incorporated. They consider characterizations of such claims as more realistic representations of the manner in which idealized theory elements are actually applied. Therefore, this part of their considerations will be discussed here more thoroughly.

For the exact characterization of approximative empirical claims Balzer, Moulines and Sneed define some special concepts. The most elementary of these concepts is that of a blur, which is simply a set of pairs of models. By means of this concept two more complicated concepts are defined. The first, the uniformity on a set of potential models, is a set of blurs constituted out of the respective potential models so that every blur includes all possible pairs of identical potential models. The second, the class of admissable blurs in a uniformity on a set of potential models, is a special kind of subset of the given uniformity.

The respective subset of the uniformity has to meet four requirements for being a class of admissable blurs: firstly, it has to be non-empty; secondly, membership in this class has to be invariant when the order of elements within the pairs of the blurs is changed; i.e. if a certain blur belongs to the class, then the blur in which the order of models within the pairs is reversed must also belong to the class; thirdly, blurs with equal corresponding blurs on the non-theoretical level must not be separated; i.e. if two blurs of potential models are transformed into the same blur of partial potential models by removing the theoretical functions, then both blurs of potential models have to be either both inside or both outside the class of admissable blurs; fourthly, the class of admissable blurs has to be bounded; i.e. there has to be a set of 'greatest' blurs, so that each admissable blur is a subset of one of these 'greatest' blurs [cf. Balzer, Moulines and Sneed 1987, p. 348]. In structuralist language this set of greatest blurs is refered to as the bound of the class of admissable blurs.

Balzer, Moulines and Sneed apply the concept of a class of admissable blurs to represent those pairs of potential models which can be accepted as approximations of each other. All their various characterizations of approximative empirical claims rely on this concept. In doing this, they adopt a notion of approximation which may be unnecessarily restrictive. To be specific, it is the second requirement which is criticized here. Application of this requirement implies that a model x can only be approximated by a model x' if this relationship also holds the other way round. Balzer, Moulines and Sneed formulate this demand without further extensive investigations of the empirical testing procedures which are actually performed. It will, however, be shown further below that in the case of the BTL-model the actually performed procedures do not meet this requirement. Therefore, the original concept proposed by Balzer, Moulines and Sneed will be modified here by removing the second requirement. In the resulting concept a different epistemological function is attributed to both models of the pair. That is to say, the first model is considered as the approximated and the second as the approximating model.

Balzer, Moulines and Sneed discuss three possible characterizations of an approximative empirical claim. All these three versions result from combining the idea of approximation with the exact empirical claim made by means of an idealized theory element. In the first version the approximation pertains to the empirical content of the theory element, in the second version to the data, and in the third version to both sides. Balzer, Moulines and Sneed [1987] show that the first version implies the second and that the second implies the third. From this it follows that the third version is the empirically least restrictive one. Balzer, Moulines and Sneed argue that this empirically least restrictive version is the most appropriate for reconstructing the empirical testing procedures which are actually performed in empirical sciences. In the case of the BTL-model even the second version will do.

The original structuralist considerations are concerned with empirical claims which refer to all intended application of a theory element. For the purpose of this article only the application to one single intended application is of interest. Therefore, the just selected version of an empirical claim will be correspondingly simplified in order to construct an adequate frame for reconstruction.

Now let i be a single intended application, A a class of admissable blurs for potential models (in the liberalized sense proposed above) and $B(A)$ the corresponding class of blurs for the partial potential models. Let $x \approx_{B(A)} x'$ denote that there is a $v \in B(A)$ so that $<x,x'> \in v$; this relation will be refered to as the empirical approximation relation. Furthermore let $cn_{nt}(T)$ denote the set of all partial potential models of a theory element which can be extended to a potential model so that an actual model results; this set will be refered to as the isolated empirical content. If now the second version of an approximative empirical claim is simplified so that it refers only to single intended applications the resulting formulation is[1]:

There is an $x \in M_{pp}(T)$ so that $i \approx_{B(A)} x$ and $x \in cn_{ni}(T)$.

Actual Reconstruction

With this conception as an interpretational pattern two aspects have to be discussed. Firstly, it must be determined whether the criteria which are usually applied to decide whether the BTL-model empirically holds for a given set of data can also be applied to define a class of admissable blurs. Secondly, it must be determined whether the actual empirical application procedure can be interpreted as an item of the version presented here of an approximative empirical claim.

The procedure which is actually performed for testing the BTL-model consists of two separate steps: firstly, those sensation values are estimated which fit best to the data under presupposition of the model; secondly, a check is made as to whether the probabilities predicted by the model and the estimated sensation values are sufficiently consistent with the data. In order to avoid unneccessary complications, only the case will be discussed where the respective set of data is large[2]. In this case parameters are usually estimated by means of a maximum-likelihood-procedure [Van Putten, 1982] and the final test is performed by means of the Pearson-test-function [cf. Bosch, 1976, Chap. 6]. This test-function and its application is treated here as the best candidate for a conceptual basis for defining the required class of admissable blurs.

Suppose now that the set of objects is ordered according to any arbitrarily given order. Furthermore let p_{ij} be the prefence-probability which refers to choice pair (a_i, a_j) and which is predicted by the model and the corre-

---

1 For readers who want to compare the simplified version with the original more general version of Balzer, Moulines and Sneed [1987, p. 355, (ß)] it should be noted that there seems to be an erratum in the original formulation. As far as can be judged by a reader it should be $X \in Cn(K)$ instead of $\mathbf{X} \in Cn(\mathbf{K})$.

2 In this context a large set of data can be characterized by at least four different choice-objects and by about twenty independent data for each pair of choice objects. For the exact requirements which data have to meet so that the discussed procedures can be applied see van Putten [1982] in the case of parameter estimation and Bosch [1976] in the case of the statistical test.

sponding estimated parameters. The Pearson-test-function applied to the BTL-model is then

$$f = \sum_{i=1}^{|A|} \sum_{\substack{j=1 \\ i \neq j}}^{|A|} \frac{[k(a_i,a_j) - n(a_i,a_j) \times p_{ij}]^2}{[n(a_i,a_j) \times p_{ij}]} \, .$$

Under the given presuppositions and under validity of the model this function is approximately chi-square-distributed with $|A| \times (|A|-3)$ degrees of freedom[3]. As a prerequisite for the empirical decision the acceptable probability for falsely rejecting the model is chosen and a test-function-value is determined so that the corresponding distribution function value is equal to one minus the given probability. If the empirically determined function value exceeds this critical value, the model is rejected, otherwise it is retained.

In the usual statistical jargon the p_{ij} are often refered to as theoretical and the $k(a_i,a_j)$ and $n(a_i,a_j)$ as empirical values. Hence it may be tempting to interpret the Pearson-test-function as a measure of similarity between non-theoretical and theoretical expressions which are both constructed out of function values of the same potential model. With respect to reconstructing the empirical testing procedure by means of the concept of a class of admissable blurs this interpretation, however, leads in the wrong direction. To be specific the concept of admissable blurs is concerned with similarities between two different models and not with similarities between two expressions constructed from different terms out of the same model. Therefore, it is more promising to conceive of the k- and n-values as values of one model and the p-values as values of a different model.

This interpretation, however, is still confronted with a fundamental difficulty, namely, that the third requirement in the original definition of a class of admissable blurs of potential models is violated. According to this requirement two blurs of potential models which are transformed into the same blur of partial potential models by removing the theoretical functions have to be either both inside or both outside the class of admissable blurs. If, however, − as it has just been discussed − similarity measures defined by means of theoretical functions are applied for determining the extension

---

3 The calculation of the degrees of freedom is based upon the assumption that $k(a_i,a_j)$ and $k(a_j,a_i)$ result from the same sample of presentations; i.e. that

$$k(a_i,a_j) = n(a_i,a_j) - k(a_j,a_i) = n(a_j,a_i) - k(a_j,a_i)$$

For the principles for calculating degrees of freedom for the Pearson-test-function see Bosch [1976] or Fisz [1971].

of a class of admissable blurs, this reqirement will always be violated. Therefore, only such similarity functions should be applied which are defined by means of non-theoretical terms. The Pearson-test-function can be transformed into such a non-theoretical similarity measure, when the area of possible arguments of the function is restricted to pairs of models with the same number of choice-objects and identical n-values and when then the expression $n(a_i, a_j) \times p_{ij}$ is interpreted as an estimation of the $k(a_i, a_j)$ in a different model for which the BTL-model holds exactly.

Now let α be a given accepted probability for falsely rejecting the model and let $f(\alpha, |A|)$ be the critical Pearson-test-function-value which has been determined by the procedure described above. An admissable blur of potential models can then be defined as

$$u_\alpha = \{ <x, x'> \mid x, x' \in M_p(BTL) \text{ and } |A| = |A'|$$

and the objects in both models can be ordered so that for all $a_i, a_j \in A$ and for all $a_i', a_j' \in A$: $n(a_i, a_j) = n(a_i', a_j')$ and

$$\sum_{i=1}^{|A|} \sum_{\substack{j=1 \\ i \neq j}}^{|A|} \frac{[k(a_i, a_j) - k(a_i', a_j')]^2}{k(a_i', a_j')} < f(\alpha, |A|) \}.$$

The definition for the corresponding blur for the partial potential models can be easily generated by replacing $x, x' \in M_p(BTL)$ with $x, x' \in M_{pp}(BTL)$.

As already hinted above application of the Pearson-test-function for defining a class of blurs results in a mathematical entity which does not meet the second requirement of the original definition of a class of admissable blurs. Thus if the bound of the class of admissable blurs is defined by means of a fixed α then pairs of models can be constructed so that the calculated test-function will exceed the critical value for one order of the models and so that it will not exceed this value for the revised order. If the less restrictive conception which has been proposed here is applied this aspect will cause no problems for the further reconstruction.

The process of parameter estimation for the BTL-model can now be interpreted as determining the theoretical values of that actual BTL-model which is — with respect to a theoretical approximation relation defined by means of u_α (see above) — most similar to the potential models which can be constructed by extending data with theoretical function values. Under presupposition of this interpretation the n-values of the most similar actual model are also uniquely defined; according to the approximation relation they have to be equal to the n-values of data. Consequently, the k-values of this most similar actual model are likewise uniquely defined. They can be calculated by means of the n- and p-values and by means of the fundamental law. In total this means that the process of parameter estimation can also be interpreted as indirectly determining that element of the isolated empirical content of the BTL-model which is − with respect to the corresponding empirical approximation relation − most similar to the data. The process of empirical testing can then be conceived as checking whether this similarity is strong enough for the empirical approximation relation to hold.

Concluding Remarks

Within this article one possible structuralist reconstruction of a special probabilistic model has been presented. The core idea of this reconstruction consists of conceiving probability as idealized relative frequency. Consequently, investigations have been made into whether the strategies which are actually applied for testing the empirical validity of the model can be reconstructed by means of the structuralist concept of an approximative realistic empirical claim. With the exception of one aspect this concept proved to be adequate. A possible improvement of this concept with respect to this one problematic aspect has been proposed.

The reconstruction presented here only pertains to one selected model. The applied principle of reconstructing, however, could also be applied to the reconstruction of different probabilistic models. This may be performed quite straightforwardly for all other choice models discussed in psychology [cf. Colonius 1984]. For different probabilistic models further, more fundamental considerations might be required.

Institut für Psychologie
RWTH Aachen
Jägerstrasse 17/19
52066 Aachen, Germany

ACKNOWLEDGEMENTS

I would like to thank Mrs. Ute Schmid for a critical review of an earlier version of the manuscript and Mr. Peter Bereza for revising my English.

REFERENCES

Balzer, W., Moulines, C.U. and Sneed, J.D. [1987]. *An Architectonic for Science.* Dordrecht: Reidel.

Bosch, K. [1976]. *Angewandte mathematische Statistik.* Wiesbaden: Vieweg.

Colonius, H. [1984]. Stochastische Theorien individuellen Wahlverhaltens. In: Albert, D., Pawlik, K., Stapf, K.-H. and Stroebe, W. (Eds.). *Lehr- und Forschungstexte Psychologie.* Berlin, Heidelberg: Springer.

Fisz, M. [1971]. *Wahrscheinichkeitsrechnung und mathematische Statistik.* Berlin: Deutscher Verlag der Wissenschaften.

Luce, R.D. [1959]. *Individual Choice Behavior.* New York: Wiley.

Sneed, J.D. [1971]. *The Logical Structure of Mathematical Physics.* Dordrecht.

Van Putten, W.L.J. [1982]. Maximum Likelihood Estimation for Luce's Choice Model. *Journal of Mathematical Psychology*, 25, 163−174.

Poznań Studies in the Philosophy
of the Sciences and the Humanities
Vol. 42, pp. 299–312

Reinhard Suck and Joachim Wienöbst

THE EMPIRICAL CLAIM OF PROBABILITY STATEMENTS, IDEALIZED BERNOULLI EXPERIMENTS AND THEIR APPROXIMATE VERSION

Introduction

Scientific theories in many different areas use probability statements. Furthermore, in applied science and engineering the assertion that a certain event has a high or a low probability can have important practical implications. But what is meant by a statement

(1) $P(A) = \pi$

in which A is an event, π a real number between 0 and 1, and P shorthand for "probability of ..."? Clearly every student who has learned her lesson on probability theory knows that there are some tacit assumptions; a probability space $(\Omega, \mathcal{A}, P)$ $P: \mathcal{A} \longrightarrow [0,1]$, $A \in \mathcal{A}$ not to speak of the Kolmogorov axioms.

In most cases, however, (1) has an empirical meaning, or more cautiously, refers to a situation in the real world where (1) has consequences.

In the structuralist formulation we rephrase this as: some special situations such as experiments etc. are intended applications, and the empirical claim of a theory which uses (1) is the statement that the intended application − or some of the intended applications − are models of the theory.

But as far as we know up to now no structuralist (re)construction of existing theories deals with laws which encompass or consist of statements like (1).

It should be pointed out that this lack of clarification of empirical consequences of random phenomena is not a drawback of the structuralist way of dealing with this problem. Traditionally one would "fit a model" and in case of a satisfactory fit adopt the model leaving the question of the empirical consequences to the user. Questions of the willingness to take risks in decision making are generally held to cope with this problem. In structuralism great importance is ascribed to clarification of empirical claims [cf. Balzer, Moulines and Sneed, 1987, Ch. II]. This is reason enough to extend it to probability statements.

In Suck [1992] the basic ingredients of mathematical probability are incorporated in the structuralist formulation of theories. In the present paper we try to push this project a bit further by investigating in simple cases the empirical claim of probability statements.

Different opinions on the exact meaning of probabilities go down to the foundational roots: historically as well as logically and epistemologically.

We do not deal with foundational aspects of probability and we do not want to add another paper to the discussion whether subjective probabilities are probabilities etc.

We do, however, base our reconstruction on frequencies. In this respect we are close to the frequentist group consisting of Reichenbach and others whose intention was to base the very concept of probability on relative frequency. Our point of view will be to take the usual mathematical definitions of probability as they are usually employed but to take to frequencies when it comes to *determining* probabilities and to interpreting probability statements, in short in all kinds of empirical and practical issues in which probabilities play a part.

Mathematically we base our considerations on the various laws of large numbers which provide the connection between probabilities and frequencies. These theorems are not subject to discussions or contradicting opinions. They are a safe point to start with. Relative frequencies are − in principle − easy to determine. Problems arise with the limiting process which the laws of large numbers entail and with the condition of independence.

For the first problem (limits) we draw on the topological construction of uniformities which are already used by Balzer et al. [1987] but also by Ludwig [1990] to give imprecision and approximation of theories a precise meaning. Concerning the second problem we are really at a loss. Actually what we do in part II is to push the problem around a bit until it is connected to another problem for which a practical solution is known *not* to exist.

I
Theory of Bernoulli Experiments

Throughout the following notation will be employed. The set of natural numbers is denoted by I, the real numbers by $\mathbb{R}$. A probability space $(\Omega, \mathcal{A}, P)$ consists of the elements $\omega \in \Omega$, the events $\mathcal{A}$ and the probability measure P. An infinite sequence of elements of Ω is a mapping $\xi: I \longrightarrow \Omega$, the set of such sequences is Ω^I. The product space of a probability space $(\Omega, \mathcal{A}, P)$ is $(\Omega^I, \mathcal{A}^I, P^I)$. For a given event $A \in \mathcal{A}$ the *indicator function* $I_A(\omega)$ is defined by

$$I_A(\omega) := \begin{cases} 1 & \text{if } \omega \in A \\ 0 & \text{otherwise .} \end{cases}$$

Definition 1 *Let* $x = \, <A, (\Omega, \mathcal{A}, P), \pi>$.

　　a) x *is a potential model of a* zero-one-structure *or ZO-structure iff*

　　　　(ZO_1) Ω *is a non-empty set.*

　　　　(ZO_2) $A \subseteq \Omega$.

　　　　(ZO_3) $(\Omega, \mathcal{A})$ *is a measure space, i.e.* $\mathcal{A}$ *is a σ-algebra of subsets of* Ω.

　　　　(ZO_4) $A \in \mathcal{A}$.

　　　　(ZO_5) π *is a real number satisfying* $0 \leq \pi \leq 1$.

　　b) x *is a model of a ZO-structure iff*

　　　　(ZO_6) $\exists P: \mathcal{A} \longrightarrow [0,1]$ *such that* $(\Omega, \mathcal{A}, P)$ *is a probability space.*

　　　　(ZO_7) $P(A) = \pi$.

Theorem 1 *If* $x = \, <A, (\Omega, \mathcal{A}, P), \pi>$ *is a model of a ZO-structure then for the product measure* P^I *of* P *on* $(\Omega^I, \mathcal{A}^I)$ *holds*

$$P^I\left(\{\xi \in \Omega^I; f(A,\xi) = \pi\}\right) = 1$$

where

$$(2) \quad f(A,\xi) := \begin{cases} \lim_{n \to \infty} \dfrac{1}{n} \cdot \sum_{i=1}^n I_A(\xi(i)) & \text{if the limit exists} \\ 2 & \text{otherwise} \end{cases}$$

Remark: The number 2 in (2) is arbitrary. Any value not in [0,1] would also do.

Proof. The strong law of large numbers yields for $n \longrightarrow \infty$

$$P\left(\bigcup_{i=1}^{\infty} \left\{\omega; \left|\frac{S_{n+i}(\omega)}{n+i} - P(A)\right| \geq \varepsilon\right\}\right) \longrightarrow 0$$

where $S_k(\omega) := \sum_{i=1}^{k} I_A(\omega)$.

The Theorem follows now from the definition of the product measure P^I and of the frequency function $f(A,\xi)$. ☐

Theorem 1 provides us with a tool to "measure" probabilities at least in an approximate sense. Moreover, we can exploit it to construct a theory element in the structuralist sense whose models are the experiments conventially used to gain data to estimate probabilities. The main idea is to interpret the function

(3) $\xi: I \longrightarrow \Omega$

restricted to $I_n := \{1,...,n\}$ for some large $n \in I$ as the "counter" of a Bernoulli experiment. To be more precise another structure of potential models is introduced in the following definition.

Definition 2 *Let* $x = \ <A, I,(\Omega, \mathcal{A}, P), \pi, \xi>$.
   a) x *is a potential model of a* Bernoulli experiment, $x \in M_p(B)$ *iff*
   (B_1) $<A,(\Omega, \mathcal{A}, P), \pi>$ *is a model of a ZO-structure.*
   (B_2) $I = \{1,2,3,...\}$ *is the set of natural numbers.*
   (B_3) $\xi : I \longrightarrow \Omega$, *i.e.*, $\xi \in \Omega^I$.
   b) x *is a model of a* Bernoulli experiment, $x \in M(B)$ *iff*
   (B_4) *The sequence* $(I_A(\xi(i)))$ *satisfies independence*
   (B_5) $f(A,\xi) = \pi$.

If it comes to identifying models of Bernoulli-experiments i.e., to verify that an intended application is a model or not, we are in a fix. (B_4) and (B_5) are — for different reasons — neither liable to an exact verification nor to a falsification. (B_5) entails a limiting process and for (B_4) an infinite number of equalities has to be checked, each of them consisting of a probability the determination of which is in fact impossible.

Actually the case of (B_5) is not that puzzling. We can estimate $f(A,\xi)$ from a large though finite number of values of ξ and accept or reject (B_5) with a certain amount of confidence or imprecision. More accurately, we form an (N_0,N,ε)-blur, i.e., an element of a uniformity [cf. Balzer et al. 1987, Ch. VII] by selecting a number $N,N_0 \in I$ (N the length of the experiment $> N_0$) and defining for $\varepsilon > 0$ $u_{\varepsilon,N_0,N}$ by

$$
(4) \quad u_{\varepsilon,N_0,N} := \Big\{ (x,x') \in M_p(B) \times M_p(B); \ \forall n \, N_0 \le n \le N,
$$
$$
\frac{1}{n} \sum_{i=1}^{n} | \, I_A(\xi(i)) - I_A(\xi'(i) \, | < \varepsilon \Big\}
$$

where ξ belongs to the potential model x and ξ' to x'. Furthermore, let U be the set of all blurs defined by (4), i.e.,

$$
U = \{ u_{\varepsilon,N_0,N}; \ \varepsilon > 0, \ N_0, \ N \in I, \ N_0 < N \}.
$$

It is easy to demonstrate that U satisfies the conditions of a uniformity and the identification of admissable approximations in the sense of Balzer et al. 1987, Chap. VII, 2.2. is possible by this procedure.

If ξ is one of the P^I-almost all elements of Ω^I with $f(A,\xi) = \pi$ then for all $n \ge N_0(\varepsilon)$

$$
(5) \quad \left| \frac{1}{n} \sum_{i=1}^{n} I_A(\xi(i)) - \pi \right| < \varepsilon \, .
$$

We are, however, forced to decide on the validity of (5) from finitely many n. The following lemma describes how models and elements of a blur $u_{\varepsilon,N_0,N}$ are related.

Lemma 1 *If x' is a model of a Bernoulli experiment then for all $\varepsilon > 0$ exists $N_0 = N_0(\varepsilon)$ such that for all $N > N_0$ all $x \in M_p(B)$ with $(x,x') \in u_{\varepsilon/2,N_0,N}$ satisfy Eq. (5).*

Proof. Clearly, the estimation

$$\left| \frac{1}{n} \sum_{i=1}^{n} I_A(\xi(i)) - \pi \right| \leq \left| \frac{1}{n} \sum_{i=1}^{n} I_A(\xi(i)) - \frac{1}{n} \sum_{i=1}^{n} I_A(\xi'(i)) \right| +$$
$$\left| \frac{1}{n} \sum_{i=1}^{n} I_A(\xi'(i)) - \pi \right|$$

holds yielding

$$\left| \frac{1}{n} \sum_{i=1}^{n} I_A(\xi(i)) - \pi \right| < \varepsilon .$$

□

As a summary of the results above it may suffice to note that models of a Bernoulli experiment are approximately characterized if the condition of independence is assumed to hold or can be imported from another theory element. The latter alternative is the subject of the next section.

II
Independence

The property of stochastic independence of a set of events of a probability space has often been claimed to be the most important concept of probability theory. Many theorems of this theory and of statistics invoke i.i.d. − identically independently distributed − random variables.

To stipulate this condition and to guarantee it in an experimental setting are, of course, too different things and it seems that taking it for granted is in almost all cases a kind of idealization.

In structuralist terminology we can perhaps occasionally formulate it as a law in a theory element but how are we to verify that an intended application satisfies it at least approximately? The description of the experiment must provide the means to decide on this question. But how can this be accomplished? The example of roling a die though conceptually easy shows the difficulty: If we describe the die (its material, the homogeneity of the distribution of the material, the geometry of the cube) and the process of throwing, e.g. the forces that are at work, the side conditions in the laboratory (temperature etc.), and if we are very accurate in this: by which consideration do we known that this implies stochastic independence?

Before suggesting our solution(s) to this very general question we mention in passing a related problem which was of some importance in the development of probability theory: *Bertrand's paradoxon*. What is the probability of a randomly chosen secant of a given circle exceeding in length the side of a equilateral triangle inscribed in the same circle? This seemingly innocuous question has three different and superfically equally suggestive answers because there are three approaches for calculating the probability.

First approach: A point within the circle is randomly chosen; this point uniquely determines a secant (the one with the chosen point as midpoint). Since any point whose distance from the center does not exceed $r/2$ gives rise to a secant of the required sort and all other points not, and if we assume a uniform distribution over the circle the probability equals the proportion of the areas of the circles with radius $r/2$ and r, i.e., $1/4$.

Second approach: Because of symmetry it suffices to pick a radius and choose randomly a point on this radius. The straight line perpendicular to the radius through this point is a secant, its length is greater than the side of the triangle if, and only if, the distance of the chosen point from the center exceeds $r/2$. Assuming uniform distribution over the radius yields the probability $1/2$.

Third approach: This time we fix a point on the circumference and choose randomly a second point on the circumference (again with uniform distribution). The ensuing secant is greater than the side of the triangle if, and only if, the second point is on the part between BC of the circumference of the circle in Fig. 1 where A is the chosen point. Thus, the probability is $1/3$ in this case.

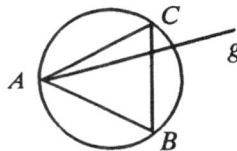

Figure 1: The intuition underlying the third approach to Bertrand's paradox

All three solutions to Bertrand's problem are physically realizable; at least idealizations of physically realizable processes seem to be solutions with different results.

This state of things becomes less paradoxical if one becomes aware of the weak point of the original question: what is a randomly chosen secant. In this respect the three proposed solutions differ. Three different experimental procedures of random choice — none of them can at first sight be evaluated as inadequate — lead to different conclusions. To be more specific, they lead to different probability measures over the same σ-algebra. To resume the issue of independence, for differing probabilities the concept of stochastic independence is, naturally, different, too. It is not difficult to exhibt sets A and B of secants which are independent in one of the approaches and dependent in the other. One should note that by introducing further requirements the measure can be made unique. For example by a classical result of integral geometry the only probability which is invariant under rigid motions is obtained from the second approach.

The conclusion we draw from the consideration of Bertrand's Paradoxon is that, *if the description of the experiment yields experimental independence and this should be reflected by stochastic independence then the description must be specific to the extent that the probability measure is uniquely determined.*

If for some reason or other in the setting of Bertrand's paradoxon the resulting probability should be motion invariant (because that fits best to the physical realisation of selecting a secant randomly) then the question of independence is also settled.

However, in experiments with repeated trials the product probability used in the preceeding section which suggests itself for describing independent repeated trials is not an adequate tool because the product probability of $A \times A$ (i.e., A occurs in the first trial and in the second) satisfies *by definition*

$$P^2(A \times A) = P(A) \cdot P(A).$$

What is needed is a criterion or a probability by which independent and dependent cases can be told apart.

In the sequel we present techniques which can be applied to test for independence in the actual sequence of events generated by an experimental series. Then we introduce a theory element based on an investigation of Luce and Narens [1978] in which the aforementioned considerations are to some extent included. However, we do not see how to apply it directly to the problem of (B_4) in Definition 2.

Independence Tested by the Binomial Distribution

The question of independence of the sequence $(I_A(\xi(i)))$ is much more involved than the problem of the limit. Essentially it is the experimental design which has to guarantee that the value of $I_A(\xi(i))$ is not influenced by $I_A(\xi(1)),\ldots,I_A(\xi(i-1))$. Mathematically this is transformed into stochastic independence and expressed by

$$P(I_A(\xi(i)) = 1 \,|\, (I_A(\xi(1)),\ldots,I_A(\xi(i\text{-}1)))) = P(I_A(\xi(i)) = 1)$$

Condition (B_5) of Definition 2 can be fulfilled by highly dependent sequences. However, according to Theorem 1, there is at least theoretically the chance of detecting deviations from independence by investigation particularly chosen events A_i (e.g. the event 00011 in the sequence $(I_A(\xi(i)))$. This consideration is the idea of the usual randomization tests. However for each test there exist sequences (with highly sequential dependence) which pass this test. Nevertheless, one can select such a test, apply it to the sequence of the particular experiment and decide for or against independence within the usual error margin, where the probability of error can again be reconstructed as a blur chosen from a uniformity.

A Probability Space on the Positive Integers to Reflect Independent Trials

Next we introduce a conditional probability space on the set of positive integers on which the function $\xi\colon I \longrightarrow \Omega$ is a random variable. By this technique we can transform the critical condition of independence into the asymptotic behavior of certain functions providing us with a means of approximately testing for independence. Thus with the usual restrictions concerning blurs and approximation we can "reconstruct" independence, i.e., describe a theory-element which includes a function ξ satisfying the conditions required for repeated trials. The laws which describe models in the class of potential models can be formulated.

Let $\xi\colon I \longrightarrow \Omega$ and $<I, 2^I, \mathcal{F}_\xi, P_\xi(.\,|D)>$ be defined as follows:

$$A \subseteq \Omega$$
$$p_i := P(I_A(\xi(i)) = 1)$$
$$\mathcal{F}_\xi := \left\{ D \subseteq I; 0 < \sum_{i \in D} p_i < \infty \right\}$$

For $D \in \mathcal{F}_\xi$ let

$$P_\xi(.\,|D)\colon 2^I \longrightarrow \mathbb{R}$$

(6) $\quad P_\xi(B|D) := \dfrac{\displaystyle\sum_{i \in B \cap D} p_i}{\displaystyle\sum_{i \in D} p_i}.$

In this way a conditional probability algebra is defined which is generated by a usual probability if $\sum_{i=1}^{\infty} p_i$ converges. The possible failure to meet this condition makes the more complicated construction necessary. Probabilities like the preceding one are often used in number theory.

Let us calculate $P_\xi(B|D)$ for some finite $D \in \mathcal{F}_\xi$ with $B \subset D$ under the assumption of ξ generating independent trials. Equation (6) yields

$$P_\xi(B|D) = \frac{\sum\limits_{i \in B \cap D} P(I_A(\xi(i) = 1)}{\sum\limits_{i \in D} P(I_A(\xi(i) = 1)}$$

(7) $$\approx \frac{|B|}{|D|} \, .$$

(7) follows from the fact that the expected number of occurrences of an event in a Bernoulli-experiment of length n is $n \cdot p$.

Now, we assume a random sequence (ϵ_i) i.e., $\epsilon_i \in \{0,1\}$ such that (ϵ_i) is a model of the theory RS. We use this sequence (ϵ_i) to determine a random sequence (K_n) for $n \to \infty$ of sets $K_n \subset I$ and subsets κ_n of K_n in the following way:

1. For convinience let $n = 2^m$ with $m \in \mathbb{N}$.
2. $i_1 = \sum_{i=1}^{m} \epsilon_i \cdot 2^{i-i}$.
3. $j = \sum_{i=m+1}^{2m} \epsilon_i \cdot 2^{i-m-1}$.
4. If $j \neq i_1$ then $i_2 = j$. If $j = i_1$ then calculate a new j with $\epsilon_{m+2}, \ldots, \epsilon_{2m+1}$ and test for $j \neq i_1$. Repeat this until i_2 is constructed.
5. Construct analogously $i_3, \ldots, i_n$. Define $K_n := \{i_1, \ldots, i_n\}$.
6. Evaluate $S := \sum_{i \in K_n} p_i$. If $S = 0$ repeat the procedure until a K_n is constructed with $S > 0$.
7. Let $k = 2^{m-1}$.
8. Delete from (ϵ_i) the initial segment of indices already used in the preceding procedure.
9. Construct with the new (ϵ_i) a set $\{m_1, \ldots, m_k\}$ using the steps 2. to 5.
10. Define $\kappa_n := \{i_{m_1}, \ldots, i_{m_k}\}$.

We suppose that for each n this process terminates after a finite number of steps. After going through this procedure we end up with two sets κ_n, $K_n \subset I$ which enjoy the properties:

$$|K_n| = n, \quad |\kappa_n| = \frac{n}{2}$$

$$\kappa_n \subset K_n .$$

If (ϵ_i) is a random sequence then (7) reduces to $P_\xi(\kappa_n|K_n) = 1/2$ if the sequence $(\xi(i))$ satisfies independence. Therefore we introduce

Definition 3 ξ *is an* independence generating function relative to (ϵ_i) *iff for sets κ_n and K_n selected according to (ϵ_i) holds*

(8) $$\lim_{n \to \infty} P_\xi(\kappa_n|K_n) = \frac{1}{2} .$$

With these concepts a test for independence consists of an approximate verification of (8).

Thus, with Definition 3 it is feasible to replace (B_4) of Definition 2 by

(B_4) ξ is an independence generating function relative to some random sequence (ϵ_i).

Obviously, at this point of the development a theory of random sequences (or of random number generators) is called for. We shall neither provide such a theory nor a reconstruction of an existing concept, but refer to Chaitin [1987] or related developments of Kolmogorov or Martin-Löf (cf. Fine [1973]). Chaitin gives a definition which seems theoretically satisfactory, but is − for reasons connected to Gödels incompleteness theorem − not a practical instrument to determine for a given sequence whether it is random or not.

Thus it may seem that little is gained in reducing the concept of independence to the idea of randomness. However, if a random process is at our disposal (e.g. we have a coin which can be regarded as unbiased and which, when used repeatedly produces independently 'heads' and 'tails') then we might use it to produce (ϵ_i) and test the procedure under scrutiny (i.e., the function ξ).

Qualitative Independence of Experiments

Another way of dealing with the independence assumption is to involve a qualitative structure in which stochastic independence is a primitive. Luce and Narens [1978] investigated this idea in the context of axiomatic measurement theory. They define what they call an independent joint qualitative probability structure. Here the independence comes in via a condition which is known in conjoint measurement as an independence condition (see Theorem 2 below) which is at first sight quite different from stochastic independece. However, given the other axioms, a numerical representation is proved to exist which maps pairs, triples,..., n-tupels of events multiplicatively into the reals. This property is reminiscent of stochastic independence, at least, if the representing function is a probability (and, of course, the whole structure is a probability space). But this property is established by an embeddability argument: The structure is embedded in a probability space, and the numerical representation is extended to a probability such that multiplicativity carries over to the usual independence.

The n-tupels of events can be interpreted as repeated trials in each of which the "same" algebra of set is the relevant algebra of events and multiplicativity is tantamount to independence. By this procedure experimental independence is captured by qualitative axioms. Furthermore, it is possible to formulate the conditions for this probability to coincide with the product probability P^l on $(\Omega^l, \mathcal{A}^l)$.

In our structuralist reconstruction the theory element looks somewhat different because we introduce the result of the representation theorem as a law distinguishing potential models and models.

Definition 4 Let $x = <\Omega, \mathcal{A}, \precsim, P>$.

   a) *The system x is a potential model of an* independent experiment, *$x \in M_p(IE)$ iff*

   (IE_1) Ω *is a non-empty set.*

   (IE_2) $\mathcal{A}$ *is an algebra of subsets of Ω.*

   (IE_3) $\precsim$ *is a weak order on $\mathcal{A} \times A$.*

   (IE_4) $(\Omega, \mathcal{A}, P)$ *is a finitely additive probability space.*

   b) *x is a model of an independent experiment iff*

   (IE_5) *For all $A,B,C,D \in \mathcal{A}$*

   $$(A,B) \precsim (C,D) \text{ iff } P(A) \cdot P(B) \leq P(C) \cdot P(D).$$

It can be shown [cf. Luce and Narens 1978, Theorem 5] that (IE_5) holds true if some qualitative axioms are fulfilled which are quite natural in conjoint measurement structures. We mention only three of them:

Theorem 2 Let $<\Omega, \mathcal{A}, \precsim>$ *be a system satisfying IE_1, IE_2, and IE_3 of Definition 4. If it satisfies for all $A,B,C,D \in \mathcal{A}$*

independence, *i.e., $(A,C) \precsim (B,C)$ iff $(A,D) \precsim (B,D)$.*

symmetry, *i.e., $(A,B) \sim (B,A)$.*

distribution, *i.e., for $A \cap B = C \cap D = \emptyset$, $(A,E) \preceq (C,F)$, and $(B,E) \preceq (D,F)$ hold then $(A \cup A, E) \preceq (C \cup D, F)$.*

and a few technical conditions then there is a probability P on $\mathcal{A}$ such that $x = <\Omega, \mathcal{A} \preceq, P>$ is a model of an independent experiment, i.e., x satisfies IE_4 and IE_5.

The significance of a result of this kind lies in the fact that the multiplicativity of P (for stochastic independent events) need not be postulated as in (IE_5), but can be inferred from other axioms which do not refer to a real-valued probability function. Once P is obtained from a measurement theoretic representation theorem (which represents x in the sense that (IE_5) is satisfied), the concept of independence is also fixed.

To accomplish this aim, Luce and Narens [1978] prove another theorem which guarantees that a model of an independent experiment can be related to a probability space in which those events are stochastic independent which one expects to be experimentally independent.

Discussion

Seventy years on the dust has settled whirled up by the debate on the foundations of probability in the twenties. Relative frequencies which von Mises and Reichenbach employed to define probability are now linked to probabilities by the Laws of Large Numbers. In the preceding sections we used relative frequencies and these laws as a means − as the only means − of measuring probabilities and to link statements involving probabilistic parts witsh phenomena. In this respect we are close to frequentist interpretations.

The same or at least a similar point of view ist held by van Fraassen [1979] in a philosophical context and by Ludwig [1990] for physical theories.

In a sense the statement $P(A) = \pi$ is an idealization and real phenomena and data can verify or falsify it only approximately. A law of large numbers characteristically involves a limiting process (the various laws differ with respect to this limit) which is translated in the approximation by a blur of a uniformity. This shows the discrepancy between the ideal and its approximation. We want to emphasize that in the present context probabilities are not used to model neglected or not controlled or not controllable chance effects as a source of imprecision. That is the usual way statistics is employed in social sciences. We dealt with a situation where the probabilistic statement is a congeneric part of the theory under scrutiny.

312

The problem of independence and the approach to deal with it presented in part II seems to be extremely hard, when it comes to a precise structuralist reconstruction. Our intention to base it on Chaitin's randomness definition is apparently not satisfactory because Chaitin [1987] has related it to Gödel's incompleteness result. From this interrelation he concludes the impossibility of exhibiting randomness. Despite the criticism raised by Lambagden [1989] against this point of view we hold that this is a "natural" feature of randomness. The impossibility result transfers to our independence reconstruction.

Does that mean that we cannot recognize models under the intended applications? Yes. But here a similar remark applies as with respect to the limit. Independence is the ideal, we can detect violations of independence and if we do not find any in an intended application we can think of it as an approximation to the ideal of independence.

Finally, we want to point out that the independence question is not restricted to Bernoulli experiments and the context of probability statement like (1). Many theories in physics but also in social sciences and psychology use devices with repeated trials which have to satisfy experimental independence which is in most cases mathematically described by stochastic independence. In all these cases an analysis similar to ours is warranted.

Department of Psychology
University of Osnabrück
D-49069 Osnabrück, FRG.
e-mail: Suck at DOSUNI1.BITNET

REFERENCES

Balzer, W., Moulines, C.U., Sneed, J.D. [1987]. *An Architectonic for Science*. Dordrecht: Reidel.

Chaitin, G.J. [1987]. *Information, Randomness, and Incompleteness — Papers on Algorithmic Information Theory*. Singapore: World Scientific.

Fraassen, B.C. van [1979]. Relative Frequencies. In: Salmon, W.C. (Ed.). *Hans Reichenbach: Logical Empiricist*. Dordrecht: Reidel.

Lambagden, M. van [1989]. Algorithmic Information Theory. *The Journal of Symbolic Logic*, **54**,4, 1389–1400.

Luce, R.D., Narens, L. [1978]. Qualitative Independence in Probability Theory. *Theory and Decision*, **9**, 225–239.

Ludwig, G. [1990]. *Die Grundstrukturen einer physikalischen Theorie*. Berlin: Springer.

Suck, R. [1992]. Probabilistic Components of a Theory in the Structuralistic Reconstruction. In: Westmeyer, H. (Ed.). *The Structuralist Program in Psychology: Foundations and Applications*. Toronto: Hogrefe & Huber.

Poznań Studies in the Philosophy
of the Sciences and the Humanities
Vol. 42, pp. 313–323

Pekka J. Lahti

IDEALISATIONS IN QUANTUM THEORY OF MEASUREMENT

Introduction

Quantum theory of measurement is the part of quantum mechanics which investigates the measurement process as a physical process subject to the laws of quantum physics. This theory has both practical and conceptual dimensions, ranging from a study of the measurement accuracy limitations in instrumentation technology to an investigation of the fundamental issues of quantum mechanics.

Recent advances in ultrahigh technology has brought these two divergent subjects nearer to each other, even to a quite surprising extent. Extremely controlled experimentation on individual objects, like atoms, neutrons, electrons, or photons is getting a daily enterprise in experimental quantum physics. To say the least then, this progress has made it highly wishful to develop an interpretation of quantum mechanics which goes beyond the phenomenological level of a statistical interpretation.

Difficulties entountered in developing an interpretation of quantum mechanics as a theory of individual objects and their properties are known to be fundamentally rooted in the irreducible nature of the measurement outcome probabilities predicted by the theory.

Progress on mathematical and conceptual foundations of quantum mechanics, especially on its theory of measurement, together with some new ideas on the interpretation of the theory, like the modal interpretations, have, however, made it possible to formulate in a systematic fashion an interpretation of quantum mechanics which exceeds a purely statistical level, and which comes closer to the present day experimental practise.

In this paper I shall attempt to review some of that development and to show, in particular, how an increase in the degree of idealisation in measurement leads to a further structural specification of a measurement, opening thereby perspectives for enriching the statistical interpretation of quantum mechanics to its realistic interpretation as a theory of individual objects and their properties.

I
The Universe of Discourse
Quantum Mechanics with its Minimal Interpretation

In the ordinary Hilbert space formulation of quantum mechanics the description of a physical system is based on a complex separable Hilbert space $\mathcal{H}$, with the inner product $< \cdot \,|\, \cdot >$. In their most common representation *states* and *observables* of the system are represented as unit vectors φ of $\mathcal{H}$ and as self-adjoint operators A acting in $\mathcal{H}$, respectively. If $A = \int_{\mathbf{R}} a\,dP^A(a)$ is the spectral decomposition of A, then the *probability measure* defined by this observable and a state φ obtains the explicit form

(1) $\quad p_{\varphi}^{A}(X) := \; <\varphi\,|\,P^A(X)\varphi>$

where $P^A(X)$ is the spectral projection of A associated with the (Borel) subset X of the real line $\mathbf{R}$. According to the *minimal* (Born) *interpretation* these numbers are probabilities for measurement outcomes: $p_{\varphi}^{A}(X)$ *is the probability that a measurement of the observable A leads to a result in the set X when performed on the system in a state φ*. Clearly, any two unit vectors which differ only by a phase give rise to the same probabilities for all observables, a fact which amounts to representing states as one-dimensional projection operators $P[\varphi]$ rather than as their generating unit vectors φ.

For our subsequent needs we recall that in addition to the vector states φ, or rather $P[\varphi]$, there are the mixed states T represented as density operators, or positive trace-one operators. If T_1 and T_2 are any two such states and λ is a real number between 0 and 1, then the concex combination, or the mixture, $T = \lambda T_1 + (1-\lambda)T_2$ is also a state. We say that T_1 is *a convex component* of T if $\lambda \neq 0$. In that representation the vector states are exactly those states which have no other convex component than the state itself. An equivalent algebraic characterisation of such states is their idempotency: $T^2 = T$. We recall that any state can be decomposed even to vector states, but such a decomposition is never unique unless the state is a vector state. Much of the difficulties in the interpretation of quantum mechanics is due to this nonuniqueness. (For a further discussion, see, for instance, the of monographs of Beltrametti and Cassinelli [1981], Busch *et al.* [1991], and van Fraassen [1991]).

For arbitrary states T the measurement outcome probabilities have the structure $p_T^A(X) = tr[TP^A(X)]$, which for the vector states $T = P[\varphi]$ reduces to the above formula (1).

We remark also that the representation of an observable as a self-adjoint operator is unnecessarily restrictive, the most appropriate representation being the one given by a positive-operator-valued measure. However, for the present purpose the above formulation of an observable is sufficient. Moreover, for most part of this paper it is sufficient to consider only discrete observables, that is, observables which are represented as discrete self-adjoint operators $A = \sum a_i P_i$, where a_i are the eigenvalues of A and P_i are the corresponding eigenprojections. Then $p_\varphi^A(a_i) = <\varphi | P_i \varphi>$ is the probability that a measurement of A leads to the result a_i when performed on the system in a state φ.

Some further concepts and results are still in due. First of all, let me recall that for any state T there is the smallest projection operator P_T such that

$$(2) \quad T = P_T T = T P_T .$$

This is the support projection of T, and it is the projection on the closure of the range of T, $\overline{ran}(T)$. As concerns the measurement outcome probabilities, the role of this projection operator appears in the fact that $p_T^A(X) = tr[TP^A(X)] = tr[TP_T P^A(X)P_T]$ showing that nonzero probabilities may occur only for those sets X for which $P_T P^A(X) P_T$ is nonzero.

There are various important options to introduce an order among states. Here we need to recall only the following one: for any two states T_1 and T_2, T_1 is *possible relative to* T_2, in symbols $T_1 \prec T_2$, if the support projection of T_1 is contained in that of T_2, or equivalently, if the closure of the range of T_1 is contained in the closure of the range of T_2, that is

$$(3) \quad T_1 \prec T_2 \text{ iff } P_{T_1} \leq P_{T_2} \text{ iff } \overline{ran}(T_1) \subset \overline{ran}(T_1).$$

This defines, indeed, an order on the set of states, the vector states being minimal with respect to that order whereas any T with $P_T = I$ is maximal. Let me remark that the condition $T_1 \prec T_2$ does not yet imply that T_1 were a convex component of T_2. Clearly, if a state is a convex component of another state, then it is also possible with respect to that state.

Let $\mathcal{P}(\mathcal{H})$ denote the set of projection operators on $\mathcal{H}$. For any state T we define the following sets:

$$(4) \quad P_1(T) := \{P \in \mathcal{P}(\mathcal{H}) : tr[TP] = 1\}$$

316

(5) $\mathcal{P}_1(T) := \{P \in \mathcal{P}(\mathcal{H}) :$ there is a $T' \prec T$ such that $tr[T'P] = 1\}$

so that for all states T

(6) $P_1(T) \subset \mathcal{P}_1(T)$.

Moreover, we observe that

(7) $P_1(T) = \mathcal{P}_1(T)$

if and only if T is a vector state. If $T = P[\varphi]$ we write $\mathcal{P}_1(\varphi)$ instead of $\mathcal{P}_1(P[\varphi])$.

The sets $P_1(T)$ and $\mathcal{P}_1(T)$ play an important role in our considerations. So do they also in the Copenhagen variant of the modal interpretation of quantum mechanics as developed by van Fraassen [1991].

For a first orientation, consider an observable $A = \sum a_i P_i$. If the system is in a state T, then $p_T^A(a_i) = tr[TP_i]$ are the measurement outcome probabilities for this observable. If $P_k \in \mathcal{P}_1(T)$ for some k then a measurement of this observable is certain to yield the result a_k. We take this condition as being *sufficient* for A having the value a_k:

(8)   if $p_T^A(a_k) = 1$, then A *has* the value a_k.

Recalling that properties of the system are represented as projection operators, we may then say that the set $P_1(T)$ contains properties which the system has in that state.

The problems would immediately be manifold if we were to follow von Neumann (1955) and consider the condition $p_T^A(a_k) = 1$ also as necessary for A having the value a_k. To avoid these difficulties, let me therefore emphasise, that for us condition (8) is *sufficient but not necessary* for an observable having a value. Indeed, we wish to keep it open that if the system is in a mixed state T, then A still could have the value a_k even though $p_T^A(a_k) \neq 1$. The set $\mathcal{P}_1(T)$ is then taken to contain the properties that the system *could have* in state T, or, equivalently, it is taken to specifiy the values that an observable could have in that state. The spelling out of the details of the relevant characterisations requires the use of the measurement theory which shall be discussed subsequently. But, let me recall, that from a probabilistic point of view the identification of the set $\mathcal{P}_1(T)$ as the set of properties that the system could have in that state rests on the following fact: $P \in \mathcal{P}_1(T)$ if and only if there is a state T' which is possible relative to T such that $tr[T'P] = 1$.

In closing this introductory part, let me note still that the set $\mathcal{P}_1(T)$ can be characterised as follows [Cassinelli et al, 1993a]:

(9) $\mathcal{P}_1(T) = \cup (P_1(\varphi) : P[\varphi] \leq P_T)$

Thus, indeed, the set $\mathcal{P}_1(T)$ equals to $P_1(T)$ exactly when $T = P[\varphi]$ for some φ. Therefore, if the system is in a pure state, then all the properties which the system then could have are exactly those which it actally has. We note also that if a property P is in $\mathcal{P}_1(T)$, then its measurement outcome probability $tr[TP]$ is nonzero, or, equivalently, for $A = \sum a_i P_i$, if $P_i \in \mathcal{P}_1(T)$, then $p_T^A(a_k) \neq 0$. However, the converse is not true: $p_T^A(a_k) \neq 0$ does not necessarily imply that $P_i \in \mathcal{P}_1(T)$.

II
Measurements as Calibrations

Consider an observable $A = \int_R a dP^A(a)$. To model a measurement of A one usually fixes a measuring apparatus, with its Hilbert space $\mathcal{K}$, an initial state Φ of the apparatus, a pointer observable $Z = \int_R z dP^Z(z)$, and a (unitary) measurement coupling $U: \mathcal{H} \otimes \mathcal{K} \to \mathcal{H} \otimes \mathcal{K}$. If φ is the initial state of the measured system, then $U(\varphi \otimes \Phi)$ is the system-apparatus state after the measurement. Denoting the corresponding reduced states of the measured system and the measuring apparatus as $T(\varphi)$ and $W(\varphi)$, respectively, we have the schematic representation of the state transformations associated with a measurement:

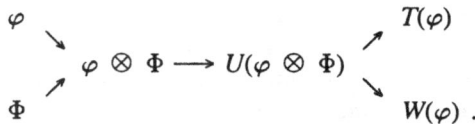

A minimal requirement for $\mathcal{K}$, Z, Φ, and U to constitute a measurement of A is the *calibration condition* [Busch et al, 1991]: for any X and φ,

(10) $p_\varphi^A(X) = 1$, then $p_{W(\varphi)}^Z(\overline{X}) = 1$

where $\overline{X}$ is the value set of the pointer observable corresponding to that of the measured observable. Using the notations of Section 1 we may express the calibration condition also as follows: for any X and φ,

(11)   if $P^A(X) \in P_1(\varphi)$, then $P^Z(\bar{X}) \in P_1(W(\varphi))$.

The calibration condition is equivalent to the apparently stronger *probability reproducibility condition*: for any X and φ,

(12) $p^A_\varphi(X) = p^Z_{W(\varphi)}(\bar{X})$.

For a discrete observable $A = \sum a_i P_i$ these three equivalent conditions read: for any i and φ,

(13a) if $p^A_\varphi(a_i) = 1$, then $p^Z_{W(\varphi)}(z_i) = 1$
(13b) if $P_i \in P_1(\varphi)$, then $Z_i \in P_1(W(\varphi))$
(13c) $p^A_\varphi(a_i) = p^Z_{W(\varphi)}(z_i)$

where now $Z = \sum z_i Z_i$. Using the terminology of Section 1 we may now conclude that the basic requirement of a measurement is the following: *if a measurement of an observable is certain to yield a particular result, then the pointer observable has the corresponding value after the measurement.*

III
Measurements Leading to Results

A lot a work has gone to characterise measurements in terms of the calibration condition, or the probability reproducibility condition. A number of properties of measurements are therefore known in that general level. However, it is obvious that this condition does not exhaust the physics of a measurement process, some further specifications are in due. From the conceptual point of view there remains the question of analysing the obvious requirement that "a measurement leads to a result". In attempting to approach this question one has to keep well in mind the basic problem of quantum mechanics: the theory does not allow one in any obvious way to explain the occurence of a measurement result. Therefore, some care is needed.

Weak Modality: Possible Pointer Values

As a first attempt to study the above question we follow the idea expressed by van Fraassen [1991] that *to assume that a measurement leads to a result is to assume that the pointer observable has a value after the measurement.*

In the present context this assumption leads to a further specification of a measurement process. Indeed, consider a measurement $<\mathcal{K}, Z, \Phi, U>$ of an observable A of the object system. If φ is the initial state of the system, then $W(\varphi)$ is the state of the apparatus after the measurement. According to the point of view followed here, the set $\mathcal{P}_1(W(\varphi))$ contains all the properties which the apparatus could have in that state, or, to put it in the other way, this set specifies the values that an apparatus-observable could now have. Therefore, the above identification leads to the following requirement on the measurement:

WEAK MODALITY. Consider a measurement $<\mathcal{K}, Z, \phi, U>$ of an observable $A = \sum a_i P_i$. For any possible value a_i of A and for any initial state φ of the measured system,

(14)   if $p_\varphi^A(a_i) \neq 0$, then $Z_i \in \mathcal{P}_1(W(\varphi))$.

Recalling that $p_\varphi^A(a_i) = p_{W(\varphi)}^Z(z_i)$ for all φ and i, one may supplement this condition with the following assumption: *the pointer observable Z has the value z_i in the state $W(\varphi)$ with the probability $p_\varphi^A(a_i)$.* This assumption poses, however, no further structural constrains on the measurement.

The weak modality condition (14) together with the given association of probabilities for possible pointer values constitutes the basic assumption of the Copenhagen variant of the modal interpretation of quantum mechanics. Since $p_\varphi^A(a_i) = p_{W(\varphi)}^Z(z_i) \neq 0$ does not necessarily imply that $Z_i \in \mathcal{P}_1(W(\varphi))$, condition (14) is, indeed, a specification of the measurement which goes beyond the calibration condition. It is easy to give sufficient conditions for a measurement to satisfy this condition. Indeed, the final apparatus state can always be decomposed as $W(\varphi) = \sum_{ij} Z_i W(\varphi) Z_j$. It may now happen that the "off-diagonal elements" $Z_i W(\varphi) Z_j$ do not contribute to $W(\varphi)$ so that, in fact, $W(\varphi) = \sum_i Z_i W(\varphi) Z_i$. In such a case $Z_i \in \mathcal{P}_1(W(\varphi))$ whenever $p_\varphi^A(a_i) \neq 0$. For instance, the value reproducible measurements of part IV of the paper are such. Also necessary and sufficient conditions for (14) to hold are now known. In order not to get involved with too many technical questions here I shall content to give only a reference where an interested reader may find these results with further details, see Cassinelli and Lahti [1993a].

Strong Modality: Possible Values

In addition to the above ideas, one may consider the possibility that also the measured observable has a value after the measurement, and that it would have a particular value with the same probability as the pointer observable has the corresponding value after the measurement. Again, this idea lends itself readily to a formal expression in our universe of discourse.

Consider a measurement $<\mathcal{K}, Z, \phi, U>$ of an observable $A = \sum a_i P_i$. If φ is the initial state of the measured system, then $T(\varphi)$ is its final state. The set $\mathcal{P}_1(T(\varphi))$ contains the properties that the measured system could have after the measurement. The question at issue now is whether $P_i \in \mathcal{P}_1(T(\varphi))$ whenever $p_\varphi^A(a_i) \neq 0$. It may easily be demonstrated that neither the calibration condition nor the weak modality condition implies, in general, that this were the case. Therefore, to assume that the measured observable has a particular value after the measurement is a further condition on measurement. We formulate the involved set of ideas as follows:

STRONG MODALITY. Consider a measurement $<\mathcal{K}, Z, \Phi, U>$ of an observable $A = \sum a_i P_i$. For any possible value a_i of A and for any initial state φ of the measured system,

(15)   if $P_i \in \mathcal{P}_1(\varphi)$, then $P_i \in \mathcal{P}_1(T(\varphi))$.

It might be natural to supplement this condition with the following two assumptions: for each i and φ

(16) $p_\varphi^A(a_i) \neq 0$, then $P_i \in \mathcal{P}_1(T(\varphi))$;
(17)   the measured observable A has the value a_i in the state $T(\varphi)$ with the probability $p_\varphi^A(a_i)$.

However, the innocent looking condition (15) is very strong and it already contains (16) and also essentially (17). Therefore, they can be dropped from our considerations. The strong modality condition specifies the structure of a measurement to a large extent, but not yet completely. This will become clear below.

IV
Measurements Reproducing Values

It has been a subject of much discussions whether a measurement should be value reproducible or even value determinative, that is, whether the measured observable should also posses a value after the measurement, and whether, for instance, upon a repetion the measurement should lead to the same result. There are various formulations of these intuitively quite different ideas. But they all are formally equivalent in the present framework. I shall now discuss some of them here.

Consider a measurement $<\mathcal{K}, Z, \Phi, U>$ of an observable $A = \sum a_i P_i$. We say that such a measurement is *value reproducible* if the following implication always holds true:

(18) if $p_\varphi^A(a_i) = 1$ then $p_{T(\varphi)}^A(a_i) = 1$.

The value reproducibility of a measurement is known to be equivalent to a seemingly stronger property, the measurement being of the *first kind*: for any i and φ

(19) $p_\varphi^A(a_i) = p_{T(\varphi)}^A(a_i)$.

On its turn, this condition is equivalent to the fact that the final state of the measured system is a *mixture of eigenstates* of the measured observable:

(20) $T(\varphi) = \sum P_i T(\varphi) P_i$.

But this means that $P_i \in P_1(T(\varphi))$ whenever $p_\varphi^A(a_i) = 1$. Thus we may conclude that a measurement $<\mathcal{K}, Z, \Phi, U>$ of an observable A satisfies the strong modality condition if and only if it is value reproducible, or, equivalently, if it is of the first kind. There are even further equivalent characterisations of such measurements in terms of strong correlations and conditional probabilities [Cassinelli and Lahti, 1993b].

To close this section, let me note that both the value reproducibility condition as well as the first kind property can be expressed in a natural way not only for discrete observables $A = \sum a_i P_i$ but for arbitrary observables $A = \int a dP^A(a)$, as well. The two properties of a measurement remain equivalent and any of them implies that the measured observable is, in fact, a *discrete* one. Clearly, the same holds also true for the strong modality condition. It can be formulated for arbitrary observables with the implication that the measured observable is a discrete one and that the measurement in question is of the first kind. This already shows clearly that the strong modality condition is, unlike its weak counterpart, a very strong condition excluding, for instance, any measurement of a continuous observable. Such observables are basic in constituting Galilei or Poincare invariant objects in quantum physics.

V
Measurements Preserving Properties

To close this discussion I shall formulate the strongest possible form of the ideality assumptions on measurements. This assumption is known simply as the ideality condition. Apart from its weak appearance, it determines completely the structure of the measurement to that of the paradigm of a measurement in quantum mechanics: the von Neumann−Lüders measurement.

Consider any observable represented by a self-adjoint operator A, but do not assume that it is discrete. Let $<\mathcal{K}, Z, \Phi, U>$ be any of its measurements. When performed this measurement changes the state of the system according to $\varphi \longmapsto T(\varphi)$. In general, the state of the system after the measurement is different from its state before the measurement. Therefore, also the properties which the system could have before and after the measurement are, in general, different. We say that a measurement is ideal if the state of the system is changed only to the extent that is necessary: any property which pertains to the system before the measurement and which is compatible with the measured observable should also pertain to the system after the measurement. Recalling that the compatibility of a property $P \in \mathcal{P}(\mathcal{H})$ with an observable A, $P \sim A$, is properly expressed as their commutativity, we arrive at the following formal definition: a measurement $<\mathcal{K}, Z, \Phi, U>$ of A is *ideal* if for any initial state φ of the system and for any property $P \in \mathcal{P}(\mathcal{H})$ the following implication holds true:

(21)   if $P \in P(\varphi)$ and $P \sim A$, then $P \in P_1(T(\varphi))$.

The following result is then obtained: if an observable represented by a self-adjoint operator admits an ideal measurement, then this observable is discrete and the measurement in question is equivalent to a von Neumann−Lüders measurement of this observable [Lahti *et al.*, 1991].

Concluding Remarks

We have discussed a number of idealising assumptions on measurements in quantum mechanics, starting with the minimal calibration condition and ending up with the ideality assumption. In between there are the weak and strong modality conditions, and the value reproducibility condition, with its many equivalent forms. Any of these assumptions is a further structural specification of the measurement, the ideality assumption implying a complete specification of the measurement. It appears plausible that in addition to the calibration condition also the weak modality condition is a common property of any measurement. It is an open question to what extent the other discussed properties are to be considered as general features of measurements. Their feasibility is subject to empirical investigation. In any case, they all contribute to the understanding of quantum mechanics as a theory of individual objects and their properties.

Department of Physics
University of Turku
20500 Turku, Finland

REFERENCES

Beltrametti, E. and G. Cassinelli [1981]. *The Logic of Quantum Mechanics*. Addison-Wesley, Reading, Massachusetts.

Busch, P., P. Lahti, and P. Mittelstaedt [1991]: *The Quantum Theory of Measurement*, *LNP* m2, Springer-Verlag, Berlin.

Cassinelli, G. and P. Lahti [1993a]. *Foundations of Physics Letters*, 6, 553.

Cassinelli, G. and P. Lahti [1993b]. *Nuovo Cimento* 108B, 45.

van Fraassen, B.C. [1991]. *Quantum Mechanics: an empiricist view*. Clarendon Press, Oxford.

Lahti, P., P. Busch, and P. Mittelstaedt [1991]. *Journal of Mathematical Physics* 32, 2770.

von Neumann, J. [1955]. *Mathematical Foundations of Quantum Mechanics*. Princeton UP, Princeton.

ABSTRACTS

Theo A.F. Kuipers
Department of Philosophy
University of Groeningen

THE REFINED STRUCTURE OF THEORIES

The paper gives a systematic introduction to the basic ideas of the structuralist reconstruction of empirical theories. It starts from a number of global ideas about the nature and structure of empirical theories.

According to the structuralist view an axiomatizad theory defines a class of structures, and the conditions imposed on the components of the structures are the axioms of the theory. The link with reality is made by the claim, associated with the theory, that the set of set-theoretic representations of the so-called intended applications forms a subset of the class of structures of the theory.

The main advantage of the structuralist approach is that its representations and analyses of theories and their relations are as close to their actual presentations in textbooks as is formally possible. However, even theoretical questions concerning for instance idealization and concretization as a truth approximation strategies can be treated relatively easily in structuralist terms.

I will present the main general aspects and I will not go into technical details which are not of primary importance for actual practice. My main goal is to make clear what kind of entities one may be looking for in theory formation and how standard questions about these entities can be explicated. Moreover, in passing I will explicate some standard Popperian concepts in structuralist terms.

C. Ulises Moulines and Reinhold Straub
Institut für Philosophie, Logik und Wissenschaftstheorie
University of Munich

APPROXIMATION AND IDEALIZATION
FROM THE STRUCTURALIST POINT OF VIEW

There are at least three things in philosophy of science we have learnt since the heroic times of logical positivism: 1) there are no such things as brute empirical facts upon which a theory may be built or with which it may be confronted, 2) a scientific theory is a cultural product which essentially contains irreducibly pragmatic components and 3) in general, no theory works unless a certain measure of "idealization" with respect to its "outer world" is allowed.

We call these results the principles of "theory-ladenness", of "praxis-ladenness", and of "approximation-ladenness" of science. The first two are, it seems, widely accepted and taken into account in present-day philosophy of science. The third one is not so popular, though it has increasingly become a matter of study from different perspectives in the last years.

The three principles mentioned are tied together in a deep, though not entirely obvious way. To make the point very briefly, our thesis is that, on the one hand, empirically meaningful approximations and idealizations are only possible because of theory-ladenness and, on the other hand, the concepts of approximation and idealization we need for empirical theories are essentially constituted by some irreducibly pragmatic components in addition to semantic ones.

Wolfgang Balzer and Gerhard Zoubek
Institut für Philosophie, Logik und Wissenschaftstheorie
University of Munich

STRUCTURALIST ASPECTS OF IDEALIZATION

Recently, the scheme of idealization and concretization has become the subject of properly methodological investigation, mainly through the work of Nowak and Krajewski. Attempts are made to explicate the concepts, to apply them to cases from the history of science, and to evaluate them in a context of other methodological issues.

In this paper we investigate the notions of idealization and concretization as intertheoretical relations in structuralist terms. Krajewski's and Nowak's explications use a rather narrow syntactic format. We generalize their concept of idealization by using the structuralist format. Our generalization makes these notions applicable to various reconstructed examples as they can be found in the literature.

On the other hand, we do *not* generalize the treatment of concretization for this would lead us to a general notion which has been studied in detail under the label of approximative reduction. On our account, idealization and concretization are seen as special cases of approximative reduction which can be clearly distinguished from the latter, and may be studied on its own.

In addition to this general point, we address three more special ones. First, we elaborate on a feature neglected by Nowak and Krajewski, namely that the laws of the two theories (the idealized and the concretized one) have the same form. Second, we stress that in the examples from the natural sciences an additional requirement of continuity is satisfied. Third, we suggest that *quasi-metrical* spaces are to be used for the treatment of 'real-life' examples of idealization and concretization.

Andoni Ibarra
Departemento de Lógica y Filosofía de las Ciencas
Universidad del Pais Vasco, UPV/EHV, Donostia/San Sebastian
Thomas Mormann
Institut für Philosophie, Logik und Wissenschaftstheorie
University of Munich

COUNTERFACTUAL DEFORMATION AND IDEALIZATION IN A STRUCTURALIST FRAMEWORK

One possible way to explain the idealized character and indirect applicability of scientific theories is provided by possible worlds semantics. The framework of possible world semantics enables us to understand the counterfactual character of scientific laws. However, it remains incomplete as long as the relation between the actual and the "ideal" worlds is not elucidated.

Another approach that has contributed to a deeper understanding of the problem of the applicability of empirical theories is structuralism. This has been done by developing a highly sophisticated description of the structure of empirical theories. However, until today, structuralism has hardly taken any notice of possible world semantics.

Finally, there is a third approach to philosophy of science which has explicitly dealt with the problem of idealization, viz., the "Poznan School". Although Nowak repeatedly emphasizes the counterfactual character of economic laws, he never refers to any kind of possible world semantics or to any other account of modal logic.

In this paper we introduce counterfactual (or idealizing) deformation procedures following some recent ideas of Nowak. We then study counterfactual deformation operators in the structuralist approach, and the complementary concepts of idealization and concretization, are introduced. Finally, we apply the framework of structuralism cum idealization structure to the elucidation of the counterfactual character of empirical laws.

Ilkka A. Kieseppä
Department of Philosophy
University of Helsinki

ASSESSING THE STRUCTURALIST THEORY
OF VERISIMILITUDE

I shall discuss the structuralist theory of verisimilitude developed by T.A.F. Kuipers. I concentrate mostly on applying the definition of verisimilitude to the values of a finite number of quantitative variables. A theorem is proved which shows that, in an important special case, the definition of Kuipers only makes use of a small part of the information contained in the hypotheses or theories compared. I shall also discuss in detail a physical system which provides us with several examples of counterintuitive results of another kind that the definition leads to.

Leszek Nowak
Department of Philosophy
University of Poznań

REMARKS ON THE NATURE OF GALILEO'S METHODOLOGICAL REVOLUTION

The Galilean revolution consisted in making evident the misleading nature of the world image which senses produce. We only see phenomena which are the joint effects of the relevant forces. As a result, senses do not contribute in the slighest to the understanding of the facts. To understand phenomena we must take into account the work of reason for reason is needed to select some features of the objects through idealization. These idealized models differ a great deal from their sensory prototypes. What is more, they present images of hidden relationships which could not be captured in experience at all.

This gap between the abstract world of laws and the world of senses can be filled with the aid of concretization which takes into account what has been previously abstracted from. Because of this, abstract laws become more and more realistic and the distance between them and the actual facts diminishes. Idealization and concretization constitute the essence of the method whose adoption in physics Galileo initiated. This method had been systematically applied by Newton. Also our understanding of it had been deepened in Newton's *Principia*.

Ilkka Niiniluoto
Department of Philosophy
University of Helsinki

APPROXIMATION IN APPLIED SCIENCE

Applied science exists in two forms: predictive and design science. The former tries to establish dynamic regularities that help to predict the future state of a natural or social system; the latter attempts to establish technical norms or conditional rules of action. It is typical of both cases that idealized theoretical descriptive models are combined with empirical information. When the idealized model is concretized in Nowak's sense, the predictions derived and technical norms can be likewise improved, so that their degree of approximate truth or truthlikeness increases. These methodological ideas can be illustrated by the history of exterior ballistics.

Elke Heise, Peter Gerjets and Rainer Westermann
Institut für Psychologie
University of Goettingen

IDEALIZED ACTION PHASES: A CONCISE RUBICON THEORY

If human goal-directed behavior is to be analyzed from an action-theoretical perspective, a theory is required which comprises all relevant processes between the deliberation of potential action goals and the final evaluation of action outcomes. The Rubicon theory of action phases offers an adequate framework for a sequential description of goal-directed activities.

A concise version of the Rubicon theory will be outlined in structuralist terms, comprising only those assumptions that are essential for either theoretical or empirical reasons. An assumption is classified as essential if it uses fundamental explanatory concepts of the Rubicon theory, if it has been empirically corroborated, or if it is at least expected to underlie future applications. All other assumptions, especially those which refer only to different operationalizations of higher-order concepts will be omitted. On the basis of the concise structuralist version of the theory, it is possible to illustrate the idealizing assumptions of the theory in more detail.

Klaus G. Troitzsch
Social Science Informatics Institute
Koblenz—Landau University

MODELLING, SIMULATION, AND STRUCTURALISM

During the last twenty years, the interdisciplinary research efforts of synergetics have invented methods that make new and powerful tools available for the social scientist, too. It would be fruitful to study these theoretical tools from a structuralist point of view.

We distinguish between two approaches: in the first one we have models in which individuals do not interact directly, but change the state of the whole system by their behaviour and react to changes of this collective with individual changes in their behavioural states. In the beginning we have an "aggregate" of initially unconnected individuals which turns by itself into a system. This process is due to the fact that the individuals are endowed with the ability to move in a potential which is built up as a result of the sheer existence of these individuals.

In the second approach we have models in which individuals interact directly, mostly in a network or, as it were, in a cellular automaton. Here too, the population is taken to be homogeneous in the beginning. Then stochastically influenced interactions change the states of the interacting individuals as well as the relations between them. As a result we find stable clusters, groups, subnets, or strata in the whole population.

The aim of the paper is to give a logical reconstruction of such approaches to self-organization in the social sciences. I shall try to show that computer simulation supports the structuralist reconstruction of these approaches if it is done in a certain way.

Veikko Rantala and Tere Vadén
Department of Mathematical Sciences
University of Tampere

IDEALIZATION IN COGNITIVE SCIENCE
A Study in Counterfactual Correspondence

It is suggested by Smolensky that in cognitive science the symbolic is an idealization of the subsymbolic. They may be seen as complementary ways to understand and explain cognition, and therefore we should not try to eliminate one of them in favour of the other. Rather, we should think of their relationship as providing a cognitive correspondence principle, a principle analogous to the correspondence principle much discussed in the philosophy of physics. But Smolensky's proposals are vague, and there seem to be no attempts in the literature to give them a more definite form. For this reason it is somewhat difficult to see their real significance.

In this paper, we shall evaluate the proposals by relating them to the work recently done in the philosophy of science, and offer a case study of the relationship between symbolic and subsymbolic representations. It turns out that good sense can be given to the suggestion that a particular kind of symbolic representation (or a theory of symbolic representation) is a limiting case of a particular kind of subsymbolic representation (or a theory). Due to the method used, the case study also sheds some light on the notion of idealization in cognitive science.

Matti Sintonen and Mika Kiikeri
Department of Mathematical Sciences
University of Tampere

IDEALIZATION IN EVOLUTIONARY BIOLOGY

The basic ideas of the Poznań school are highly suggestive, and the motivation laudable. However, we shall argue for some amendations. First, it is not clear that a theory as a whole has a structure which is amenable to treatment along Poznań lines. Secondly, its notion of idealized laws and their application process does not do full justice to the model-constructing activities found in many sciences. We shall single out one crucial feature of model building, viz. the interplay between general laws on the one hand and model-specific assumptions on the other. We shall try to show that this is more complex than Poznań philosophers have assumed.

The example of theory construction and model building we have chosen comes from the domain of evolutionary biology. We shall start with a brief account of evolutionary theory and the Poznań view and distinguish several senses in which a theory may be idealized and needs to be concretized. We shall show that evolutionary theory at large does not fit the view in which theories have a unique core, and that if we want to adopt a realistic picture of model building in evolutionary subtheories such as population genetics we have to make alterations in the Poznań account.

Timo Tuomivaara
Department of Philosophy
University of Helsinki

ON IDEALIZATIONS IN ECOLOGY

There has been much dispute concerning the use of the method of idealization in ecology. Its use has been defended especially by those ecologists who, following the exemplar of mechanistic physical science, believe that the quantitative, mathematical and analytical methods of physical sciences are also applicable in ecology. On the other hand its use has been criticized by those who have abandoned the mechanistic approach, arguing that it is incompatible with the view of the holistic, unique, and historically changing nature of ecological entities. I believe, however, that a more viable intermediate position between the extremities of mechanism and anti-mechanism exists. I call it the approach of *interactive particularism.*

Interactive particularism, developed in detail in the paper, is my ontological and methodological starting point. Two important additional theses are presupposed: *the realistic conception of theory* and *the theoryladenness of all data.* First, the *theoretical* and the *empirical contents* of a theory are separated. Second, I claim that all observation, measurement and experimentation in ecology is founded on the *theoretical ideas* of *the processes* of data generation.

Theory construction is analysed as a *process* in which the theoretical content of theory is gradually developed and explicated by using the *methods of isolation*, *idealization*, and *concretization* or *specification*. Theory construction proceeding in this way typically results in a hierarchy of theoretical models with varying degrees of generality and realism.

For empirical tests, the empirical content of a theory in some designed test situation must be specified via a *theory of data generation.* The relevant properties of the process of data generation are also defined step by step by using the methods of isolation, idealization, and concretization.

Martti Kuokkanen and Matti Häyry
The Academy of Finland and
Department of Philosophy, University of Helsinki

EARLY UTILITARIANISM AND ITS IDEALIZATIONS FROM A SYSTEMATIC POINT OF VIEW

We shall reconstruct some basic ideas of Early Utilitarian Theorizing using the framework of the structuralist theory of science. We first divide the core assumptions of Early Utilitarianism into three groups: the Greatest Happiness Principle, the Impartiality Principle and assumptions about the quantitative and qualitative aspects of pleasures.

Historically several substantially different versions of the Greatest Happiness Principle have been differentiable. We consider only two of them explicitly; *Strict Universal Altruism* and *Classical Utilitarianism*.

There are also several formulations of the Impartiality Principle. We formulate two of them, a weak and a strong one. We study four different positions concerning the assumptions about the qualitative and quantitative aspects of pleasures.

It is argued that the three core assumptions of *Early Utilitarianism* constitute three *mutually independent (and compatible) sets*. They make it possible to *specialize* the core assumptions in the *structuralist sense*. There are two basic theory-elements, a *quantitative* and a *qualitative* one, both constituting a theory-net.

It is shown that three of the four positions on the relation of the qualitative and quantitative aspects of pleasures contain idealizing assumptions. Second, it is shown that *applying utilitarian* theorizing presupposes idealizations. The *general form* of idealizations can be crystallized using the Poznań School Theory of Idealizations. It turns out that the idealizations about the qualitative and quantitative aspects of pleasures are special, stronger cases of the general form of idealizations presupposed in applications of utilitarian theorizing.

Rainer Westermann
Institut für Psychologie
University of Goettingen

MEASUREMENT-THEORETICAL IDEALIZATIONS AND EMPIRICAL RESEARCH PRACTICE

In this paper I will discuss the question how substantial theories can be connected with measurement theories in order to adequately specify the theoretical basis of scores and numbers used in empirical research.

The main problem dealt with here is the fact that, although there is no empirical research without some form of measurement, in analyzing and reconstructing actual scientific theories and research processes it becomes apparent that there is little or no manifest connection to a theory of measurement.

The most basic problem lies in the fact that measurement theories usually involve idealizations that make it difficult to apply them in actual empirical research practice. Measurement theory primarily deals with an ideal form of measurement, the "representational" measurement, whereas the vast majority of variables are, at least in the social and behavioral sciences, amenable only to non-ideal forms of measurement which are called "derived" and "quasi-representational".

These three types of measurement are explained with the use of specific examples and some suggestions are given as to how they can be incorporated into theory-holons to form an adequate basis for deriving scale values for non-theoretical terms and assessing their scale type.

In addition, most measurement-theoretical structures are strictly deterministic and pertain to ideal error-free situations. This idealization can be dealt with by probabilistic formulations and specific error theories or, more simply, by testing derived statistical hypotheses. A short explanation of this latter approach is also given analyzing quasi-representational measurement.

Uwe Konerding
Institut für Psychologie
RWTH Aachen

PROBABILITY AS AN IDEALIZATION OF
RELATIVE FREQUENCY:
A CASE STUDY BY MEANS OF THE BTL-MODEL

It is common scientific experience that deterministic empirical laws are seldom exactly valid. In nearly every non-trivial empirical application there are some data which are not in complete correspondence with the law.

The whole conceptual framework of structuralism has been developed with regard to deterministically formulated theories. The formulation and empirical application of probabilistic models has hardly ever been treated within the structuralist framework. In order to counter this deficit, the attempt is made within this article to apply the structuralist conception for the reconstruction of a probabilistic model together with common strategies of its empirical application. The probabilistic model which is to be reconstructed is the so-called BTL-model, which is of great importance in psychology.

Two alternative reconstructions are presented. The first reconstruction is nearer to usual expositions of the model in textbooks of mathematical psychology. It will, however, be argued that this reconstruction provides no adequate conceptual basis for reconstructing the empirical application of the model. In contrast, the second reconstruction is less similar to usual expositions; but on the other hand it is adequate for giving an account of how the model is empirically applied. The core idea of the latter reconstruction consists of interpreting probability as an idealized relative frequency.

Reinhard Suck and Joachim Wienöbst
Fachbereich Psychologie
University of Osnabrück

THE EMPIRICAL CLAIM OF PROBABILITY STATEMENTS, IDEALIZED BERNOULLI EXPERIMENTS AND THEIR APPROXIMATE VERSION

In an earlier paper of Suck the basic ingredients of mathematical probability are incorporated in the structuralist formulation of theories. In this paper we investigate the empirical claims of probability statements in some simple cases.

We base our reconstruction on frequencies in the following special sense: the standard mathematical definitions of probability are used on the theoretical level. However, frequencies are used in *determining* probabilities and *interpreting* probability statements, i.e., in all kinds of empirical and practical issues in which probabilities play a part.

Mathematically we base our considerations on the various laws of large numbers which provide the connection between probabilities and frequencies. These theorems are not subject to discussions or contradicting opinions, and are therefore safe to start with. Relative frequencies are — in principle — easy to determine. Problems arise with the limiting process which the laws of large numbers entail and with the condition of independence.

For the limit problem we draw on the topological construction of uniformities. As to the independence problem, we push the problem around a bit until it is connected to another problem for which there exists no known practical solution.

Pekka J. Lahti
Department of Physics
University of Turku

IDEALISATIONS IN QUANTUM THEORY OF MEASUREMENT

Quantum theory of measurement is the part of quantum mechanics which investigates the measurement process as a physical process subject to the laws of quantum physics. This theory has both practical and conceptual dimensions, ranging from a study of the measurement accuracy limitations in instrumentation technology to an investigation of the fundamental issues of quantum mechanics.

Recent advances in ultrahigh technology has brought these two divergent subjects nearer to each other, to a quite surprising extent. Extremely well-controlled experimentation on individual objects, like atoms, neutrons, electrons, or photons is a daily practise in experimental quantum physics. To say the least then, this progress has made it highly wishful to develop an interpretation of quantum mechanics which goes beyond the phenomenological level of a statistical interpretation.

Progress on the mathematical and conceptual foundations of quantum mechanics, especially on its theory of measurement, together with some new ideas on the interpretation of the theory, like the modal interpretations, have made it possible to formulate in a systematic fashion an interpretation of quantum mechanics which exceeds a purely statistical level, and which comes closer to present day experimental practise.

In this paper I shall attempt to review some of that development and to show, in particular, how an increase in the degree of idealisation in measurement leads to a further structural specification of a measurement, thereby opening perspectives for enriching the statistical interpretation of quantum mechanics to its realistic interpretation as a theory of individual objects and their properties.

POZNAŃ STUDIES IN THE PHILOSOPHY OF THE SCIENCES AND THE HUMANITIES

Contents of back issues

VOLUME 1 (1975)

VOLUME 2 (1976)

No. 1 (Sold out)

Articles – W. Mejbaum, *Value and Conceptualization*; J. Kmita, *On Axiological Heterogeneity of Evaluating Predicates*; P. Buczkowski, *The Marxian Category of Bourgeois Scientist*; L. Nowak, *Evaluation and Cognition*; T.M. Jaroszewski, *The Peculiarity of Man*. **Discussions** – M. Franklin, *Professor Topolski on Lenin's Theory of History*; F.J. Fleron, *Comments on "Lenin's Theory of History" by Jerzy Topolski*.

No. 2 (Sold out)

Articles – M. Przełęcki, *On Possibility and Possible Worlds*; Z. Augustynek, *Relational Becoming*; T. Kubiński, *On Foundations of Deontic Logic*; Z. Ziembiński, *Factual Assumptions of Normative Utterances*; J. Agassi, *On Spontaneity in the Arts*; B. Tuchańska, *A Phenomenon, the Essence of the Phenomenon, a Theory*; T. Pawłowski, *Indicators in Comparative Research. The Problems. Introductory Remarks*. **Discussions** – F. J. Vandamme, *Theory Change, Incompatibility and Non-Deducibility*; D.H. DeGrood, *Dialectical Method and "Idealizations"*; I. Supek, *Boscovich's Philosophy of Nature*.

No. 3 (Sold out)

Idealizational Concept of Science – L. Nowak, *Essence – Idealization – Praxis. An Attempt at a Certain Interpretation of the Marxist Concept of Science*; B. Tuchańska, *Factor versus Magnitude*; J. Brzeziński, *Empirical Essentialist Procedures in Behavioral Inquiry*; J. Brzeziński, J. Burbelka, A. Klawiter, K. Łastowski, S. Magala, L. Nowak, *Law and Theory. A Contribution to the Idealizational Interpretation of Marxist Methodology*; P. Chwalisz, P. Kowalik, L. Nowak, W. Patryas, M. Stefański, *The Peculiarities of Practical Research*. **Discussions** – T. Batóg, *Concretization and Generalization*; R. Zielińska, *On Inter-Functional Concretization*; L. Nowak, *A Note on Simplicity*; L. Witkowski, *A Note on Implicational Concept of Correspondence*.

No. 4 (Sold out)

Categorial Interpretation of Dialectics – L. Nowak, *On the Categorial Interpretation of History*; I. Nowakowa, *The Principle of Universal Nexus in the Categorial Interpretation of Dialectics*; S. Magala, *The Phenomenon of Complementarity in the Categorial Dialectics*; I. Nowakowa, *Partial Truth – Relative Truth – Absolute Truth. An Attempt at a Construction of the Ordering Concept of Essential Truth*. **Discussions** – J. Witt-Hansen, *Reflections on Marxian Dialectics*; A. Pałubicka, J. Kmita, *Some Remarks on the Dialectical Method of Karl Marx*; D.W. Felder, *Contradiction and Conflict*; M.A. Amer, *Contradiction: Dialectical and Logical*; O.A. Wojtasiewicz, *The Concept of Effect in Autodynamic World*; A. Miś, *Remarks on a Certain Interpretation of the Marxist Methodology*.

VOLUME 3 (1977)

Nos. 1-4 (Sold out)

Aspects of the Production of Scientific Knowledge (Edited by J. Witt-Hansen) – J. Witt-Hansen, *Marx's Method in Social Science, and Its Relationship to Classical and Modern Physics and Mathematics*; S.A. Pedersen, *Logic and Ontology in Study of Theory Change*;

H.S. Jensen, *On the Production of Scientific Knowledge*. **Articles** – R. Arthur, *The Empiricist Account of Scientific Knowledge – A Polemic Evaluation*; S. Shibata, *Die Moderne Wissenschaft und marxistische Philosophie. Zur Problematik der Methodologie der Wissenschaft*; W. Patryas, *The Sense of Empirical Testing*; K. Łastowski, *The Method of Idealization in the Populational Genetics*; B. Tuchańska, *An Idealizational View on Measurement and Indicator-Based Reasoning*; J. Topolski, *On the Class Approach to History*. **Discussions** – W. Patryas, *When do Two Theorems Correspond?*; E. Paszkiewicz, *Context of Discovery and Context of Justification – Opposition or Complementarity?*; J. Woleński, *Jörgensen's Dilemma and the Problem of the Logic of Norms*.

VOLUME 4 (1978)

Nos. 1-4 (Sold out)

Aspects of the Growth of Science (Edited by W. Krajewski) – W. Krajewski, *Preface*; I. Szumilewicz, *The Postulate of Simplicity*; E. Pietruska-Madej, *Anomalies and the Dynamics of Scientific Theories*; A. Motycka, *Kuhn's Sociological Principle of Demarcation*; A. Lewenstam, *Models of Atom. Remarks on the Development of Science*. **Articles** – J. Agassi, *Logic and Logic of*; E. Köhler, *Observation Established by Artificial Intelligence and the Solution to Goodman's Paradox*; M. Przełęcki, *Some Approach to Inexact Measurement*; S. Bartlett, *Lower Bounds of Ambiguity and Redundancy*; J. Such, *Idealization and Concretization in Natural Sciences*; J. Pogonowski, J. Wiśniewski, *A Formal Approach to Causality*; S. Richmond, *Polanyi's Critique of Methodology and Liberalism*; M. Nowakowska, *Towards a Formal Theory of Group Action*; K. Zamiara, *The Problem of Psychologistic Reduction*; M.A. Finocchiaro, *Theory and Practice in the Philosophy of Science. Comments on Galileo and Clavelin*; M.J. Siemek, *Lukács' Dialectical Epistemology*; I. Nowakowa, L. Nowak, *Marxism and Positivism: The Idea of a Scientific Philosophy*. **Discussions** – E. Kronthaler, *Induction und Gesetzsartigkeit. Zu einer Theorie von Nelson Goodman*; E. Kronthaler, *Marginalie zu G.H. von Wright's* The Logical Problem of Induction.

VOLUME 5 (1979)

Nos. 1-4 (Sold out)

Methodological Problems of Historical Research (Edited by J. Topolski) – J. Topolski, *The Basic Problems of the Methodology of History*; G. Castellan, *Histoire et Psychologie*; G.G. Iggers, *History as an Historical Science*; L. Canfora, *Epic and Historical Claim to Totality*; A. Pałubicka, *The Positivist and Instrumentalist Concepts of the So-called Archeological Culture*; T. Kostyrko, *On the Concept of Historical Sources*. **Articles** – W. Marciszewski, *Epistemological Foundations of Democratism in Cartesian Philosophy*; G. di Bernardo, *Thetic and Prohairetic Normative Systems*; J. Wróblewski, *The Theory of Law – Multilevel, Empirical or Sociological?*; T. Maruszewski, *On Humanist Interpretation in Psychology: The Case of Investigations of the Defence Mechanisms*; K. Nielsen, *On Marx and Moralizing Social Science*; H. Skolimowski, *Space in Architecture: A Phenomenological Analysis*; B. Dziemidok, *On Cathartic-Compensatory Function of Art*; W. Ławniczak, *On Different Concepts of a Model*; K. Łastowski, *On Possibility of Adaptive Interpretation of Dialectics*; I.S. Narski, *The Functions and Structure of Dialectical Logic*. **Discussions** – K. Berka, *Critical Remarks on the Axiomatized Expected Utility Theory*; J. Ławniczak, *Anti-Individualism, Scientific Discovery and Third World*; R. McCleary, *Remarks on Historical Materialism and Dialectics*.

VOLUME 6 (1982)

SOCIAL CLASSES ACTION & HISTORICAL MATERIALISM

On Classes – I. Fetscher, *Some Theories on the Role of Intellectuals in the Revolutionary Labour-Movement. Brecht and Sartre*; R.P. Sieferle, *Class Consciousness in Marx's Theory of Revolution: Some Remarks*; M. Franklin, *On the Magic of Meritocracy*; O.A. Wojtasiewicz, *Status and Social Structures*. **On Action** – J. Ritsert, *"Wertbeziehung" to the Societal Basis – Two Lines of Argument in Max Weber's Theory of Action?*; A. Klawiter, *Structure and Praxis*; W. Patryas, *The Form of the Principle of Rationality*. **The Adaptive Interpretation of Historical Materialism** – L. Nowak, *The Theory of Socio-Economic Formations as a Theory of Adaptation Processes*; K. Łastowski, *The Theory of Development of Species and the Theory of Motion of Socio-Economic Formation*; P. Buczkowski, *Toward a Theory of Economic Society. An Attempt at the Adaptive Interpretation*; J. Burbelka, *Historical Materialism: General Theory and Forms*; P. Buczkowski, A. Klawiter, L. Nowak, *Historical Materialism as a Theory of Social Whole*; A. Klawiter, *The Theory of Social Formation in Historical Materialism*; P. Buczkowski, *International Relations in the Adaptive Interpretation of Historical Materialism*. **Contributions to Historical Materialism** – J. Ritsert, *Totality, Theory and Historical Analysis. Remarks on Critical Sociology and Empirical Research*; L. Nowak, *Adaptation and Revolution. The Problem of Motion of Socio-Economic Formations in the Adaptive Interpretation of Historical Materialism*; J. Ritsert, *On Horkheimer's and Adorno's Concept of Ideology*; P. Buczkowski, *Imperialism: On the Possibility of Reinterpretation of Lenin's Concept*.

VOLUME 7 (1982)

DIALECTICAL LOGICS FOR THE POLITICAL SCIENCE
(Edited by Hayward R. Alker, Jr.)

L. Nowak, *Marxian Methodology Leads to the Generalization of Historical Materialism*; D. Mefford, *A Comparison of Dialectical and Boolean-Algebraic Models of the Genesis of Interpersonal Relations*; C. Roig, *Leninist Dialectic Seen From the Point of View of Kenneth Burke's Dialectic: Theory, Analysis and Ideology in Lenin's Speech*; H.R. Alker Jr., *Logic, Dialectics, Politics: Some Recent Controversies*; F. Loeser, *What is Dialectical Logic?*

VOLUME 8 (1985)

CONSCIOUSNESS: METHODOLOGICAL AND PSYCHOLOGICAL APPROACHES
(Edited by Jerzy Brzeziński)

Editor's Introduction; A. Klawiter, *A Scientific Theory versus Theoretical Program. A Contribution to the Problem of a Social Character of Cognition*; S. Magala, L. Nowak, *The Problem of Historicity of Cognition in the Idealizational Concept of Science*; J. Brzeziński, *The Protoidealizational Model of the Investigative Process in Psychology*; M. Gaul, *The Problem of Interaction in the Protoidealizational Model of Investigative Process. Modifications and Developments*; J. Kmita, *A Humanistic Coefficient of Activity and the Psychology-Humanities Relation*; K. Zamiara, *In Support of Psycho-Physical Parallelism*; M. Ziółkowski, *Some Remarks on the Notion of Social Consciousness*; S. Kowalik, *The Psycho-*

Social Problem and Modern Psychology; J. Topolski, On the Concept and Structure of Methodological Consciousness of Historians; T. Maruszewski, Are the Idealizational Procedures Used within the Scope of Common Sense Knowledge?; T. Zgółka. Stylistic and Sociolinguistics. **Discussions** – T.A.F. Kuipers, The Paradigm of Concretization: The Law of van der Waals; N. Ziv, The Paradoxical Nature of Bloor's Strong Program in the Sociology of Knowledge.

VOLUME 9 (1986)

THEORIES OF IDEOLOGY AND IDEOLOGY OF THEORIES
(Edited by Piotr Buczkowski and Andrzej Klawiter)

I. Theories of Ideology – P. Buczkowski, The Levels of Consciousness: Some Remarks on Sociology of Knowledge; L. Nowak, Ideology versus Utopia. A Contribution to the Analysis of the Role of Social Consciousness in the Movement of Socio-Economic Formation; J. Kmita, The Antagonism of Art and Science as a Worldview Component. **II. Ideology of Theories** – J. Kmita, Scientism and Anti-Scientism; L. Nowak, Science, that is, Domination through Truth; J. Sensat, Recasting Marxism: Habermas's Proposals. **Discussions** – J. Brzeziński. A Statistical Model of Data Analysis in Interactional Psychology. Comments on the Quantitative Analysis of the Scores of the "S-R" Inventory of Anxiousness.

VOLUME 10 (1987)

WHAT IS CLOSER-TO-THE-TRUTH ?
A PARADE OF APPROACHES TO TRUTHLIKENESS
(Edited by Theo A.F. Kuipers)

Introduction. **Part I: Four Approaches to Truthlikeness** – I. Niiniluoto, How to Define Verisimilitude; G. Oddie, The Picture Theory of Truthlikeness; G. Schurz, P. Weingartner. Verisimilitude Defined by Relevant Consequence-Elements; T.A.F. Kuipers, A Structuralist Approach to Truthlikeness. **Part II: Eight Aspects of Truthlikeness** – J. van Benthem. Verisimilitude and Conditionals; J. Cohen, Verisimilitude and Legisimilitude; R. Festa, Theory of Similarity, Similarity of Theories, and Verisimilitude; T.A.F. Kuipers, Truthlikeness of Stratified Theories; I. Niiniluoto. Verisimilitude with Indefinite Truth; G. Oddie, Truthlikeness and the Convexity of Propositions; G. de Vries, Explaining 'Truth' in a Relativist Way; H. Zandvoort, Verisimilitude and Novel Facts.

VOLUME 11 (1988)

NORMATIVE STRUCTURES OF THE SOCIAL WORLD
(Edited by Giuliano di Bernardo)

Editor's Introduction; H.-N. Castañeda, Ought, Reasons, Motivation, and the Unity of the Social Sciences; G.H. von Wright, Action Logic as a Basis for Deontic Logic; S. Galvan. Über den Begriff von möglicher Welt in den Anwendungen der Modallogik; G. di Bernardo. The Formal Model M of the Game of Chess as a Type of Social Context; R. Guastini, Constitutive Rules and the Is-Ought Dichotomy; H.J. Heringer, Not by Nature nor by Intention. The Normative Power of the Language Signs; A. Kasher, Justification of Speech, Acts and Speech Acts; Z. Ziembiński. Le contenu et la structure des normes concédant les *compétences*; K. Opałek, *Directives, Norms, and Performatives*; G. Kalinowski, *Les performatifs en droit. Sur la distinction entre le langage prescriptif et le langage performatif*; J. Wróblewski, *Cognition of Norms and Cognition through Norms*; A.G. Conte. *Eidos. An Essay on Constitutive Rules*. **Discussions** – A. Wiśniewski, *On Some Consequences of the Postulate of an Empirical Falsifiability of Logical Theorems*; H. Gipper. *Die Sprachphilosophie Adam Schaffs*.

VOLUME 12 (1987)

POLISH CONTRIBUTIONS TO THE THEORY AND PHILOSOPHY OF LAW
(Edited by Zygmunt Ziembiński)

Editor's Introduction. **Part I** – S. Czepita, *Leon Petrażycki and Czesław Znamierowski – Founders of the Polish Theory of Law*; K. Opałek, *Normativism against the Background of the Methodological Inquiries in Polish Legal Theory*; C. Znamierowski, *The Basic Concepts of Theory of Law. Introductory Remarks*. **Part II** – Z. Ziembiński, *The Methodological Problems of Theory and Philosophy of Law: a Survey*; J. Wróblewski, *Paradigm of Legal Dogmatics and the Legal Sciences*; J. Woleński, *Jurisprudence, Science, Philosophy*; T. Gizbert-Studnicki, *Is an Empirical Theory of Language of Law Possible?* **Part III** – L. Nowak, S. Wronkowska, M. Zieliński. Z. Ziembiński, *Conventional Acts in Law*; L. Nowak, *A Concept of Rational Legislator*; S. Wronkowska, *The Rational Legislator as a Model for the Real Lawmaker*; M. Zieliński, *Decoding Legal Text*. **Part IV** – S. Wronkowska. M. Zieliński, Z. Ziembiński. *The Methodological Problems Concerning the "Principles of Law"*; S. Pałczyński, *The Concept of Human Act in the Legal Sciences*. **Discussions** – Fr. Castellani, *Some Remarks on the Semantic Interpretation of Intensional Logic*.

VOLUME 13 (1988)

DIMENSIONS OF THE HISTORICAL PROCESS
(Edited by Leszek Nowak)

Editor's Introduction. **I. Analyzing the Dimensions of the Historical Process** – A. Klawiter, *Historical Materialism and the Visions of Social Development. A Study of Transformations of Marxian Periodization Formula*; L. Kreisberg. *The Role of Consensus in Social Conflicts*; I.C. Jarvie, *Objective versus Mentalist Conceptions of Social Class*. **II. Individual/Civiliza- tional/Transformational Approach** – A. Honneth, *Work and Instrumental Action*; P. Buczkowski, *Rationality and the Levels of Organization of Society*; A. Bartoszek. E. Bogalska-Czajkowska, W. Czajkowski, A. Małkiewicz. *The Social Division of Labor*. **III. Globalistic/Civilizational/Adaptive Approach** – A. Klawiter, K. Łastowski. L. Nowak, W. Patryas, *Adaptation, Learning, Praxis. Some Applications of the Adaptive Conceptual Apparatus*; A. Klawiter, *Adaptation and Competition. A Contribution to the Classification of Adaptive Relationships*; M. Witkowski, *On Adaptive and Functional Dependencies. An Attempt at Categorial Approach*; K. Niedźwiadek, *The Structure and Development of the Society's Modes of Spiritual Production*. **IV. Globalistic/Conflict/Transformational Approach** – T. Söderqvist, *The Macro- and Micro-Conflict Structure of the Information- and Knowledge Society*; D. Pels, *Towards a Non-Hegelian Conception of the "New Class"*; L. Nowak, *An Idealizational Model of Capitalist Society*; G. Tomczak, *The Economic Collapse in Two Models of Socio-Economic Formation*: M. Niewiadomski, *Toward a Model of Economic Institutions*; A. Falkiewicz, *The Anatomy of Utopia*: K. Paprzycka, L. Nowak. *On the Social Nature of Colonization*.

VOLUME 14 (1990)

ACTION AND PERFORMANCE: MODELS AND TESTS. CONTRIBUTIONS TO THE QUANTITATIVE PSYCHOLOGY AND ITS METHODOLOGY

(Edited by Jerzy Brzeziński and Tadeusz Marek)

VOLUME 15 (1989)

VISIONS OF CULTURE AND THE MODELS OF CULTURAL SCIENCES

(Edited by Jerzy Kmita and Krystyna Zamiara)

Notion of Pedagogical Culture; P. Ozdowski, *Logical Semantics and Psychology of Cognitive Processes*; T. Zgółka, *Study of Language as a Form of Symbolic Culture: Methodological Foundations*; G. Banaszak, *Contemporary Musical Consciousness in the Perspective of the Socio-pragmatic Theory of Culture*; A. Zeidler, *Valorization of Tradition as a Means of Preserving the Continuity of Culture*.

VOLUME 16 (1990)

IDEALIZATION I: GENERAL PROBLEMS
(Edited by Jerzy Brzeziński, Francesco Coniglione, Theo A.F. Kuipers and
Leszek Nowak)

Introduction – I. Niiniluoto, *Theories, Aproximations, and Idealizations*. **Historical Studies** – F. Coniglione, *Abstraction and Idealization in Hegel and Marx*; B. Hamminga, *The Structure of Six Transformations in Marx's Capital*; A.G. de la Sienra, *Marx's Dialectical Method*; J. Birner, *Idealization and the Development of Capital Theory*. **Approaches to Idealization** – L.J. Cohen, *Idealization as a Form of Inductive Reasoning*; C. Dilworth, *Idealization and the Abstractive-Theoretical Model of Explanation*; R. Harré, *Idealization in Scientific Practice*; L.Nowak, *Abstracts Are Not Our Constructs. The Mental Constructs Are Abstracts*. **Idealization and problems of the philosophy of science** – M. Gaul, *Models of Cognition or Models of Reality?*; P.P. Kirschenmann, *Heuristic Strategies: Another Look at Idealization and Concretization*; T.A.F. Kuipers, *Reduction of Laws and Theories*; K. Paprzycka, *Reduction and Correspondence in the Idealizational Approach to Science*.

VOLUME 17 (1990)

IDEALIZATION II: FORMS AND APPLICATIONS
(Edited by Jerzy Brzeziński, Francesco Coniglione, Theo A.F. Kuipers and
Leszek Nowak)

Forms of Idealization – R. Zielińska, *A Contribution to the Characteristic of Abstraction*; A. Machowski, *Significance: An Attempt at a Variational Interpretation*; K. Łastowski, *On Multi-Level Scientific Theories*; A. Kupracz, *Concretization and the Correction of Data*; E. Hornowska, *A Certain Approach to Operationalization*; I. Nowakowa, *External and Internal Determinants of the Development of Science: Some Methodological Remarks*. **Idealization in Science** – H. Rott, *Approximation versus Idealization: The Kepler-Newton Case*; J. Such, *The Idealizational Conception of Science and the Law of Universal Gravitation*; G. Boscarino, *Absolute Space and Idealization in Newton*; M. Sachs, *Space, Time and Motion in Einstein's Theory of Relativity*; J. Brzeziński, *On Experimental Discovery of Essential Factors in Psychological Research*; T. Maruszewski, *On Some Elements of Science in Everyday Knowledge*.

VOLUME 18 (1990)

STUDIES ON MARIO BUNGE'S TREATISE
(Edited by Paul Weingartner and Georg J.W. Dorn)

Section I: Semantics – M. Dillinger, *On the Concept of 'a Language'*; J. Ferrater-Mora, *On Mario Bunge's Semantical Realism*; A. Bartels, *Mario Bunge's Semantical Realism. An*

Antidote against Incommensurability?; J.-P. Marquis, *Partial Truths about Partial Truth*; I. Nowakowa and L. Nowak, *Approximation and the Two Ideas of Truth*. **Section II: Ontology** – J. M. Bocheński, *On the System*; J Agassi, *Ontology and Its Discontent*; Y. Wand and R. Weber, *Mario Bunge's Ontology as a Formal Foundation for Information Systems Concepts*; D. Blitz, *Emergent Evolution and the Level Structure of Reality*; M. Espinoza, *The Four Causes*; B. Dubrovsky, *A Comment on Three Topics in Volume 7 of the* Treatise: *Teleology, the Mind-Body Problem, and Health and Disease*; W. R. Shea, *Tackling the Mind*; H. Kurosaki, *Mario Bunge on the Mind-Body Problem and Ontology. Critical Examinations.* **Section III: Epistemology** – R. Boudon, *On Relativism*; G. Vollmer, *Against Instrumentalism*; M. Kary, *Information Theory and the* Treatise: *Towards a New Understanding.* **Section IV: Philosophy of Science and Technology** – F. Miró-Quesada-Canturias, *Mario Bunge's Philosophy of Logic and Mathematics*; M. Paty, *Reality and Probability in Bunge's* Treatise; M. García-Sucre, *On the Relationship Between Mathematics and Physics*; B. Kanitscheider, *Does Physical Cosmology Transcend the Limits of Naturalistic Reasoning?*; M. Stöckler, *Realism and Classicism, or Something More? Some Comments on Mario Bunge's Philosophy of Quantum Mechanics*; M. E. Burgos, *Projections Are a Law of Nature*; A. Barceló, *Are there Economic Laws?*; R. Mattessich, *Mario Bunge's Influence on the Administrative and Systems Sciences*; Y. Gingras and J. Niosi, *Technology and Society: A View from Sociology*; D. A. Seni, *The Sociotechnology of Sociotechnical Systems: Elements of a Theory of Plans*; P. Weingartner, *The Non-Statement View. A Dialogue between Socrates and Theaetetus.* **Section V: Ethics** – E. Garzón-Valdes, *Basic Needs, Legitimate Wants and Political Legitimacy in Mario Bunge's Conception of Ethics*; F. Forman, *Virtue, The Missing Link to the Last Volume of the* Treatise: G. Zecha. *Which Values Are Conductive to Human Survival? A Bunge-Test of Bunge-Ethics.* **Section VI: The Treatise as a Philosophy** – A. Cupani, *The Significance of the* Treatise *in the Light of the Western Philosophical Tradition*; E. Rosenthal, *Scientific Philosophical and Sociopolitical Aspects of the Work of Mario Bunge.* **Section VII: Mario Bunge Replies. Section VIII: Mario Bunge's Life and Work** – M. Bunge, *Instant Autobiography*; D. Blitz, *Bibliography of Mario Bunge's Publications 1939-1989.*

VOLUME 19 (1990)

NARRATION AND EXPLANATION.
CONTRIBUTION TO THE METHODOLOGY OF THE HISTORICAL RESEARCH
(Edited by Jerzy Topolski)

G. Iggers, *Historical Studies in the 1980's. Some Observations*; I. Nowakowa, *Historical Narration and Idealization*; J. Pomorski, *On Historical Narration. A Contribution to the Methodology of a Research Programme*; G. Zalejko, *On Cognitive and Extra-Cognitive Components of the Historical Narrative*; J. Topolski, *Historical Explanation in Historical Materialism*; A. Zybertowicz, *Technological Determinism and Historical Materialism Revisited. Some Preliminary Notes*; J. Burbelka, *Types of Historical Processes in Ethnology*; W. Wrzosek, *In Search of Historical Time. An Essay on Time, Culture and History*; A. Ponzio, *Humanism, Philosophy of Language and Theory of Knowledge in Adam Schaff*; J. Rüsen, *Was ist historische Methode?*; A. Constanzo, *Several Remarks on Principles of Law.*

VOLUME 20 (1990)

Jürgen Ritsert
MODELS AND CONCEPTS OF IDEOLOGY

Chapter 1: *Preface – Enterprise and Experiment*; Chapter 2: *Some Theoretical Problems of Theories of Ideology*; Chapter 3: *Base and Superstructure – On the Generality of Some Marxian Models*; Chapter 4: *Action, Interest and Reflection – Weber and Weberian Models of Scientific Knowledge*; Chapter 5: *Analysis of Social Content and the Debate about Internalism and Externalism*; Chapter 6: *French Connections in Models of Ideology – Or: Des idées Franco-allemandes*; Chapter 7: *Critical Theory of Ideology – On Adorno's Dialectical Model of Appearance and Ideology*.

VOLUME 21 (1991)

PROBABILITY AND RATIONALITY
STUDIES ON L. JONATHAN COHEN'S PHILOSOPHY OF SCIENCE
(Edited by Ellery Eells and Tomasz Maruszewski)

Part I: Setting the Stage – L. J. Cohen, *From a Historical Point of View*; **Part II: The Method of Relevant Variables and Other Modes of Scientific Reasoning** – L. Nowak, *The Method of Relevant Variables and Idealization*; M. Fisch, *Learning from Experience*; M. A. Finocchiaro, *Induction and Intuition in the Normative Study of Reasoning*. **Part III: Subjective Probability: Pascalian and Baconian Conceptions** – D.A. Schum, *Jonathan Cohen and Thomas Bayes on the Analysis of Chains of Reasoning*; J. Logue, *Weight of Evidence, Resiliency and Second-Order Probabilities*; C.S. Nosal, *Neurobiology of Subjective Probability*; R. Stachowski, *On Standard and Non-Standard Models in Theories of Psychological Measurement*. **Part IV: Rationality and Methodological Pluralism** – L.L. Lopes and G.C. Oden, *The Rationality of Intelligence*; G. Gigerenzer, *On Cognitive Illusions and Rationality*; J.E. Adler, *An Optimist's Pessimism: Conversation and Conjuction*; T. Maruszewski, *Human Rationality – Fact or Idealizational Assumption*: I.T. Ścigała, *Are People Programmed to be Normal? Psychological Determinants of Human Normality*. **Part V: Conclusion** – L. Jonathan Cohen, *Some Comments by L.J.C.*

VOLUME 22 (1991)

THE SOCIAL HORIZON OF KNOWLEDGE
(Edited by Piotr Buczkowski)

J. Ritsert, *The Wittgenstein-Problem in Sociology or: The "Linguistic Turn" as a Pirouette*; D. Jary, *Beyond Objectivity and Relativism: Feyerabend's "Two Argumentative Chains" and Sociology*; L. Nowak, *The Defence of a Social System Against its Ideology. A Case Study*; P. Buczkowski, *Remarks on the Structure of Social Consciousness*; A. Falkiewicz, *The Individual's Horizon and Valuation*; F. Lövenich, *Heiligsprechung des Imaginären. Das Imaginäre in Cornelius Castoriadis' Gesellschaftstheorie*; T. Maruszewski, *Everyday Knowledge as Representation of Reality*.

VOLUME 23 (1991)

ETHICAL DIMENSIONS OF LEGAL THEORY
(Edited by Wojciech Sadurski)

W. Sadurski, *Introduction*; J. Wróblewski, *Moral Values and Legal Reasoning: Some Aspects of Their Mutual Relations*; H. Tapani-Klami, J. Sorvettula and M. Hatakka, *Moral Reasoning and Evidence in Legal Decision-Making*; T. Gizbert-Studnicki, *Conflict of Values in Adjudication*; M. Krygier, *Thinking Like a Lawyer*; A. Peczenik, *Prima-facie Values and the Law*; R. Wacks, *Judges and Moral Responsibility*; Z. Ziembiński, *The Concept of Morality in Philosophy of Law*; G. Haarscher, *Law, Reason and Ethics in the Philosophy of Human Rights*; T.D. Campbell, *Unlawful Discrimination*; H. Ph. Visser' Hoft, *Intergenerational Justice: Some Reflections on Methodology*; Ch. Sampford and D. Wood, *Tax, Justice and Priority of Property*.

VOLUME 24 (1991)

ADVANCES IN SCIENTIFIC PHILOSOPHY
ESSAYS IN HONOUR OF PAUL WEINGARTNER ON THE OCCASION OF THE 60TH ANNIVERSARY OF HIS BIRTHDAY
(Edited by Gerhard Schurz and Georg J.W. Dorn)

Section I: Advances in Philosophical Logic – R. Sylvan and R. Nola, *Confirmation without Paradoxes*; B. Smith, *Relevance, Relatedness and Restricted Set Theory*; G. Schurz, *Relevant Deductive Inference: Criteria and Logics*; M.L. Dalla Chiara, *Epistemic Logic without Logical Omniscience*; P. Gochet and E. Gillet, *On Professor Weingartner's Contribution to Epistemic Logic*; H. Festini, *An Application of Weingartner's Logical Proposal for Rational Belief, Knowledge and Assumption*; L. Åquist, *Deontic Tense Logic: Restricted Equivalence of Certain Forms of Conditional Obligation and a Solution to Chisholm's Paradox*; E. Orlowska, *Relational Formalization of Temporal Logics*; W. Lenzen, *What Is (or At Least Appears to Be) Wrong with Intuitionistic Logic?* **Section II: Current Challenges in Philosophy of Science** – Ch.P. Enz, *Quantum Theory in the Light of Modern Experiments*; P. Mittelstaedt, *An Inconsistency between Quantum Mechanics and Its Interpretation: The "Disaster" of Objectification*; E. Scheibe, *Substances, Physical Systems, and Quantum Mechanics*; M. Stöckler, *Reductionism and the New Theories of Self-Organization*; B. Kanitscheider, *Unification as an Epistemological Problem*; R. Haller, *Atomism and Holism in the Vienna Circle*; M. Przełęcki, *Is the Notion of Truth Applicable to Scientific Theories?*; T.A.F. Kuipers, *Structuralist Explications of Dialectics*; C.U. Moulines, *Pragmatics in the Structuralist View of Science*; K.R. Popper, *A World of Propensities. Two New Views of Causality*; G.J.W. Dorn, *Inductive Support.* **Section III: Recent Debates in Semantics and Ontology** – F. von Kutschera, *Kripke's Doubts about Meaning*; I. Bellert, *A Linguistic Approach to Frege's Puzzles*; P.M. Simons, *Inadequacies of Intension and Extension*; E. Morscher, *Inadequacies of Peter and Paul*; J. Brandl, *Some Remarks on the "Slingshot" Argument*; A. Hieke, *Real Facts.* **Section IV: Epistemological and Ethical Problems with Society** – E. Klevakina, *The Case against Value-Free Belief*; R. Tuomela, *Mutual Beliefs and Social Characteristics*; H. Lenk nad M. Maring, *A Pie-Model of Moral Responsibility? Remarks Concerning Ethical Dilutionism*; O. Neumaier, *Are Collectives Morally Responsible?*; E. Bencivenga, *The Electronic Self.* **Section V: Analytical Philosophy of Religion** – J.M. Bocheński, *Faith and Science. A Logical Commentary on the First Question of the*

Summa; E. Nieznański, *The Beginnings of Formalization in Theology*; H. Ganthaler, *Some Comments on Paul Weingartner's Concept of Scientific Theology*. **Section VI: Methods of Philosophy** – G. Kreisel, *Suitable Descriptions for Suitable Categories*; M. Bunge, *Why We Cherish Exactness?*; L. Koj, *Exactness and Philosophy*. **Section VII: Life and Work of Paul Weingartner** – G. Zecha, *Paul Weingartner: Philosophy at Work. A Biographical Sketch*; E. Stieringer, *Bibliography of Paul Weingartner's Publications 1961-1991*.

VOLUME 25 (1992)

IDEALIZATION III: APPROXIMATION AND TRUTH
(Edited by Jerzy Brzeziński and Leszek Nowak)

Introduction – L. Nowak, *The Idealizational Approach to Science: A Survey*. **On the Nature of Idealization** – M. Kuokkanen and T. Tuomivaara, *On the Structure of Idealizations*; B. Hamminga, *Idealization in the Practice and Methodology of Classical Economics: The Logical Struggle with Lemma's and Undesired Theorems*; R. Zielińska, *The Threshold Generalization of the Idealizational Laws*; A. Kupracz, *Testing and Correspondence*; K. Paprzycka, *Why Do Idealizational Statements Apply to Reality?* **Idealization, Approximation, and Truth** – T.A.F. Kuipers, *Truth Approximation by Concretization*; I. Nowakowa, *A Notion of Truth for Idealization*; I. Nowakowa, L. Nowak, *"Truth is a System": An Explication*; I. Nowakowa, *The Idea of "Truth as a Process". An Explication*; L. Nowak, *On the Concept of Adequacy of Laws. An Idealizational Explication*; M. Paprzycki, K. Paprzycka, *Accuracy, Essentiality and Idealization*. **Discussions** – J. Sójka, *On the Origins of Idealization in the Social Experience*; M. Paprzycki, K. Paprzycka, *A Note on the Unitarian Explication of Idealization*; I. Hanzel, *The Pure Idealizational Law* – *The Inherent Law* – *The Inherent Idealizational Law*.

VOLUME 26 (1992)

IDEALIZATION IV: INTELLIGIBILITY IN SCIENCE
(Edited by Craig Dilworth)

C. Dilworth, *Introduction: Idealization and Intelligibility in Science*; E. Agazzi, *Intelligibility, Understanding and Explanation in Science*; H. Lauener, *Transcendental Arguments Pragmatically Relativized: Accepted Norms (Conventions) as an A Priori Condition for any Form of Intelligibility*; M. Paty, *L'Endoréférence d'une Science Formalisée de la Nature*; B. d'Espagnat, *De l'Intelligibilité du Monde Physique*; M. Artigas, *Three Levels of Interaction between Science and Philosophy*; J. Crompton, *The Unity of Knowledge and Understanding in Science*; G. Del Re, *The Case for Finalism in Science*; A. Cordero, *Intelligibility and Quantum Theory*; O. Costa de Beauregard, *De l'Intelligibilité en Physique. Example: Relativité, Quanta, Correlations EPR*; L. Fleischhacker, *Mathematical Abstraction, Idealization and Intelligibility in Science*; B. Ellis, *Idealization in Science*; P.T. Manicas, *Intelligibility and Idealization: Marx and Weber*; H. Lind, *Intelligibility and Formal Models in Economics*; U. Mäki, *On the Method of Isolation in Economics*; C. Dilworth, R. Pyddoke, *Principles, Facts and Theories in Economics*; J.C. Graves, *Intelligibility in Psychotherapy*; René Thom, *The True, the False and the Insignificant or Landscaping the Logos*.

VOLUME 27 (1992)

Ryszard Stachowski

THE MATHEMATICAL SOUL.
AN ANTIQUE PROTOTYPE OF THE MODERN MATEMATISATION OF PSYCHOLOGY

Chapter 1: *The Pythagorean "Number of Man"*; Chapter 2: *The Soul as a (Self-Moving) Number*; Chapter 3: *The Soul as Harmony*; Chapter 4: *The Soul as Originative of Relational Concepts*; Chapter 5: *The Faculty of the Soul as Intensive Psychological Quantity*; Chapter 6: *Soul Equals Mathematicals*; Chapter 7: *Soul to a Greater or Lesser Degree*; Epilogue; *Why Did Aristotle Call Psychology an Exact Science?*; Conclusion

VOLUME 28 (1993)

POLISH SCIENTIFIC PHILOSOPHY:
THE LVOV-WARSAW SCHOOL
(Edited by Francesco Coniglione, Roberto Poli and Jan Woleński)

Preface. **Background and Influence** – L. Albertazzi, *Brentano, Twardowski, and Polish Scientific Philosophy*; K. Schuhmann. *Husserl and Twardowski*; G. Küng. *Phenomenology and Polish Scientific Philosophy*; F. Coniglione. *Scientific Philosophy and Marxism in Poland*. **History and Systematics** – T. Kwiatkowski, *Classification of Reasonings in Contemporary Polish Philosophy*; A. Dylus, *The Problems of Ethics of Science in the Lvov-Warsaw School*; J.J. Jadacki, *Kazimierz Twardowski's Descriptive Semiotics*; P.M. Simons. *Nominalism in Poland*; T. Rzepa. R. Stachowski, *Roots of the Methodology of Polish Psychology*; K. Trzęsicki, *Łukasiewicz on Philosophy and Determinism*; A. Jedynak. *Conventionalism in Ajdukiewicz*; A. Siemianowski. *On Certain Consequences of the Radical Empiricism Hypothesis*; J. Woleński. *Tarski as a Philosopher*; R. Poli, *The Dispute over Reism: Kotarbiński – Ajdukiewicz – Brentano*.

VOLUME 29 (1993)

Zdzisław Augustynek and Jacek J. Jadacki

POSSIBLE ONTOLOGIES

Possible Ontologies – J.J. Jadacki. *Preface*; Z. Augustynek. *Point Eventism. An Outline of a Certain Ontology*; J.J. Jadacki. *Ontological Minimum*. **Discussions** – K. Paprzycka. *Carnap and Leibniz on the Problem of Being*; P. Przybysz. *Polish Discussions about Reism*.

VOLUME 30 (1993)

GOVERNMENT: SERVANT OR MASTER?
(Edited by Gerard Radnitzky and Hardy Bouillon)

G. Radnitzky, *Introduction: The Ominous Growth of the Monstrous Leviathan*. **Part I: Theory** – A. Seldon. *Politicians for or against the People*; G. Radnitzky. *Private Rights against Public Power: The Contemporary Conflict*; A. de Jasay. *Is Limited Government*

Possible?; H. Bouillon, *Mastering the Growth of Government: A Muzzle for Leviathan*; A. de Jasay, *Ownership, Agency, Socialism*. **Part II: Case Studies** – P. Bernholz, *Necessary and Sufficient Conditions to End Hyperinflations*; G. Schwarz, *The Deterioration of Switzerland's Economic System Policy*; H.O. Lenel, *Three Turns in West Germany since 1948 – Considered from the Viewpoint of the Theory of Economic Systems*; A. Flew, *Educational Services: Abuses of State Monopoly*; J.W.F. Sundberg, *Revenue-Only Taxes vs. Multipurpose Taxes: Philosophy and Implementation in Swedish High Tax Society*; S. Pejovich, *Political and Economic Reforms in Eastern Europe: The Case of Yugoslavia*. **Part III: The Limiting Case. The Totalitarian Order** – P. Bernholz, *Necessary Conditions for Totalitarianism: Supreme Values, Power and Personal Interest*.

VOLUME 31 (1993)

CREATIVITY AND CONSCIOUSNESS. PHILOSOPHICAL AND PSYCHOLOGICAL DIMENSIONS
(Edited by Jerzy Brzeziński, Santo Di Nuovo, Tadeusz Marek and Tomasz Maruszewski)

Introduction. **Part I: What Do Philosophers Say about Creativity and Consciousness?** – K. Zamiara, *Psychological Approach to Creativity. A Critical Approach*; R.L. Franklin, *Creativity and Depth in Understanding*; Z. Piątek, *Creating of Life and F. Nietzsche's Idea of Superman*; J. Janousek, *Dialogue and Joint Activity: A Psychological Approach*; K. Zamiara, *Some Remarks on Piagetian Notion of "Consciousness" and Its Importance for the Science of Culture*; A. Gałdowa, A. Nelicki, *Attitudes Toward Values as a Factor Determining Creativity*. **Part II: The Role of Creativity in Theory-Building** – L. Nowak, *Creativity in Theory-Building*; I. Nowak, *Discovery and Correspondence*; J. Brzeziński, *Factors Disturbing Validity of Experimental and Non-Experimental Psychological Research – Context's Analysis of Researcher's Psychologists Methodological Consciousness*; A. Falkowski, *Cognitive Similarity in Scientific Discovery: An Ecological Approach*. **Part III: Consciousness in Historical Perspective** – K.V. Wilkes, *Inside Insight*; F. di Maria, G. Lavanco, *Conscious/Unconscious and Group-Analysis*. **Part IV: Between Expression and Projection** – M. Stasiakiewicz, *Creativity and Projection: Paradigm Opposition and Implicit Correspondence*; A. Brzezińska, *Creative Expression versus Projection*. **Part V: The Role of Psychophysiological Components in Explanation of Phenomena of Consciousness and Creativity** – M. Bunge, *Explaining Creativity*; P. Wolski, *Hemispheric Asymmetry and Consciousness. Is there Any Relationship?*; A. Kokoszka, *A Rationale for Psychology of Consciousness*; M. Fąfrowicz, T. Marek, *Creativity and Attention*. **Part VI: Psychological Explanations of Creativity and Consciousness** – S. Di Nuovo, *Consciousness and Attention*; T. Maruszewski, *Some Remarks about Consciousness*; M. Kowalczyk, *On the Question of Functions of Consciousness*; D.K. Simonton, *From Childhood Giftedness to Creativity Genius*; M. Fąfrowicz, T. Marek, Cz. Noworol, K. Żarczyński, *Creativity and Organization*; M.A. Runco and J. R. Gaynor, *Creativity and Optimal Development*.

VOLUME 32 (1993)

FROM ONE-PARTY-SYSTEM TO DEMOCRACY
(Edited by Janina Frentzel-Zagórska)

J. Frentzel-Zagórska, *Introduction*. **Part I: Theoretical Approaches** – Z. Bauman. *A Postmodern Revolution*; L. Holmes, *On Communism, Post-communism, Modernity and Post-*

modernity; L. Nowak, *The Totalitarian Approach and the History of Socialism*; J. Pakulski, *East European Revolutions and 'Legitimacy Crisis'*. **Part II: The Transitional Period** – A. Czarnota, M. Krygier, *From State to Legal Traditions? Prospects for the Rule of Law after Communism*; M. Szabó, *Social Protest in a Post-communist Democracy: Taxi Drivers' Demonstration in Hungary*; Z. Bauman, *Dismantling Patronage State*; E. Mokrzycki, *Between Reform and Revolution: Eastern Europe Two Years after the Fall of Communism*; J. Frentzel-Zagórska, *The Road to Democratic Political System in Post-communist Eastern Europe*. **Part III: The Case of Yugoslavia** – R.F. Miller, *Yugoslavia: The End of the Experiment*.

VOLUME 33 (1993)

SOCIAL SYSTEM, RATIONALITY, AND REVOLUTION
(Edited by Marcin Paprzycki and Leszek Nowak)

Introduction. **On the Nature of Social System** – U. Preuss. *Political Order and Democracy. Carl Schmitt and his Influence*; K. Paprzycka, *A Paradox in Hobbes' Philosophy of Law*; S. Esquith, *Democratic Political Dialogue*; E. Jeliński, *Democracy in Polish Reformist Socialist Thought*; K. Paprzycka, *The Master and Slave Configuration in Hegel's System*; M. Godelier, *Lévi-Strauss, Marx and After. A Reappraisal of Structuralist and Marxist Tools for Social Logics*; K. Niedźwiadek, *On the Structure of Social System*; W. Czajkowski, *Social Being and Its Reproduction*. **On Rationality and Captivity** – M. Ziółkowski, *Power and Knowledge*. L. Nowak, *Two Inter-Human Limits to the Rationality of Man*; M. Paprzycki, *The Non-Christian Model of Man: An Attempt at Psychological Explanation*; R. Egiert. *Toward the Sophisticated Rationalistic Model of Man*. **On Social Revolution** – L. Nowak, *Revolution is an Opaque Progress but a Progress Nonetheless*; K. Paprzycka, M. Paprzycki, *How do Enslaved People Make Revolutions?*; G. Tomczak, *Is it Worth Winning a Revolution?*; K. Brzechczyn, *Civil Loops and the Absorption of Elites*; R. McCleary. *What Makes Marxist Historical Materialism Objective?*; G. Kotlarski. *Classes and Masses in Social Philosophy of Rosa Luxemburg*. **On Real Socialism** – E. Gellner, *The Civil and the Sacred*; W. Marciszewski, *Economics and the Idea of Information. Why Socialism must have Collapsed?*; L. Nowak, K. Paprzycka. M. Paprzycki, *On Multilinearity of Socialism*; A. Siegel, *The Overrepression Cycle in the Soviet Union. An Operationalization of a Theoretical Model*; K. Brzechczyn, *The State of the Teutonic Order as a Socialist Society*. **Discussions** – R. McCleary, *Socioanalysis and Philosophy*; W. Heller. *Methodological Remarks on the Public and the Private in Hannah Arendt's Political Philosophy*; K. Brzechczyn. *On Unsuccessful Conquest and Successful Subordination*.

VOLUME 34 (1994)

Izabella Nowakowa

IDEALIZATION V: THE DYNAMICS OF IDEALIZATIONS

Introduction; Chapter I: *Idealization and Theories of Correspondence*; Chapter II: *Dialectical Correspondence of Scientific Laws*; Chapter III: *Dialectical Correspondence in Science: Some Examples*; Chapter IV: *Dialectical Correspondence of Scientific Theories*; Chapter V: *Generalizations of the Rule of Correspondence*; Chapter VI: *Extensions of the Rule of Correspondence*; Chapter VII: *Correspondence and the Empirical Environment of a Theory*; Chapter VIII: *Some Methodological Problems of Dialectical Correspondence*.

VOLUME 35 (1993)

EMPIRICAL LOGIC AND PUBLIC DEBATE.
ESSAYS IN HONOUR OF ELSE M. BARTH
(Edited by E.C.W. Krabbe, R.J. Dalitz, and P.A. Smit)

Part I: Interpersonal Reasoning: Conflicts and Fallacies – T. Govier, *Needing Each Other for Knowledge: Reflections on Trust and Testimony*; J. Woods, *'Secundum quid' as a Research Programme*; G. Nuchelmans, *On the Fourfold Root of the 'argumentum ad hominem'*; F.H. van Eemeren, R. Grootendorst, *The History of the 'argumentum ad hominem' Since the Seventeenth Century*; L.S. van Epenhuysen, *Debate in a Bermuda Triangle of Medical Ethics*; E.C.W. Krabbe, *Reasonable Argument and Fallacies in the Kok-Stekelenburg Debate*; **Part II: Linguistic and Conceptual Tools** – J.D. North, *Some Weak Links in the Great Chain of Being*; A. Næss, *'You assert this?' An Empirical Study of Weight Expressions*; R. Wiche, *Gerrit Mannoury on the Communicative Functions of Negation in Ordinary Language*; J. Hoepelman, T. van Hoof, *Default and Dogma*; Ch. Goossens, *On the Logic of Nonmoral Commitment*; W. Marciszewski, *Arguments Founded on Creative Definitions*; R. Jorna, *Cognitive Science and Connectivism: Friend and Enemy or Move and Counter-Move, an Application of Empirical Logic*; **Part III: Dialectical Climates and Tempests** – R.H. Johnson, *Dialectical Fields and Manifest Rationality*; P. du Preez, *Reason Which Cannot Be Reasoned With: What Is Public Debate and How Does it Change?*; M.A. Finocchiaro, *Logic, Democracy, and Mosca*; P.A. Smit, *The Logic of Virtue and Terror*; G. van Benthem van den Bergh, *On Obstacles to Public Debate*; J.P. van Bendegem, *Real-Life Mathematics versus Ideal Mathematics: The Ugly Truth*; **Part IV: The Disempowernment of Woman: Strategies and Counter-Moves** – V. Songe-Møller, *The Road of Being and the Exclusion of the Feminine. An Analysis of the Poem of Parmenides*; R.J. Dalitz, *The Subjection of Women in the Contractual Society. An Analysis of Thomas Hobbes' Theory of Agreement*; H. Schröder, *Anti-Semitism and anti-Feminism Again: The Dissemination of Otto Weininger's 'Sex and Character' in the Seventies and Eighties*; J.R. Richards, *Traditional Spheres and Traditional Logic*.

VOLUME 36 (1994)

MARXISM AND COMMUNISM: POSTHUMOUS REFLECTIONS ON
POLITICS, SOCIETY, AND LAW
(Edited by Martin Krygier)

M. Krygier, *Introduction*; A. Flis, *From Marx to Real Socialism: The History of a Utopia*; P. Marciniak, *The Collapse of Communism: Defeat or Opportunity for Marxism in Eastern Europe*; J. Clark, A. Wildavsky, *Chronicle of a Collapse Foretold: How Marx Predicted the Demise of Communism (Although He Called It "Capitalism")*; L. Nowak, *Political Theory and Socialism. On the Main Paradigms of Political Power and Their Methodological and Historical Legitimation*; E. Mokrzycki, *Marxism, Sociology, and "Real Socialism"*; R. Bäcker, *The Collapse of Communism and Theoretical Models*; A. Zybertowicz, *Three Deaths an Ideology: The Withering Away of Marxism and the Collapse of Communism. The Case of Poland*; M. Krygier, *Marxism, Communism, and the Rule of Law*; A. Czarnota, *Marxism, Ideology, and Law*; G. Skąpska, *The Legacy of Anti-Legalism*; A. Sajo, *Law and the Legal Scholarship in the Happiest Barrack and Among the Hungry Liberated: Personal Recollections*.

VOLUME 37 (1994)

THE SOCIAL PHILOSOPHY OF AGNES HELLER
(Edited by John Burnheim)

J. Burnheim, *Introduction*; M. Vajda, *A Lover of Philosophy – A Lover of Europe*; P. Despoix, *On the Possibility of a Philosophy of Values. A Dialogue within the Budapest School*; M. Jay, *Women in Dark Times: Agnes Heller and Hannah Arendt*; J.P. Arnason, *The Human Condition and the Modern Predicament*; R.J. Bernstein, *Agnes Heller: Philosophy, Rational Utopia and Praxis*; Z. Bauman, *Narrating Modernity*; P. Beilharz, *Theories of History – Agnes Heller and R.G. Collingwood*; P. Wolin, *Heller's Theory of Everyday Life*; P. Harrison, *Radical Philosophy and the Theory of Modernity*; A.J. Jacobson, *The Limits of Formal Justice*; P. Murphy, *Civility and Radicalism*; P. Murphy, *Pluralism and Politics*; V. Camps, *The Good Life: A Moral Gesture*; L. Boella, *Philosophy Beyond the Baseless and Tragic Character of Action*; G. Márkus, *The Politics of Morals*; A. Heller, *A Reply to My Critics*; *The Bibliography of Agnes Heller*.

VOLUME 38 (1994)

IDEALIZATION VI: IDEALIZATION IN ECONOMICS
(Edited by Bert Hamminga and Neil B. De Marchi)

Introduction – B. Hamminga, N. De Marchi, *Preface*; B. Hamminga, N. De Marchi, *Idealization and the Defence of Economics: Notes Toward a History*. **Part I: General Observations on Idealization in Economics** – K.D. Hoover, *Six Queries about Idealization in an Empirical Context*; B. Walliser, *Three Generalization Processes for Economic Models*; S. Cook, D. Hendry, *The Theory of Reduction in Econometrics*; M.C.W. Janssen, *Economic Models and Their Applications*; A.G. de la Sienra, *Idealization and Empirical Adequacy in Economic Theory*; I. Nowakowa, L. Nowak, *On Correspondence between Economic Theories*; U. Mäki, *Isolation, Idealization and Truth in Economics*. **Part II: Case Studies of Idealization in Economics** – N. Cartwright, *Mill and Menger: Ideal Elements and Stable Tendencies*, W. Balzer, *Exchange Versus Influence: A Case of Idealization*; K. Cools, B. Hamminga, T.A.F. Kuipers, *Truth Approximation by Concretization in Capital Structure Theory*; D.M. Hausman, *Paul Samuelson as Dr. Frankenstein: When an Idealization Runs Amuck*; H.A. Keuzenkamp, *What if an Idealization is Problematic? The Case of the Homogeneity Condition in Consumer Demand*; W. Diederich, *Nowak on Explanation and Idealization in Marx's* Capital; G. Jorland, *Idealization and Transformation*; J. Birner, *Idealizations and Theory Development in Economics. Some History and Logic of the Logic Discovery*. **Discussions** – L. Nowak, *The Idealizational Methodology and Economics. Replies to Diederich, Hoover, Janssen, Jorland and Mäki*.

VOLUME 41 (1994)

HISTORIOGRAPHY BETWEEN MODERNISM AND POSTMODERNISM.
CONTRIBUTIONS TO THE METHODOLOGY OF
THE HISTORICAL RESEARCH
(Edited by Jerzy Topolski)

HISTORIOGRAPHY BETWEEN MODERNISM AND POSTMODERNISM CONTRIBUTIONS TO THE METHODOLOGY OF THE HISTORICAL RESEARCH

Ed. by Jerzy Topolski

Amsterdam/Atlanta, GA 1994. 221 pp.
(Poznań Studies 41)

ISBN: 90-5183-721-6	Bound Hfl. 110,-/US-$ 68.-
ISBN: 90-5183-744-5	Paper Hfl. 35,-/US-$ 21.50

Contents: JERZY TOPOLSKI: A Non-postmodernist Analysis of Historical Narratives. FRANK R. ANKERSMIT: The Origins of Postmodernist Historiography. DAVID CARR: Getting the Story Straight: Narrative and Historical Knowledge. WOJCIECH WRZOSEK: The Problem of Cultural Imputation in History. Cultures versus History. JACQUES TACQ: Causality and Virtual Finality. GWIDON ZALEJKO: Soviet Historiography as a "Normal Science". HENRYK MAMZER AND JANUSZ OSTOJA-ZAGÓRSKI: Deconstruction of the Evolutionist Paradigm in Archaeology. NICOLE LAUTIER: At the Crossroads of Epistemology and Psychology: Prospects of a Didactic of History. TERESA KOSTYRKO: Remarks on "Aesthetization" in Science on the Basis of History.

USA/Canada: Editions Rodopi, 233 Peachtree Street, N.E., Suite 404, Atlanta, GA 30303-1504, Telephone (404) 523-1964, Call toll-free 1-800-225-3998 (U.S. only), Fax (404) 522-7116

And Others: Editions Rodopi B.V., Keizersgracht 302-304, 1016 EX Amsterdam, The Netherlands. Telephone ++ (0) 20 622 75 07, Fax ++ (0) 20 638 09 48

the review of
metaphysics

a philosophical quarterly

ISSN 0034-6632

JUNE 1994 | **VOL. XLVII, No. 4** | **ISSUE No. 188** | **$11.00**

articles

books received

recent titles in philosophy

philosophical abstracts

announcements

index

Individual Subscriptions $25.00   Institutional Subscriptions $42.00   Student/Retired Subscriptions $15.00

Jude P. Dougherty, Editor

The Catholic University of America, Washington, D.C. 20064

FRANK G. FORREST

Valuemetrics ℵ :
The Science of Personal
and Professional Ethics

Amsterdam/Atlanta, GA 1994. XVIII,179 pp.
(Value Inquiry Book Series 11)
ISBN: 90–5183–683–X   Hfl. 60,–/US–$ 35.–

Valuemetrics is an elaboration of Robert S. Hartman's innovative development in the application of an abstract system to the study of ethical
problems. The system used for this purpose is a branch of logic called set
theory. Set theory fulfills this role because goodness, the fundamental
phenomenon of ethics, is defined axiomatically in terms of sets. The
similarity of structure between certain elements of set theory and the
various types and degrees of goodness makes mathematical accounting of
goodness phenomena possible. In the valuemetrics context, value judgments
are considered as an assessment of the goodness of something. Therefore,
the mathematical system for the accounting of goodness serves as a tool for
objectively making many kinds of value judgments and possible attendant
ethical decisions. One of the results of this conception, attributable to
Hartman, is the birth of the science of ethics.
The first half of the book elucidates the theory, terminology, and
mathematical system used in valuemetrics, known as Hartmanean algebra.
The second half is devoted to the application of this system to the
measurement and development of a person's value vision, and the solution
of various problems in ethics using the case study technique. Hartmanean
algebra will resolve several types of problems such as the determination of
right and wrong, good and bad; determining how to redress and amend
instances of wrongs and badness; and how to determine when, if ever,
wrongs and badness are justified.

USA/Canada: Editions Rodopi, 233 Peachtree Street, N.E., Suite 404,
    Atlanta, GA 30303–1504, Telephone (404) 523–1964, Call toll–free 1–
    800–225–3998 (U.S. only), Fax (404) 522–7116
And Others: Editions Rodopi B.V., Keizersgracht 302–304, 1016 EX
    Amsterdam, The Netherlands. Telephone ++ (0) 20 622 75 07, Fax ++
    (0) 20 638 09 48

International Studies in the Philosophy of Science

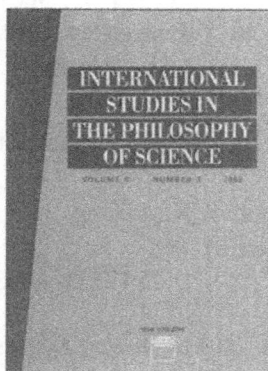

International Studies in the Philosophy of Science is an interdisciplinary journal that welcomes articles not only from philosophers, but also from historians and sociologists of science. Theoretical articles are drawn from a variety of disciplines including physics, chemistry, biology, psychology, neuroscience and mathematics. The journal has a particular interest in publishing papers from a wide range of countries around the world, thus fostering cooperation between scholars and students from a variety of backgrounds.

1994 - Volume 8 (3 issues). ISSN 0269-8595.

Subscriptions
Institutional rate: £120.00/US$210.00, post free.
Personal rate: £46.00/US$80.00, post free.

ORDER FORM

Please invoice me at the ▨ institutional ▨ personal rate

▨ Please send me an inspection copy of *International Studies in the Philosophy of Science*

Name _____

Address _____

_____

_____

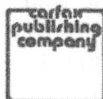

CARFAX PUBLISHING COMPANY
P O Box 25, Abingdon, Oxfordshire OX14 3UE, UK
PO Box 2025, Dunnellon, Florida 34430-2025, USA
UK Tel: +44 (0) 235 521154. Fax: +44 (0) 235 553559.

REALITÄT UND GEWIßHEIT

Tagung der Internationalen J.–G.–Fichte–Gesellschaft (6.–9. Oktober 1992) in Rammenau in Zusammenarbeit mit dem Istituto per gli Studi Filosofici (Neapel)

Hrsg. von Helmut Girndt, Wolfgang H. Schrader

Amsterdam/Atlanta, GA 1994. XI,448 pp.
(Fichte–Studien 6)
ISBN: 90–5183–739–9   Hfl. 135,–/US–$ 84.–

Inhalt: KLAUS HAMMACHER: Nachruf auf Dr. Richard Schottky. EDIT DÜSING: Johannes Schurr zum Gedenken. TEIL I. JÜRGEN STOLZENBERG: Fichtes Satz »Ich bin«. Argumentanalytische Überlegungen zu Paragraph 1 der *Grundlage der gesamten Wissenschaftslehre* von 1794/95. KUNIHIKO NAGASAWA: Intellektuelle Anschauung und Dialektik. SVEN JÜRGENSEN: Die Unterscheidung der Realität in Fichtes *Wissenschaftslehre* von 1794. WILHELM METZ: Fichtes genetische Deduktion von Raum und Zeit in Differenz zu Kant. PETER ROHS: Über die Zeit als das Mittelglied zwischen dem Intelligiblen und dem Sinnlichen. DANIEL BREAZEALE: Philosophy as the Divided Self: On the »Existential« and »Scientific« Tasks of the Jena *Wissenschaftslehre*. ENDRE KISS: Zwischen Apriorismus und Empirismus im Kontext der Isomorphie zweier Apriorismen. Zur Rekonstruktion von Fichtes philosophischer Konzeption. TEIL II. ERICH HEINTEL: Gewißheit und Wahrheit bei Fichte. H.M. EMRICH: Identität und Versprechen. HANS GEORG VON MANZ: Selbstgewißheit und Fremdgewißheit. Fichtes Konzeption des Anderen als Konstituens der Selbsterfassung unter Berücksichtigung der Perspektive Lévinas'. ALOIS K. SOLLER: Die Unbegreiflichkeit der Wechselwirkung der Geister. Das Problem einer »Interpersonalitätslehre« bei Fichte. CARLA DE PASCALE: Die Trieblehre bei Fichte. KAREN GLOY: Die Naturauffassung bei Kant, Fichte und Schelling. ALBERT MUES: Der Grund der Dualität der Materie und des Indeterminismus in der physikalischen Natur. Die Lösung eines quantenphysikalischen Rätsels. MARCO IVALDO: Zur Geschichtserkenntnis nach der Transzendentalphilosophie. FAUSTINO ONCINA COVES: Geheimnis und Öffentlichkeit bei Fichte. FRANK ASCHOFF: Rückkehr zur Metaphysik? Reinholds Abkehr von der Philosophie Fichtes. IVES RADRIZZANI: Der Übergang von der *Grundlage* zur *Wissenschaftslehre nova methodo*. L. de Vos: Die Realität der Idee. TEIL III. WILHELM LÜTTERFELDS: Fichtes Konzept absoluter Einheit (1804)– ein performativer Selbstwiderspruch? URS RICHLI: »Ich aber fordere Sie auf, absolute Genesis ins Auge zu fassen!« Realität und absolute Negativität in Fichtes *Wissenschaftslehre* von 1804 und in Hegels *Wissenschaft der Logik*. HARTMUT TRAUB: Realität und System. Das Realitätsproblem in Fichtes Theorie der Fünffachheit.

USA/Canada: Editions Rodopi, 233 Peachtree Street, N.E., Suite 404, Atlanta, GA 30303–1504, Telephone (404) 523–1964. Call toll–free 1–800–225–3998 (U.S. only), Fax (404) 522–7116

And Others: Editions Rodopi B.V., Keizersgracht 302–304, 1016 EX Amsterdam, The Netherlands. Telephone ++ (0) 20 622 75 07, Fax ++ (0) 20 638 09 48

JENS KULENKAMPFF

Kants Logik des ästhetischen Urteils

2., erweiterte Auflage 1994. 252 Seiten.
Ln DM 88.– ISBN 3-465-02646-2
Philosophische Abhandlungen Band 61

Im Mittelpunkt der Untersuchung steht die *Analytik des Schönen* in Kants *Kritik der Urteilskraft.* Es wird gezeigt, was Kant zu einer „Geschmackskritik in transzendentaler Absicht" motiviert und wie er in Form einer entdeckenden Analyse die Erklärung dafür zu finden sucht, warum reine ästhetische Urteile mit einem Allgemeingültigkeitsanspruch auftreten können, auch wenn sie ihn nicht einlösen können. Kritische Betrachtungen der Einleitungen zur *Kritik der Urteilskraft* und Analysen von Kants Theorie der schönen Form und seiner Theorie der „anhängenden Schönheit" runden die Untersuchung ab. – Dieser typographisch verbesserten, druckfehlerbereinigten, im übrigen aber unveränderten *zweiten* Auflage der Abhandlung sind ein Register und ein längeres Nachwort beigegeben, das auf einige Titel der neueren Forschungsliteratur eingeht (Ginsborg, Fricke, Henrich, Otto).

VITTORIO KLOSTERMANN · FRANKFURT AM MAIN

SHELDON RICHMOND

Aesthetic Criteria
Gombrich and the Philosophies
of Science
# of Popper and Polanyi

Amsterdam/Atlanta, GA 1994. 152 pp.
(Schriftenreihe zur Philosophie Karl R. Poppers
und des Kritischen Rationalismus 6)
ISBN: 90–5183–618–X   Hfl. 45,–/US–$ 26.–

Dr. Richmond's book examines deftly the aesthetic theory of leading art historian and critic, Sir Ernst Gombrich. Though in the psychology of art and in related matters Gombrich is an avowed follower the rationalist philosophy of science of Sir Karl Popper, in aesthetics proper he follows ideas first propounded in the irrationalist philosophy of science of Michael Polanyi. Dr. Richmond presents succinctly the ideas of these three great thinkers and finds here an unexpected irrationalist streak in the rationalistic works of Gombrich. (Joseph Agassi)

USA/Canada: Editions Rodopi, 233 Peachtree Street, N.E., Suite 404, Atlanta, GA 30303–1504, Telephone (404) 523–1964, Call toll–free 1–800–225–3998 (U.S. only), Fax (404) 522–7116
And Others: Editions Rodopi B.V., Keizersgracht 302–304, 1016 EX Amsterdam, The Netherlands. Telephone ++ (0) 20 622 75 07, Fax ++ (0) 20 638 09 48

Philosophia

Philosophical Quarterly of Israel
Editor: Asa Kasher

Articles from Volume 23, 1993-1994

Editorial addresses:
PHILOSOPHIA, Bar-Ilan University, Ramat-Gan 52100, Israel
or: Prof. Asa Kasher,
L. Schwarz-Kipp Chair of Professional Ethics and Philosophy of Practice
Tel-Aviv University, Tel-Aviv 69978, Israel
FAX: +972-3-6409457 (Attn.: Prof. Asa Kasher)
ASA0425@TAUNIVM.BITNET or ASA0425@TAUNIVM.TAU.AC.IL

BELIEF AND UNBELIEF
Psychological perspectives

Ed. by Jozef Corveleyn & Dirk Hutsebaut

Amsterdam/Atlanta, GA 1994. 246 pp.
(International Series in the Psychology of Religion 3)
IBN: 90–5183–673–2   Hfl. 75,–/US–$ 46.50

Contents: **1. Historical and theoretical approaches.** JACOB A. BELZEN: An early effort in the psychology of belief and unbelief: Critical reflections on a Dutch classic in the psychology of religion. REINDER RUARD GANZEVOORT: Crisis experiences and the development of belief and unbelief. **2. Empirical studies.** *Children and adolescents.* FRITZ K. OSER, K. HELMUT REICH & ANTON A. BUCHER: Development of belief and unbelief in childhood and adolescence. HELENA HELVE: The development of religious belief systems from childhood to adulthood: A longitudinal study of young Finns in the context of the lutheran church. GEOFFREY E.W. SCOBIE: Belief, unbelief and conversion experience. MARGO ROOIJACKERS: Ethnic identity and islam: The results of an empirical study among young Turkish immigrants in The Netherlands. JAN VAN DER LANS & MARGO ROOIJACKERS: Attitudes of second generation Turkish immigrants towards collective religious representations of their parental culture. *Adults.* GERALDO JOSÉ DE PAIVA: Religious iteneraries of academics: A psychological discussion. JAMES M. DAY: Moral development, belief and unbelief: Young adult accounts of religion in the process of moral growth. MARINUS H.F. VAN UDEN: On saints: Case–studies between belief and unbelief. JORDI BACHS: Belief, unbelief and religious experience. **Clinical studies.** JOZEF CORVELEYN & HUGO LIETAER: Religion and mental health in the eighties: A survey and critical review of the literature. OWE WIKSTRÖM: Psychotic (a–)theism? The cognitive dilemmas of two psychiatric episodes.

USA/Canada: Editions Rodopi, 233 Peachtree Street, N.E., Suite 404, Atlanta, GA 30303–1504, Telephone (404) 523–1964, Call toll-free 1–800–225–3998 (U.S. only), Fax (404) 522–7116
And Others: Editions Rodopi B.V., Keizersgracht 302–304, 1016 EX Amsterdam, The Netherlands. Telephone ++ (0) 20 622 75 07, Fax ++ (0) 20 638 09 48

ΛΝΛLOGΙΛ

Revista de Filosofía.

ANALOGIA es una revista de investigación y difusión filosóficas del Centro de Estudios de la Provincia de Santiago de México de la Orden de Predicadores (Dominicos). ANALOGIA publica artículos de calidad sobre las distintas áreas de la filosofía.

Director: Mauricio Beuchot. Consejo editorial: Ignacio Angelelli, Tomás Calvo, Roque Carrión, Gabriel Chico, Marcelo Dascal, Gabriel Ferrer, Jorge J. E. Gracia, Klaus Hedwig, Ezequiel de Olaso, Lorenzo Peña, Philibert Secretan, Enrique Villanueva.

Colaboraciones (artículos, notas, reseñas) y pagos enviarse a:
Apartado postal 23-161
Xochimilco 16000 México, D.F.
MEXICO

Peridiocidad semestral. Suscripción anual (2 números): 35 US dls.

RECONSTRUCTING FOUCAULT
ESSAYS IN THE WAKE OF THE 80s

Ed. by Ricardo Miguel–Alfonso and Silvia Caporale–Bizzini

Amsterdam/Atlanta, GA 1994. 304 pp.
(Postmodern Studies 10)
ISBN: 90–5183–708–9   Bound Hfl. 125,–/US–$ 73.–
ISBN: 90–5183–710–0   Paper Hfl. 40,–/US–$ 23.50

Contents: SILVIA CAPORALE–BIZZINI AND RICARDO MIGUEL–ALFONSO: Introduction. PART I: FOUCAULT AND POST-STRUCTURALISM. PATRICIO PEÑALVER: "Archaeology, History, Deconstruction: Foucault's Thought and the Philosophical Experience". CRISTINA DE PERETTI: "Foucault The Twofold Games of Language". PART II: FOUCAULT AND CRITICAL THEORY. CHRISTOPHER NORRIS: "What is Enlightenment? Kant according to Foucault". DANIEL T. O'HARA: "Why Foucault No Longer Matters". MICHAEL RYAN: "Foucault's Fallacy". PART III: FOUCAULT AND THE SUBJECT. ALAN D. SCHRIFT: "Reconfiguring the Subject: Foucault's Analytics of Power". KATH RENARK JONES: "Modernity, Ethics and Irony: The Return of the Subject in the Later Works of Michel Foucault. PART III: FOUCAULT IN FEMINISM: SUSAN BORDO: "Feminism, Foucault, and the Politics of the Body". ROSA MA. RODRÍGUEZ: "The Female Subject after the Death of Man". PART IV: FOUCAULT, HISTORY AND AESTHETICS. ARTHUR KROKER: "Cynical Aesthetics: The Games of Foucault. ANTONIO CAMPILLO: "On War: The Space of Knowledge, Knowledge of Space".

USA/Canada: Editions Rodopi, 233 Peachtree Street, N.E., Suite 404, Atlanta, GA 30303–1504, Telephone (404) 523–1964, Call toll–free 1–800–225–3998 (U.S. only), Fax (404) 522–7116

And Others: Editions Rodopi B.V., Keizersgracht 302–304, 1016 EX Amsterdam, The Netherlands. Telephone ++ (0) 20 622 75 07, Fax ++ (0) 20 638 09 48

THE JOURNAL OF THE BRITISH SOCIETY FOR PHENOMENOLOGY

An International Review of Philosophy and the Human Sciences
EDITOR: WOLFE MAYS

Volume 25, May 1994

Gadamer, Sartre, and Deleuze

Articles

The *JBSP* publishes papers on phenomenology and existential philosophy as well as contributions from other fields of philosophy. Papers from workers in the Humanities and human sciences interested in the philosophy of their subject will be welcome. All papers and books for review to be sent to the Editor: Dr. Wolfe Mays, Institute of Advanced Studies, The Manchester Metropolitan University, Manchester M15 6HB, England. Subscription and advertisement enquiries to be sent to the publishers: Haigh and Hochland Ltd., The Precinct Centre, Oxford Road, Manchester 13, England.

HEINRICH GOMPERZ, KARL POPPER UND DIE 'ÖSTERREICHISCHE PHILOSOPHIE'

Beiträge zum internationalen Forschungsgespräch aus Anlaß des 50. Todestages von Heinrich Gomperz (1873–1942) und des 90. Geburtstages von Karl R. Popper (1902–)

Hrsg. von Friedrich Stadler/Martin Seiler

Amsterdam/Atlanta, GA 1994. XI,227 pp.
(Studien zur österreichischen Philosophie 22)
ISBN: 90–5183–632–5   Hfl. 75,–/US–$ 44.–

Leben und Werk zweier geistesverwandter österreichischer Denker werden im Kontext österreichischer Philosophie dargestellt und deren Oeuvre miteinander verglichen.
Neben der Rekonstruktion des umfangreichen und vergessenen Lebenswerkes von *Heinrich Gomperz* (1873–1942) geht es in dieser erstmaligen Bestandsaufnahme um Gomperz spezifische Beiträge zur Philosophiegeschichte, Methodologie sowie zur praktisch orientierten Rechts– und Sozialphilosophie aus aktueller Sicht. Vor dem Hintergrund einer fast unübersichtlichen Forschungsliteratur zum Werk von *Karl Popper* wird im vorliegendem Band vor allem auf neuere Aspekte von Poppers Philosophie und Wissenschaftstheorie im Zusammenhang mit der österr. Philosophiegeschichte eingegangen.

USA/Canada: Editions Rodopi, 233 Peachtree Street, N.E., Suite 404, Atlanta, GA 30303–1504, Telephone (404) 523–1964, Call toll–free 1–800–225–3998 (U.S. only), Fax (404) 522–7116
And Others: Editions Rodopi B.V., Keizersgracht 302–304, 1016 EX Amsterdam, The Netherlands. Telephone ++ (0) 20 622 75 07, Fax ++ (0) 20 638 09 48

the review of

metaphysics

a philosophical quarterly

ISSN 0034-6632

SEPTEMBER 1994 | VOL. XLVIII, No. 1 | ISSUE No. 189 | $11.00

articles

books received

philosophical abstracts

doctoral dissertations

visiting philosophers

retiring philosophers

announcements

Individual Subscriptions $25.00   Institutional Subscriptions $42.00   Student/Retired Subscriptions $15.00

Jude P. Dougherty, Editor
The Catholic University of America, Washington, D.C. 20064

WILLIAM GERBER

The Meaning of Life
Insights of the World's Great Thinkers

Amsterdam/Atlanta, GA 1994. XVII,282 pp.
(Value Inquiry Book Series 12)
ISBN: 90–5183–691–0   Bound Hfl. 125,–/US–$ 73.50
ISBN: 90–5183–680–5   Paper Hfl. 45,–/US–$ 26.–

The book aims to present the wisdom of sages, great thinkers, renowned writers, and philosophers, of many countries and time periods, in their own words, regarding life. The book also aims to place the numerous quotations from these sources in a structured organization, with introductory and explanatory comments and comparisons.

Main Topics or Fields – See Organization or Principal Parts.

Organization or Principal Parts – Part 1, The Deepest Questions About Life in General (whence *came* life? what *is* life? what is the nature of plants and animals? etc.); Part 2, The Deepest Questions About Human Life in Particular (what is unique about humanity? who am I and what am I? do humans have free will?); Part 3, The Deepest Questions About the Quality, Meaning, and Span of Our Lives (are we more like angels or like devils? what meaning if any do our lives have? how do we march from infancy to old age and death?); and Part 4, The Deepest Questions About Living in Society (what are the basic truths about sex, love, marriage, civilization, and religion?).

Special Features – An appendix shows the source of every quotation included in the book. An index of authors lists the quotations taken from each author.

USA/Canada: Editions Rodopi, 233 Peachtree Street, N.E., Suite 404, Atlanta, GA 30303–1504, Telephone (404) 523–1964, Call toll–free 1–800–225–3998 (U.S. only), Fax (404) 522–7116

And Others: Editions Rodopi B.V., Keizersgracht 302–304, 1016 EX Amsterdam, The Netherlands. Telephone ++ (0) 20 622 75 07, Fax ++ (0) 20 638 09 48

www.ingramcontent.com/pod-product-compliance
Lightning Source LLC
Chambersburg PA
CBHW021027210326
41598CB00016B/928